Y0-BZT-036

Photoproteins in Bioanalysis

Edited by
Sylvia Daunert and
Sapna K. Deo

Related Titles

M. Goeldner, R. Givens (Eds.)

Dynamic Studies in Biology

Phototriggers, Photoswitches and Caged Biomolecules

2005

ISBN 3-527-30783-4

W. R. Briggs, J. L. Spudich (Eds.)

Handbook of Photosensory Receptors

2005

ISBN 3-527-31019-3

M. Chalfie, S. Kain (Eds.)

Green Fluorescent Protein

Properties, Applications and Protocols (Methods of Biochemical Analysis)
2nd Edition

2005

ISBN 0-471-73682-1

A. Pingoud, C. Urbanke, J. Hoggett, A. Jeltsch

Biochemical Methods

A Concise Guide for Students and Researchers

2002

ISBN 3-527-30299-9

B. Valeur

Molecular Fluorescence

Principles and Applications

2002

ISBN 3-527-29919-X

Photoproteins in Bioanalysis

Edited by
Sylvia Daunert and Sapna K. Deo

WILEY-VCH Verlag GmbH & Co. KGaA

The Editors

Dr. Sylvia Daunert
College of Arts & Sciences
Gill Eminent Professor of Analytical and Biological Chemistry
Department of Chemistry
University of Kentucky
Lexington, KY 40506-0055
USA

Sapna Deo
Department of Chemistry
Indiana University
Purdue University – Indianapolis
402 N. Blackford Street, LD 326
Indianapolis, IN 46202-3274
USA

Library of Congress Card No.: applied for
British Library Cataloguing-in-Publication Data
A catalogue record for this book is available from the British Library

Bibliographic information published by Die Deutsche Bibliothek
Die Deutsche Bibliothek lists this publication in the Deutsche Nationalbibliografie; detailed bibliographic data is available in the Internet at <http://dnb.ddb.de>

Printed in the Federal Republic of Germany
Printed on acid-free paper

Typesetting Manuela Treindl, Laaber
Printing betz-druck GmbH, Darmstadt
Bookbinding Litges & Dopf Buchbinderei GmbH, Heppenheim
Cover 4T Matthes + Traut, Darmstadt

ISBN-13: 978-3-527-31016-6
ISBN-10: 3-527-31016-9

Contents

Photoproteins in Bioanalysis. Edited by Sylvia Daunert and Sapna K. Deo

ISBN: 3-527-31016-9

Preface

Bioluminescence is a spectacular natural phenomenon where light is emitted from a living organism due to an internal biochemical reaction. Perhaps, the most well-known case of bioluminescence involves the "flashes of light" that we observe on summer nights emitted by the different species of terrestrial fireflies. While this may be the most common avenue of initial exposure to bioluminescence for millions of people, bioluminescence is even more prevalent in a variety of organisms that inhabit our seas and oceans. In the early 1960s, the discovery of two proteins, namely, aequorin and green fluorescent protein (GFP) from the jellyfish *Aequorea victoria*, by Shimomura and co-workers opened a new era in the study of bioluminescence. Specifically, this discovery spurred the interest of scientists to focus on the marine world, where bioluminescence is truly abundant; sunlight cannot penetrate efficiently in the deep oceans, and therefore, bioluminescence becomes a major source of illumination. This emission of light aids marine organisms in mating, fetching food, and scaring predators, among other functions. The quest for the identification of new marine bioluminescent organisms and for the understanding of the nature of their bioluminescence continues and is beautifully described and illustrated in our book in Chapter 2. There is no question that with the discovery of aequorin, and, later on, of the green fluorescent protein, GFP, Dr. Osamu Shimomura re-invigorated the field of bioluminescence. Dr. Shimomura reviews the history and properties of aequorin in Chapter 1.

Although bioluminescence is common in nature, only a few photoproteins have been isolated and characterized. Certain properties of photoproteins such as superior detection sensitivity, hazard-free handling, lack of cellular toxicity, and low background noise have positioned them as an excellent alternative analytical reagents to radiolabels. The cloning and recombinant production of these proteins has provided a major thrust for their applications in a variety of fields. These applications encompass the fields of biochemical, analytical, medical, and environmental science, as well as drug discovery and diagnostics. Among the proteins isolated from bioluminescent organisms, GFP and luciferases have become important tools in molecular and cell biology, whereas aequorin has proven to be an excellent label for analytical assays and intracellular calcium measurements. It is difficult to compile the enormous amount of work performed with these three proteins and others like obelin in a single book. Hence, we have narrowed the focus of our book to mainly the applications of photoproteins in bioanalysis.

Photoproteins in Bioanalysis. Edited by Sylvia Daunert and Sapna K. Deo

ISBN: 3-527-31016-9

The choice of contributed material in this book covers a variety of topics related to photoproteins. The book begins with a chapter narrating the discovery of photoproteins and their properties. The subsequent chapters discuss the phylogenetic tree of bioluminescent organisms, novel luciferases, and their applications in biotechnology. The sections following these chapters cover applications of photoproteins in bioanalysis including protein-protein interactions, nucleic acid analysis, bioluminescence resonance energy transfer assays, *in vivo* imaging, biomedical assays, whole cell sensing, and binding assays. The book also discusses the applications of photoproteins in advances in instrumentation, as well as their incorporation into micro-total analysis systems (μ-TAS). Commercial sources and description of photoproteins and their substrates compiled in a tabular format should serve as a quick, easy-to-use guide for readers interested in working with these proteins.

The aim of this book is to provide the readers with an overview of the current state-of-the-art in photoprotein-based bioanalysis, as well as to encourage a growing number of investigators interested in photoproteins to explore new avenues of research. In that regard, it was our goal to make the content of this book stimulating to a broader scientific community.

With advances in biotechnology, the creation of designer proteins with unprecedented properties is easier than ever before. Photoproteins will continue to provide an excellent scaffold for "tailor-made" proteins with applications in a variety of research fields. As bio-nanotechnology and nanoscale analysis become more prevalent, the demand will increase to develop techniques with high sensitivity that can detect biomolecules in biological and environmental samples. High throughput and multi-analyte detection will play an increasingly important role, as well. Photoproteins have a great potential to be a solution to this new challenge given their low detection limits and virtual lack of background interferences, along with their ability to supply a "palette of colors" that can be employed in parallel analysis. Moreover, as we voyage to uncharted regions of our oceans and rainforests, the relevance of photoproteins will be enhanced by the discovery of new proteins with unique and unknown properties.

Finally, we should mention that photoprotein research has been a focal point of work in our laboratories since 1994, and that this topic of study has always been rewarding and stimulating. Therefore, we are delighted to have the opportunity to assemble the work of outstanding scientists in the fields of photoproteins and bioanalysis for our book. In that regard, it is important to emphasize that completion of this book would not have been possible without the contribution from all of the authors and the support from our publisher, Wiley-VCH. We would like to extend our most sincere gratitude and appreciation to all of them, as well as to the granting agencies that support our work, namely the National Science Foundation, the National Institutes of Health, the National Institute of Environmental Health Sciences, and the National Aeronautics and Space Administration.

Sylvia Daunert, Lexington, June 2006
Sapna K. Deo, Indianapolis, June 2006

List of Contributors

Stephanie Bachas-Daunert
Department of Chemistry
University of Kentucky
Lexington
KY 40506-0055
USA

Theodore K. Christopoulos
Department of Chemistry
University of Patras
Greece
26500 Petras
Greece
and
Foundation for Research and Technology Hellas (FORTH)
Institute of Chemical Engineering and High Temperature Processes
26504 Patras
Greece

Christopher H. Contag
Departments of Pediatrics
Radiology and Microbiology & Immunology
Stanford School of Medicine
Clark Center
East Wing E150
318 Campus Drive
Stanford
CA 94305
USA

Sylvia Daunert
Department of Chemistry
University of Kentucky
Lexington
KY 40506-0055
USA

Logan Davies
Department of Chemistry
University of Kentucky
Lexington
KY 40506-0055
USA

Sapna K. Deo
Department of Chemistry & Chemical Biology
Indiana University –
Purdue University Indianapolis
Indianapolis
IN 46202
USA

Emre Dikici
Department of Chemistry & Chemical Biology
IUPUI
Indianapolis
IN 46202-3274
USA

Photoproteins in Bioanalysis. Edited by Sylvia Daunert and Sapna K. Deo

ISBN: 3-527-31016-9

Leslie Doleman
Department of Chemistry
University of Kentucky
Lexington
KY 40506-0055
USA

Jessika Feliciano
Department of Chemistry
University of Kentucky
Lexington
KY 40506-0055
USA

Massimo Guardigli
Department of Pharmaceutical Sciences
University of Bologna
Via Belmeloro 6
40126 Bologna
Italy

Steven H.D. Haddock
Monterey Bay Aquarium Research Institute
7700 Sandholdt Rd.
Moss Landing
CA 95039
USA

Penelope C. Ioannou
Department of Chemistry
University of Athens
15771 Athens
Greece

Eric Karplus
ScienceWares
30 Sandwich Road
East Falmouth
MA 02538
USA

Elisa Michelini
Department of Pharmaceutical Sciences
University of Bologna
Via Belmeloro 6
40126 Bologna
Italy

Mara Mirasoli
Department of Pharmaceutical Sciences
University of Bologna
Via Belmeloro 6
40126 Bologna
Italy

Elizabeth A. Moschou
Department of Chemistry & Chemical Biology
IUPUI
Indianapolis
IN 46202-3274
USA

Yoshihiro Ohmiya
Cell Dynamics Group
National Institute of Advanced Science and Technology (AIST)
Ikeda
Osaka 536-8577
Japan

Patrizia Pasini
Department of Chemistry
University of Kentucky
Lexington
KY 40506-0055
USA

Aldo Roda
Department of Pharmaceutical Sciences
University of Bologna
Via Belmeloro 6
40126 Bologna
Italy

Anna Rothert
Department of Chemistry & Chemical Biology
IUPUI
Indianapolis
IN 46202-3274
USA

Laura Rowe
Department of Chemistry & Chemical Biology
IUPUI
Indianapolis
IN 46202-3274
USA

Osamu Shimomura
The Photoprotein Laboratory
324 Sippewissett Road
Falmouth
MA 02540
USA

Rajesh Shinde
Departments of Pediatrics
Radiology and Microbiology & Immunology
Stanford School of Medicine
Clark Center
East Wing E150
318 Campus Drive
Stanford
CA 94305
USA

Suresh Shrestha
Department of Chemistry & Chemical Biology
Indiana University –
Purdue University Indianapolis
Indianapolis
IN 46202
USA

Yoshio Umezawa
Department of Chemistry
School of Science
The University of Tokyo
Hongo
Bunkyo-ku
Tokyo 113-0033
Japan

Monique Verhaegen
Department of Dermatology
University of Michigan
Cancer and Geriatrics Center
Ann Arbor
MI 48109
USA

Vadim R. Viviani
Departamento de Biologia Celular e Molecular
Instituto de Biociências
Universidade Estadual Paulista (UNESP)
Rio Claro
SP 13506-900
Brazil

Hui Zhao
Departments of Pediatrics
Radiology and Microbiology &
Immunology
Stanford School of Medicine
Clark Center
East Wing E150
318 Campus Drive
Stanford
CA 94305
USA

1
The Photoproteins

Osamu Shimomura

1.1
Discovery of Photoprotein

In 1961, we found an unusual bioluminescent protein in the jellyfish *Aequorea* and named it "aequorin" after its genus name (Shimomura et al. 1962). The protein had the ability to emit light in aqueous solutions merely by the addition of a trace of Ca^{2+}. Surprisingly, it luminesced even in the absence of oxygen. After some studies, we discovered that the light is emitted by an intramolecular reaction that takes place inside the protein molecule, and that the total light emitted is proportional to the amount of protein luminesced. At that time, we simply thought that aequorin was an exceptional protein accidentally made in nature.

In 1966, however, we found another unusual bioluminescent protein in the parchment tubeworm *Chaetopterus* (Shimomura and Johnson 1966). This protein emitted light when a peroxide and a trace of Fe^{2+} were added in the presence of oxygen, without the participation of any enzyme. The total light emitted was again proportional to the amount of the protein used. These two examples were clearly out of place in the classic concept of the luciferin–luciferase reaction of bioluminescence, wherein luciferin is customarily a relatively heat stable, diffusible organic substrate and luciferase is an enzyme that catalyzes the luminescent oxidation of a luciferin.

Considering the possible existence of many similar bioluminescent proteins in luminous organisms, we have introduced the new term "photoprotein" as a convenient, general term to designate unusual bioluminescent proteins such as aequorin and the *Chaetopterus* bioluminescent protein (Shimomura and Johnson 1966). Thus, "photoprotein" is a general term for the bioluminescent proteins that occur in the light organs of luminous organisms as the major luminescent component and are capable of emitting light in proportion to the amount of protein (Shimomura 1985). The proportionality of the light emission makes a clear distinction between a photoprotein and a luciferase. In a luciferin–luciferase luminescence reaction, the total amount of light emitted is proportional to the amount of luciferin, not to the amount of luciferase. If a luciferin is a protein,

Photoproteins in Bioanalysis. Edited by Sylvia Daunert and Sapna K. Deo

ISBN: 3-527-31016-9

Table 1.1 Photoproteins that have been isolated.

Source	*Name*	M_r	*Requirements for luminescence*	*Luminescence maximum (nm)*
Protozoa				
Thalassicola sp.[a]	Thalassicolin		Ca^{2+}	440
Coelenterata				
Aequorea aequorea[b]	Aequorin	21 500	Ca^{2+}	465
Halistaura sp.[c]	Halistaurin		Ca^{2+}	470
Phialidium gregarium	Phialidin[d], clytin[e]	23 000 21 600	Ca^{2+} Ca^{2+}	474
Obelia geniculata	Obelin	21 000[f]	Ca^{2+}	475[h]
Obelia geniculata	Obelin	21 000		485[i]
Obelia longissima	Obelin	22 200[g]		495[i]
Ctenophora				
Mnemiopsis sp.[j]	Mnemiopsin-1 Mnemiopsin-2	24 000 27 500	Ca^{2+} Ca^{2+}	485 485
Beroe ovata[j]	Berovin	25 000	Ca^{2+}	485
Annelida				
Chaetopterus variopedatus[k]		120 000 184 000	Fe^{2+}, hydro-peroxide, and O_2	455
Hamothoe lunulata[l]	Polynoidin	500 000	Fe^{2+}, H_2O_2, and O_2	510
Mollusca				
Pholas dactylus[m]	Pholasin	34 600	Peroxidase or Fe^{2+}, plus O_2	490
Symplectoteuthis oualaniensis[n]	Symplectin	60 000	Alkaline pH? and O_2	470[o]
Symplectoteuthis luminosa[p]		50 000	Catalase, H_2O_2, and O_2	
Diplopoda				
Luminodesmus sequoia[q]		60 000	ATP, Mg^{2+}, and O_2	496
Echinodermata				
Ophiopsila californica[r]		45 000	H_2O_2	482

a) Campbell et al. 1981
b) Shimomora 1986b
c) Shimomura et al. 1963
d) Levine and Ward 1982
e) Inouye and Tsuji 1993
f) Stephanson and Sutherland 1981
g) Illarionov et al. 1995
h) Morin and Hastings 1971
i) Markova et al. 2002
j) Ward and Seliger 1974
k) Shimomura and Johnson 1969
l) Nicolas et al. 1982
m) Michelson 1978
n) Fujii et al. 2002
o) Tsuji et al. 1981
p) Shimomura unpublished
q) Shimomura 1981, 1984
r) Shimomura 1986a

the luciferin is a photoprotein, regardless of the existence or nonexistence of a specific luciferase.

A photoprotein could be an extraordinarily stable form of enzyme–substrate complex, more stable than its dissociated forms, an enzyme and a substrate. Because of its high stability, a photoprotein, rather than its dissociated forms, occurs as the primary light-emitting component in the light organs. For example, the light organs of the jellyfish *Aequorea* contain aequorin, which is highly stable in the absence of Ca^{2+}, but its components coelenterazine and apoaequorin, both unstable, are hardly detectable in any part of the jellyfish. In the cells of luminous bacteria, the bacterial luciferase forms an intermediate by reacting with $FMNH_2$ and O_2, and this intermediate emits light when a fatty aldehyde is added (Hastings and Gibson 1963; Hastings and Nealson 1977). However, this intermediate is unstable and short-lived; thus, it does not fit the definition of photoprotein.

Presently, there are about 30 different types of bioluminescent systems for which substantial biochemical knowledge is available. About half of these types involve a photoprotein (Table 1.1). These photoproteins include the Ca^{2+}-sensitive type from various coelenterates (aequorin, obelin, etc.); the superoxide-activation types from a scale worm (polynoidin) and the clam *Pholas* (pholasin); the H_2O_2-activation type from a brittle star (*Ophiopsila*); and the ATP-activation type from a Sequoia millipede (*Luminodesmus*). For analytical applications, the photoproteins of the Ca^{2+}-sensitive type and the superoxide-sensitive type (pholasin) have been utilized, and the photoprotein aequorin has been in extensive use in various biological studies for the past 35 years. Each of the various photoproteins are briefly described in the next section, followed by a discussion on the extraction and purification of photoproteins and a more detailed account on the photoprotein aequorin.

1.2 Various Types of Photoproteins Presently Known

1.2.1 Radiolarian (Protozoa) Photoproteins

A Ca^{2+}-sensitive photoprotein that resembles coelenterate photoproteins was isolated from the radiolarian *Thalassicola* sp., but its properties were not investigated in detail (Campbell et al. 1981). It is of interest as the only known example of a Ca^{2+}-sensitive photoprotein other than the coelenterate photoproteins.

1.2.2 Coelenterate Photoproteins

Several kinds of photoprotein, including aequorin and obelin, were isolated from hydrozoan jellyfishes and hydroids. All of them emit blue light when Ca^{2+} is added, regardless of the presence or absence of oxygen. The coelenterate photoproteins are suitable for use in the detection and measurement of trace amounts of Ca^{2+}, and aequorin has been widely used in the studies of Ca^{2+} in various biological systems, including single cells (Blinks et al. 1976; Ashley and Campbell 1979). The overwhelming popularity of this type of photoprotein compared with the other types sometimes leads to the misconception that the photoproteins are Ca^{2+}-sensitive bioluminescent proteins.

Detailed studies have been made with only three kinds of photoproteins: aequorin obtained from *Aequorea*, obelin obtained from *Obelia*, and phialidin (clytin) obtained from *Phialidium*. Aequorin was isolated from *Aequorea aequorea* by Shimomura et al. (1962), and its purification was described in several papers (Shimomura and Johnson 1969, 1976; Blinks et al. 1978). The recombinant form of aequorin has been made (Inouye et al. 1985, 1986; Prasher et al. 1985, 1987). Obelin was isolated from *Obelia geniculata* (Campbell 1974) and also from *O. australis* and *O. geniculata* (Stephenson and Sutherland 1981). Its recombinant form was prepared by Illarionov et al. (2000). Phialidin was isolated from *Phialidium gregarium* by Levine and Ward (1982) and was cloned by Inouye and Tsuji (1993); the recombinant protein was named clytin.

All coelenterate photoproteins have a molecular weight close to 20 000. The concentrated solutions of purified photoproteins are slightly yellowish (weak absorption at about 460 nm) and non-fluorescent except for ordinary protein fluorescence. After Ca^{2+}-triggered luminescence, the solutions turn colorless and become fluorescent in blue (emission λ_{max} 460 nm). The intensity of the blue fluorescence is dependent on the concentrations of the spent protein and Ca^{2+}; however, the fluorescence intensity is not proportional to the concentration of the protein (Morise et al. 1974). In the case of aequorin, the emission spectrum of blue fluorescence is almost superimposable on the emission spectrum of Ca^{2+}-triggered luminescence, suggesting that the blue fluorescent chromophore formed in the luminescence reaction is probably the light emitter (Shimomura and Johnson 1970).

As a Ca^{2+} indicator, aequorin is useful at a concentration range of Ca^{2+} between $10^{-7.5}$ M and $10^{-4.5}$ M (Blinks et al. 1978), whereas obelin is useful at a range between $10^{-6.5}$ M and $10^{-3.5}$ M under similar conditions (Stephenson and Sutherland 1981). The Ca^{2+} sensitivity of phialidin is about equal to that of obelin (Shimomura and Shimomura 1985). It should be noted here that the Ca^{2+} sensitivity and certain other properties of aequorin, and probably of all coelenterate photoproteins, can be modified by replacing the coelenterazine moiety of the photoprotein with its analogues (explained later).

The chemistry of the bioluminescence reaction of aequorin has been elucidated in considerable detail and will be described later in this chapter. The reaction mechanisms of all hydrozoan photoproteins are believed to be essentially identical with that of aequorin. However, the luminescence reaction differs in luminous anthozoans, which are taxonomically closely related to hydrozoan. The luminous species of anthozoans contain a luciferin (coelenterazine) and a species-specific luciferase instead of a photoprotein, although the presence of a small amount of Ca^{2+}-sensitive photoprotein is suspected in some species, such as the sea pen *Ptilosarcus gurneyi* and the sea cactus *Cavernularia obesa* (Shimomura and Johnson 1979b).

Spent aequorin that has been luminesced with Ca^{2+} can be regenerated into the active original form by incubation with coelenterazine in the presence of O_2 and a low concentration of 2-mercaptoethanol (Shimomura and Johnson 1975a). The regenerated aequorin is indistinguishable from the original aequorin in every aspect of its properties. The yield of the regeneration is practically 100% when the protein concentration is over 0.1 mg mL^{-1} (Shimomura and Shimomura 1981). Thus, a sample of aequorin can be luminesced and recharged repeatedly. The regeneration of spent photoprotein takes place also with obelin (Campbell et al. 1981), as well as with halistaurin and phialidin (unpublished results).

1.2.3 Ctenophore Photoproteins

The photoproteins mnemiopsin and berovin were isolated from *Mnemiopsis* sp. and *Beroe lovata*, respectively (Ward and Seliger 1974). They are Ca^{2+}-sensitive photoproteins that are similar to aequorin, except that these photoproteins are photosensitive. The absorption maximum of mnemiopsin-2 is 435 nm, which is about 20 nm shorter than that of aequorin. The photosensitivity of ctenophore photoproteins is strikingly different from that of aequorin. Mnemiopsin and berovin are extremely sensitive to light (Hastings and Morin 1968), being easily inactivated by a broad spectral range of light (wavelength 230–570 nm) (Ward and Seliger 1976). Aequorin and other hydrozoan photoproteins are not affected by light.

Photoinactivated mnemiopsin, as well as spent mnemiopsin after Ca^{2+}-triggered luminescence, can be regenerated into its active form by incubation with coelenterazine in the presence of oxygen, like aequorin; however, the regeneration takes place only at a narrow pH range around 9.0 (Anctil and Shimomura 1984).

1.2.4 Pholasin (*Pholas* Luciferin)

The boring clam *Pholas dactylus* is historically important in the field of bioluminescence because it was one of the two luminous species with which Dubois first demonstrated luciferin–luciferase luminescence in 1887. Thus, the luminescence of *Pholas* was originally considered to be a luciferin–luciferase reaction involving *Pholas* luciferin and *Pholas* luciferase. However, *Pholas* luciferin is a glycoprotein with a molecular weight of 34 600 (Henry et al. 1973a, 1973b; Michelson 1978). Therefore, it is appropriate to call this luciferin a photoprotein. The name “pholasin” was first used by Roberts et al. (1987).

The ultraviolet absorption spectrum of pholasin shows a bulge at about 325 nm, in addition to the protein peak at 280 nm. Pholasin emits light (λ_{max} 490 nm) in the presence of various substances such as *Pholas* luciferase, ferrous ions, H_2O_2, peroxidase, superoxide anions, hypochlorite, and certain other oxidants, all in the presence of molecular oxygen (Henry and Michelson 1970; Henry et al. 1970, 1973a, 1973b; Müller and Campbell 1990). Thus, *Pholas* luciferase is clearly not an essential component for the luminescence of pholasin. The luminescence reaction of pholasin with *Pholas* luciferase is optimum at pH 8–9 and at an ionic strength of about 0.5 M, giving a quantum yield of 0.09 for pholasin (Michelson 1978). According to Reichl et al. (2000), the addition of horseradish peroxidase compounds I and II to pholasin induces an intense luminescence. Moreover, the addition of H_2O_2 to a mixture of myeloperoxidase and pholasin gives an intense burst of light. The chromophore of pholasin is still not chemically identified.

The cloning and expression of apopholasin was achieved by Dunstan et al. (2000), but attempts to reconstitute the recombinant apopholasin into pholasin by the addition of an acidic methanol extract of *Pholas* failed, although the mixture gave luminescence by the addition of sodium hypochlorite. Pholasin is commercially available from Knight Scientific, Plymouth, UK. The main application of pholasin is the measurement of oxygen radicals.

1.2.5
Chaetopterus Photoprotein

The photoprotein of the parchment tubeworm *Chaetopterus variopedatus* purified by chromatography has a molecular mass of approx. 120–130 kDa (Shimomura and Johnson 1966, 1968). The protein is amorphous when precipitated with ammonium sulfate, but it can be converted into a crystalline form with an increased molecular mass of 184 kDa by slow crystallization with ammonium sulfate. The photoprotein emits light in the presence of Fe^{2+}, a peroxide, and molecular oxygen. As a peroxide, H_2O_2 can be used, but an unidentified hydroperoxide existing in old dioxane or tetrahydrofuran was far more effective. Two kinds of additional activators were found to give brighter luminescence, but they were not identified. The light emission of this photoprotein is strongly affected by the pH of the medium, showing a peak at pH 7.7 with a sharp decrease at both sides (50% decreases at pH 6.5 and pH 8.3); the light intensity is not significantly influenced by the salt concentration up to 1 M when tested with NaCl. The optimum temperature for the luminescence intensity is 22 °C. With this photoprotein, a concentration of Fe^{2+} as low as 0.1 μM can be detected.

The purified photoprotein is practically colorless, and its absorption spectrum shows, in addition to the 280-nm protein absorption peak, a very slight absorption in the region of 330–380 nm, although its significance is unclear. A solution of the photoprotein is moderately blue fluorescent, with a fluorescence emission maximum at 453–455 nm and an excitation maximum at 375 nm, and these peaks do not significantly change after the luminescence reaction. The luminescence spectrum of purified photoprotein (λ_{max} 453–455 nm) closely matched with the fluorescence emission spectrum.

1.2.6
Polynoidin

A membrane photoprotein isolated from the scales of the scale worm *Harmothoe lunulata* was named "polynoidin" (Nicolas et al. 1982). The purified photoprotein (M_r 500 000) emits light in the presence of molecular oxygen (λ_{max} 510 nm) by the action of sodium hydrosulfite, the xanthine–xanthine oxidase system, Fenton's reagent (H_2O_2 plus Fe^{2+}), or other reagents that produce superoxide radicals. The photoprotein luminescence was 30% brighter in phosphate buffer than in Tris buffer, and the luminescence response was significantly increased by including a complexing agent such as EGTA. However, the injection of polynoidin solution into the mixture of H_2O_2 and Fe^{2+} failed to produce light; Fe^{2+} must be added last to initiate light emission.

The photoprotein is not fluorescent (except for usual protein fluorescence) after the bioluminescence reaction or before the reaction. The requirements for its luminescence reaction are similar to that of the bioluminescence systems of *Pholas* and *Chaetopterus*, suggesting the involvement of a common basic mechanism in these luminescence systems.

1.2.7
Symplectin

The luminescent substance of the squid *Symplectoteuthis oualaniensis* was first obtained in the form of insoluble particles by Tsuji et al. (1981). The suspension of the particles emitted light in the presence of monovalent cations such as K^+, Rb^+, Na^+, Cs^+, NH_4^+, and Li^+ (in decreasing order of effect). Molecular oxygen was needed for the luminescence. Divalent ions such as Ca^{2+} and Mg^{2+} did not trigger light emission. The light emission (λ_{max} 470 nm) was optimal in the presence of 0.6 M KCl or NaCl and at a pH of 7.8.

The soluble form of the *Symplectoteuthis* photoprotein was isolated and purified from the granular light organs of the squid and was named "symplectin" (Takahashi and Isobe 1994; Fujii et al 2002). The light organs were first extracted with a pH 6 buffer containing 0.4 M KCl to remove impurities, and then symplectin was extracted from the residue with a pH 6 buffer containing 0.6 M KCl. All solutions used in the experiments contained 0.25 M sucrose, 1 mM dithiothreitol, and 1 mM EDTA. The 0.6 M KCl extract was chromatographed by size-exclusion HPLC on a TSK G3000SW column. Symplectin was eluted as two major components of oligomers, having molecular masses of 200 kDa or more, and a minor component of monomer (60 kDa). All processes of extraction and purification were carried out at 4 °C. Warming up a solution of symplectin, adjusted to pH 8, to room temperature causes the luminescence reaction to begin, and the light emission lasts for hours.

A tryptic digestion of the KCl extract increased the content of the 60-kDa species at the expense of the two high-molecular-weight species, accompanied by the formation of 40-kDa and 16-kDa species. SDS-PAGE analysis of the two high-molecular-weight oligomers revealed that they consist mainly of the 60-kDa protein. The 60-kDa protein and the 40-kDa protein were fluorescent in the SDS-PAGE analysis. The spent protein of symplectin after luminescence (aposymplectin) could be reconstituted into original symplectin by treatment with dehydrocoelenterazine (Isobe et al. 2002).

1.2.8 *Luminodesmus* Photoprotein

This is presently the only example of a photoprotein of terrestrial origin. The millipede *Luminodesmus sequoia* (Loomis and Davenport 1951) emits light from the surface of its whole body continuously day and night. The photoprotein extracted and purified from this organism emits light (λ_{max} 496 nm) when ATP and Mg^{2+} are added in the presence of molecular oxygen (Hastings and Davenport 1957; Shimomura 1981). Thus, the luminescence system of *Luminodesmus* resembles that of the fireflies in that it requires ATP and Mg^{2+}, but it differs in that it needs only the photoprotein rather than the luciferin and luciferase required in the firefly system. The molecular weight of the photoprotein is 104 000, which is close to the molecular weight of firefly luciferase reported earlier (100 000) but larger than its newer value (62 000; Wood et al. 1984). Although it was suspected that the photoprotein might be a complex of a firefly-type luciferase and firefly luciferin, firefly luciferin itself was not detected in this photoprotein. Recently, the possible presence of a porphyrin chromophore in the photoprotein has been suggested, although the role of this chromophore in the light-emitting reaction is unclear (Shimomura 1984). Using the luminescence system of *Luminodesmus*, 0.01 μM ATP and 1 μM Mg^{2+} can be detected.

1.2.9 *Ophiopsila* Photoprotein

The brittle star *Ophiopsila californica* is abundant around Catalina Island, off the coast of Los Angeles (Shimomura 1986a). An animal of average size weighs about 3–4 g, and has five arms of about 10 cm long. The purified photoprotein luminesces in the presence of H_2O_2, emitting a greenish-blue light (λ_{max} 482 nm). Molecular oxygen is probably not needed for the luminescence reaction. The molecular weight of *Ophiopsila* photoprotein is estimated to be about 45 000 by gel filtration. The absorption spectrum of a solution of the photoprotein showed a small peak (λ_{max} 423 nm, with a shoulder at about 450 nm) in addition to the 280-nm protein peak. The 423-nm peak decreased slightly through the H_2O_2-triggered luminescence reaction, accompanied by a slight red shift of the peak. The photoprotein was fluorescent in greenish-blue (emission λ_{max} 482 nm; excitation λ_{max} 437 nm), and the fluorescence emission spectrum exactly coincided with the luminescence spectrum of photoprotein in the presence of H_2O_2, suggesting the possibility that the fluorescent chromophore might be the light emitter. However, the fluorescence emission of the photoprotein did not show any detectable change after the H_2O_2-triggered luminescence reaction; an anticipated increase in the 482-nm fluorescence did not occur.

1.3
Basic Strategy of Extracting and Purifying Photoproteins

Photoproteins are usually highly reactive, unstable substances, like luciferins. Their luminescence activities are easily lost by spontaneous light emission and various other causes. In isolating active photoproteins, it is extremely important to pay special attention to prevent the loss of the luminescence activity. Compared with the isolation of luciferins, however, techniques available for isolating photoproteins are somewhat limited because of their protein nature.

The basic principle is to extract a photoprotein in an aqueous solution and purify the photoprotein by various means of protein purification, all under conditions that prevent the luminescence and denaturation of the protein molecules. Thus, the luminescence system must be reversibly inhibited during the extraction and purification of a photoprotein. The method of reversible inhibition differs depending on the nature and cofactor requirement of the system to be isolated. For example, the calcium chelator EDTA or EGTA is used to inhibit the luminescence of the Ca^{2+}-sensitive photoproteins of coelenterates and ctenophores such as aequorin, obelin, and mnemiopsin (Shimomura et al.,1962; Campbell 1974; Hastings and Morin 1968; Ward and Seliger 1974). Before the discovery of the Ca^{2+} requirement, however, aequorin was extracted with a pH 4 buffer that reversibly inactivated the photoprotein (Shimomura et al. 1962; Shimomura 1995c). In the case of the luminescence systems of *Chaetopterus* and *Pholas*, the metal ion inhibitors 8-hydroxyquinoline and diethyldithiocarbamate, respectively, were used (Shimomura and Johnson 1966; Henry and Monny 1977).

The ionic strength and the pH of buffers are also important, and these conditions should be chosen to optimize the yield of active photoprotein. The use of acidic buffers, pH 5.6–5.8, was effective in suppressing spontaneous luminescence during the extraction of the photoproteins of euphausiids and *Luminodesmus* (Shimomura and Johnson 1967; Shimomura 1981). In the case of the membrane photoprotein polynoidin and the squid photoprotein symplectin, easily soluble impurities were all washed out and the substances that cause the luminescence of the photoprotein are completely removed before the solubilization of the photoproteins; thus, inhibitors were not needed (Nicolas et al. 1982; Fujii et al. 2002).

1.4
The Photoprotein Aequorin

1.4.1
Extraction and Purification of Aequorin

Aequorin is the best-known photoprotein and has been used widely in various applications. The first step in the extraction of aequorin from the jellyfish *Aequorea* (average body weight 50 g) is to cut off the circumferential margin of umbrella that contains light organs, making about 2-mm-wide strips commonly called "rings". This process is important because it eliminates about 99% of unnecessary body parts that do not contain aequorin. The rings can be made efficiently by using specially made cutting devices (Johnson and Shimomura 1978; Blinks et al. 1978) or, much less efficiently, with scissors. The rings (about 0.5 g each) containing light organs are kept in cold seawater. Then, about 500 rings are shaken vigorously by hand with cold, saturated ammonium sulfate solution containing 50 mM EDTA (Johnson and Shimomura 1972) or with cold seawater (Blinks et al. 1978) to dislodge the particles of light organs from the rings. Then, the rings are removed by filtering through a net of Dacron or Nylon (50–100 mesh), and the light organ particles suspended in the filtrate are collected by filtration on a Büchner funnel with the aid of some Celite. The light organ particles in the filter cake are cytolyzed and the aequorin therein is extracted by shaking with cold 50 mM EDTA (pH 6.5). After filtration, crude aequorin is precipitated by saturation with ammonium sulfate.

Of the two methods of shaking the rings mentioned above, using seawater results in much cleaner crude extracts, with a little less yield, than are obtainable by shaking in saturated ammonium sulfate containing EDTA. On the other hand, saturated ammonium sulfate strongly inhibits the luminescence response of the photogenic particles to mechanical stimulation such as shaking and stirring, and it also salts out and stabilizes aequorin, thus resulting in a better yield of aequorin and less effect on the isoform composition of aequorin extracted, compared with that obtainable by shaking in seawater.

With regard to the purification of aequorin, Blinks et al. (1978) described a well-designed method for purifying an aequorin extract that has been obtained by the "seawater shaking method". The method included gel filtration on Sephadex G-50 and ion-exchange chromatography on DEAE-Sephadex A-50 and QAE-Sephadex A-50. The ion exchangers effectively separated the green fluorescent protein from aequorin. For the purification of the extract obtained by the "saturated ammonium sulfate shaking method", gel filtration on a column of Sephadex G-75 or G-100, using buffers containing 1 M ammonium sulfate and not containing ammonium sulfate, and ion-exchange chromatography on DEAE cellulose have been used (Johnson and Shimomura 1972; Shimomura and Johnson 1969, 1976). Aequorin in 1 M ammonium sulfate aggregates to a larger size ($M_r > 50\,000$). Thus, crude aequorin is first chromatographed on Sephadex G-100 with a low-salt buffer not containing ammonium sulfate, then the aequorin fraction obtained is rechromatographed on the same column using a buffer containing 1 M ammonium sulfate to obtain purified aequorin. Using Sephadex G-50 is not recommended in this case, at least for the initial step, because of the presence of a large amount of the aggregated form of impurities.

1.4.1.1 Hydrophobic Interaction Chromatography

Butyl-Sepharose 4 Fast Flow (Pharmacia) is an excellent medium for purifying aequorin, supplementing the methods described above. Aequorin in a buffer solution containing 5–10 mM EDTA and 1.8 M ammonium sulfate is adsorbed on the column, and then aequorin is eluted with a buffer containing decreasing concentrations of ammonium sulfate and 5 mM EDTA. Aequorin elutes at an ammonium sulfate concentration between 1 M and 0.5 M.

Because apoaequorin elutes at ammonium sulfate concentrations lower than 0.1 M, aequorin is cleanly separated from apoaequorin. Thus, it is possible to prepare virtually pure samples of aequorin using a single column as follows. The aequorin sample is first luminesced by the addition of a sufficient amount of Ca^{2+}. The spent solution, after dissolving 1 M ammonium sulfate, is adsorbed on a column of butyl-Sepharose 4. The apoaequorin adsorbed on the column is eluted with decreasing concentration of ammonium sulfate starting from 1 M; apoaequorin elutes at an ammonium sulfate concentration lower than 0.1 M. The apoaequorin eluted is regenerated with coelenterazine in the presence of 5 mM EDTA and 5 mM 2-mercaptoethanol (see the section 1.4.5.3). The solution of regenerated aequorin in 1.8 M ammonium sulfate is adsorbed on the same butyl-Sepharose 4 column. Aequorin adsorbed on the column is eluted with a decreasing concentration of ammonium sulfate, yielding highly purified aequorin.

1.4.2 Properties of Aequorin

Aequorin is a conjugated protein that has a relative molecular mass of approximately 20 000–21 000 (Blinks et al. 1976), and it contains a functional chromophore corresponding to roughly 2% of the total weight. A concentrated solution of aequorin is yellowish because of its absorption peak (λ_{max} about 460 nm), in addition to a protein absorption peak at 280 nm ($A_{1\%,1cm}$ 30.0; Shimomura 1986b). Aequorin is non-fluorescent, except for a weak ultraviolet fluorescence that is due to its protein moiety. Natural aequorin is a mixture of isoforms, containing more than a dozen of them, designated aequorins A, B, C, etc. (Blinks et al. 1976; Shimomura 1986b). The isoelectric points of these isoforms lie between 4.2 and 4.9 (Blinks and Harrer 1975). The solubility of aequorin in aqueous buffers is generally greater than 30 mg mL^{-1} (Shimomura and Johnson 1979a). Aequorin can be salted out from aqueous buffers with ammonium sulfate, although the salting out is not complete even after the complete saturation of ammonium sulfate. Usually 1–2% of aequorin remains in the solution.

One milligram of aequorin emits 4.3–5.0 × 10^{15} photons at 25 °C when Ca^{2+} is added, at a quantum yield of 0.16 (Shimomura and Johnson 1969, 1970, 1979a; Shimomura 1986b). In the presence of an excess of Ca^{2+}, the luminescence reaction of aequorin has a rate constant of 100–500 s^{-1} for the rise and 0.6–1.25 s^{-1} for the decay (Loschen and Chance 1971; Hastings et al. 1969).

1.4.2.1 **Stability**

Aequorin is always emitting a low level of luminescence, spontaneously deteriorating by itself. Thus, the information concerning its stability is important when aequorin is used as a calcium probe. The stability of aequorin in aqueous solutions containing EDTA or EGTA varies widely by temperature, pH, concentration of salts, and impurities. To minimize the deterioration of aequorin, it is most important to keep the temperature as low as possible. The half-life of aequorin in 10 mM EDTA, pH 6.5, is about 7 days at 25 °C. At room temperature, aequorin is most stable in solutions containing 2 M ammonium sulfate or when it is precipitated from saturated ammonium sulfate. Freeze-dried aequorin is also stable, but the process of drying always causes a loss of luminescence activity (see below). All forms of aequorin are satisfactorily stable for many years at −50 °C or below, but all deteriorate rapidly at temperatures above 30–35 °C. A solution of aequorin should be stored frozen whenever possible because repeated freeze–thaw cycles do not harm aequorin activity.

1.4.2.2 **Freeze-drying**

A note on freeze-dried aequorin may be appropriate here, because most commercial preparations of aequorin are sold in a dried form. The process of freeze-drying aequorin always results in some loss of luminescence activity. Therefore, aequorin should not be dried if a fully active aequorin is required. The loss is about 5% at the minimum, typically about 10%. The loss can be slightly lessened by certain additives; the addition of 50–100 mM KCl and some sugar (50–100 mM) in the buffer seems to be beneficial. The buffer composition used for the freeze-drying of aequorin at the author's laboratory is as follows: 100 mM KCl, 50 mM glucose, 3 mM HEPES, 3 mM Bis-Tris, and 0.05 mM EDTA, pH 7.0.

1.4.3 Specificity to Ca^{2+}

Several kinds of cations other than Ca^{2+} elicit the light emission of aequorin. Some lanthanide ions (such as La^{3+} and Y^{3+}) trigger the luminescence of aequorin as efficiently as Ca^{2+}. In addition, Sr^{2+}, Pb^{2+}, and Cd^{2+} cause significant levels of luminescence; Cu^{2+} and Co^{2+} give some luminescence only in slightly alkaline buffer. However, Be^{2+}, Ba^{2+}, Mn^{2+}, Fe^{2+}, Fe^{3+}, and Ni^{2+} do not elicit any light from aequorin (Shimomura and Johnson 1973). In testing biological systems, however, aequorin is considered to be highly specific to Ca^{2+}, because the occurrence of a significant amount of metal ions other than Ca^{2+} is unlikely. In an *in vitro* test, all of these metal ions except Ca^{2+}, Sr^{2+}, and lanthanoids could be completely masked by including 1 mM sodium diethyldithiocarbamate in the test solution (Shimomura and Johnson 1975b).

1.4.4
Luminescence of Aequorin by Substances Other Than Divalent Cations

As already mentioned, all forms of aequorin emit photons spontaneously and constantly, regardless of its molecular status or environment conditions, even in the absence of Ca^{2+} or in the presence of a large excess of EDTA. The light emission results in a gradual deterioration of the luminescence capability of aequorin. A luminescence intensity of this type is quite low at 0 °C, though it can be easily measured with a light meter. The intensity is temperature dependent and steeply increases with rising temperature, reaching a maximum intensity at around 60 °C (Shimomura and Johnson 1979a). Such a temperature-dependent luminescence occurs with aequorin dissolved in aqueous solutions, as well as with freeze-dried aequorin and its suspension in a certain organic solvents, such as toluene, acetone, and diglyme (bis-2-methoxyethyl ether). The quantum yield of the spontaneous luminescence of dried aequorin, when warmed with or without an organic solvent, is generally in the range of 0.003–0.005, whereas that of aequorin in aqueous solutions is considerably less (about 0.001 at 43 °C).

Aequorin also emits luminescence in the presence of thiol-modification reagents such as *p*-benzoquinone, Br_2, 1_2, *N*-bromosuccinimide, *N*-ethylmaleimide, iodoacetic acid, and *p*-hydroxymercuribenzoate (Shimomura et al. 1974). The luminescence is probably caused by the conformational change of the protein that results from the modification of cysteine residues (by causing the decomposition of the coelenterazine peroxide moiety). The luminescence is weak but lasts for more than one hour. The quantum yields in this type of luminescence never exceed 0.02 (about 15% of Ca^{2+}-triggered luminescence) at 23–25 °C. To prevent this type of luminescence, any reagents that might react with an SH group should be avoided.

1.4.5
Mechanism of Aequorin Luminescence and Regeneration of Aequorin

1.4.5.1 Structure of Aequorin

Aequorin is a globular protein with three "EF-hand" domains to bind Ca^{2+}, and it accommodates a peroxidized coelenterazine in the central cavity of the protein (Head et al. 2000). The presence of a peroxy group bound to position 2 of the coelenterazine moiety was previously suggested (Shimomura and Johnson 1978) and confirmed by ^{13}C nuclear magnetic resonance spectroscopy (Musicki et al. 1986). The protein conformation of aequorin is much more compact and rigid than that of apoaequorin, consistent with the results of the fluorescence polarization studies and the papain digestion of those proteins (La and Shimomura 1982). The functional group, peroxidized coelenterazine, is shielded from outside solvent. Therefore, no reagent can react with this group without first reacting with the residues of the protein, and any reaction with the protein residues triggers the breakdown of the peroxidized coelenterazine.

1.4.5.2 Luminescence Reaction

In the case of aequorin reacting with Ca^{2+}, a conformational change of protein takes place when one molecule of aequorin is bound with two Ca^{2+} ions (Shimomura 1995b). The conformational change results in the cyclization of the peroxide of coelenterazine into the corresponding dioxetanone, which instantly decomposes and produces the excited state of coelenteramide and CO_2 (Shimomura et al. 1974; Shimomura and Johnson 1978). When the energy level of the excited state of coelenteramide falls to ground state, light is emitted. A simplified mechanism of the luminescence reaction is illustrated in Fig. 1.1.

The spent solution of the luminescence reaction of aequorin is a mixture of coelenteramide, apoaequorin, and Ca^{2+} that forms a complex called "blue fluorescent protein" (fluorescence emission maximum about 465–470 nm). The dissociation constant of the complex into coelenteramide plus apoaequorin in the presence of 0.5 mM Ca^{2+} is 7×10^{-6} M at pH 7.4 and 25 °C (Morise et al. 1974; based on the molecular weight of aequorin 21 000). Thus, the luminescence

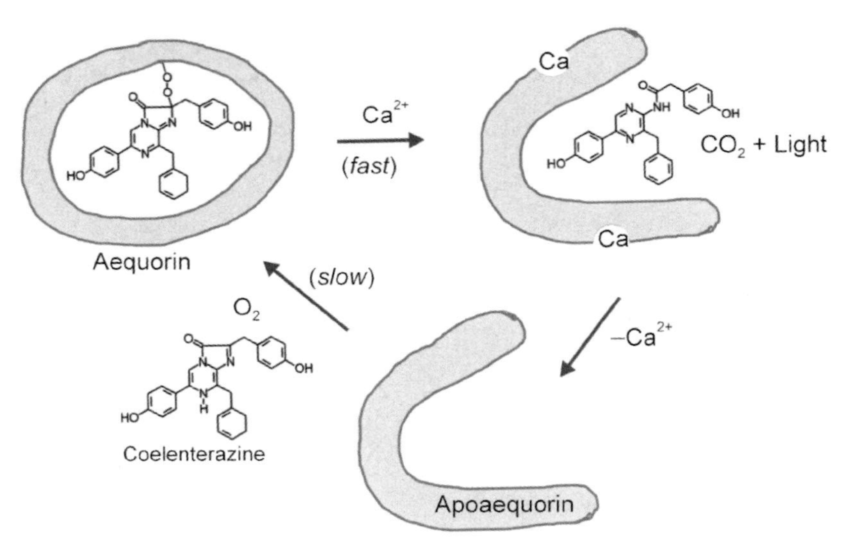

Fig. 1.1 Schematic illustration of a simplified mechanism of the luminescence and regeneration of aequorin. Aequorin (upper left) is a globular protein that contains peroxidized coelenterazine sealed in its central cavity and has three EF-hand Ca^{2+}-binding sites on the outside. When the protein is bound with two Ca^{2+} ions, an intramolecular reaction starts, resulting in the formation of coelenteramide and CO_2, accompanied by the emission of blue light (λ_{max} 460 nm) and opening of the protein shell (upper right). The protein part, apoaequorin (bottom), can be regenerated into the original aequorin by incubation with coelenterazine and molecular oxygen in the absence of Ca^{2+}. In the regeneration reaction, addition of a low concentration of 2-mercaptoethanol increases the yield of regenerated aequorin by protecting the functional cysteine residues of apoprotein.

reaction product of aequorin is usually blue fluorescent, unless the concentration of aequorin used is too low (much less than 1 μM) to form the fluorescent complex. The blue fluorescence of the complex (λ_{max} 465–470 nm) closely matches the bioluminescence emission of aequorin, giving a basis to the postulation that the fluorescent complex is the light emitter of aequorin bioluminescence (Shimomura and Johnson 1970), although it now seems an oversimplification considering that the conformation of apoaequorin continues to change for several minutes after the light emission.

When the light emission of aequorin is measured in low-ionic-strength buffers containing no inhibitor, the log–log plot of the luminescence intensity versus Ca^{2+} concentration gives a sigmoid curve having a maximum slope of about 2.0 for its middle part (Shimomura and Johnson 1976; Shimomura and Shimomura 1982), indicating that the binding of two Ca^{2+} ions to one molecule of aequorin is required to trigger the luminescence of aequorin.

1.4.5.3 **Regeneration**

Apoaequorin can be reconstituted into aequorin by incubation with coelenterazine in the presence of O_2 and 2-mercaptoethanol, which the role of the latter substance is to protect the functional sulfhydryl groups of apoaequorin during the regeneration reaction (Shimomura and Johnson 1975a). For the regeneration reaction to occur, there is no need to separate coelenteramide from apoaequorin if the material contains it. Usually, the product of the luminescence reaction is incubated at 0–5 °C in a pH 7.5 buffer solution containing 5 mM EDTA, 3 mM 2-mercaptoethanol, and an excess of coelenterazine (at least 2 μg mL^{-1} more than the calculated amount). The regeneration is usually 50% complete within 30 min and practically 100% complete after 3 h.

When the regeneration reaction of apoaequorin is carried out in the presence of an excess of free Ca^{2+}, rather than in 5 mM EDTA, the result is a continuous, weak light emission from the reaction mixture. This weak luminescence lasts many hours, differing from the short, bright flash of the Ca^{2+}-triggered luminescence of aequorin. The weak luminescence of the regeneration mixture in the presence of Ca^{2+} can be intensified several times by including 0.5% diethylmalonate in the reaction medium (Shimomura and Shimomura 1981).

During the regeneration in the presence of Ca^{2+} described above, apoaequorin appears to be acting as an enzyme that catalyzes the luminescent oxidation of coelenterazine. The mechanism involved might be a simple, straightforward one: aequorin is first formed, and then it instantly reacts with Ca^{2+} to emit light. This simple mechanism, however, has no experimental support at present; the regeneration reaction of aequorin in the presence of EDTA was not activated by diethylmalonate, suggesting either that Ca^{2+} is needed in the activation by diethylmalonate or that aequorin is not an intermediate in the luminescence reaction in the presence of Ca^{2+} (Shimomura and Shimomura 1981). Whatever the mechanism, apoaequorin must be a very sluggish enzyme if it is an enzyme. Apoaequorin has a turnover number of 1–2 per hour (Shimomura and Johnson 1976).

1.4.6
Inhibitors of Aequorin Luminescence

All thiol-modification reagents cause weak, spontaneous luminescence of aequorin in the absence of Ca^{2+}, as already mentioned. They are in effect inhibitors of the Ca^{2+}-triggered luminescence of aequorin, because the quantum yields of aequorin in the luminescence caused by these reagents (~0.008) are much lower than that of the Ca^{2+}-triggered luminescence of aequorin (Shimomura et al. 1974).

Bisulfite, dithionite, and *p*-dimethyaminobenzaldehyde are all strongly inhibitory even at micromolar concentrations (Shimomura et al. 1962). It has been found that the functional group of aequorin, i.e., a peroxide of coelenterazine, decomposes without light emission when the photoprotein is treated with bisulfite or dithionite, resulting in the formation of a corresponding hydroxy-coelenterazine or coelenterazine (Shimomura and Johnson 1978).

A number of inorganic and organic substances at high concentrations (> 50 mM) suppress the luminescence intensity of the Ca^{2+}-triggered light emission. Thus, KCl (100–150 mM) used in physiological buffers is significantly inhibitory. Magnesium ions are inhibitory at millimolar concentrations, probably by competing with Ca^{2+} (cf. Blinks et al. 1976).

EDTA and EGTA can inhibit the Ca^{2+}-triggered luminescence of aequorin in two ways: (1) when free Ca^{2+} is removed from the reaction medium by chelation, the luminescence reaction is practically stopped; and (2) when the free (unchelated) forms of these chelators directly bind with the molecules of aequorin, inhibition results (Shimomura and Shimomura 1982; Ridgway and Snow 1983). The second type of inhibition is strong in solutions of low ionic strength and in the absence of other inhibitor ions such as Mg^{2+}, but it is relatively weak in the presence of 0.1 M KCl (Shimomura and Shimomura 1984), presumably because aequorin is already inhibited by KCl. Therefore, great care must be taken if EDTA or EGTA is to be used in the calibration of the Ca^{2+} sensitivity of aequorin; this is especially important in the case of low-ionic-strength calcium buffers. It should also be noted that in usual calcium buffers, the lower the Ca^{2+} concentration, the higher the inhibitory free chelator concentration, resulting in a slope steeper than the true slope in the log–log plot of luminescence intensity versus Ca^{2+} concentration.

1.4.7
Recombinant Aequorin

The cloning and expression of apoaequorin cDNA was accomplished by two independent groups in 1985. One of these groups analyzed the cDNA clone AQ440 they obtained and reported that apoaequorin is composed of 189 amino acid residues (M_r 21 400) with an NH_2-terminal valine and a COOH-terminus proline (Inouye et al. 1985, 1986), which is consistent with the results of the amino acid sequence analysis of native aequorin reported by Charbonneau et al. (1985). In contrast, the other group reported that the cDNA AEQ1 they obtained contains the entire protein-coding region of 196 amino acid residues, which includes seven additional residues attached to the N-terminus, and the apoaequorin expressed in *Escherichia coli* showed a molecular weight of 20 600 (Prasher et al. 1985, 1987). The recombinant aequorin of the former group did not exactly match any of the isoforms of natural aequorin in the HPLC mobilities and the properties of Ca^{2+}-triggered luminescence (Shimomura et al. 1990). No detailed comparison has been made with the recombinant aequorin obtained by the latter group, although a brief test indicated that the recombinant aequorins from both sources are practically identical.

1.4.8
Semi-synthetic Aequorins

The core cavity of the aequorin molecule can accommodate various synthetic analogues of coelenterazine in place of coelenterazine. The coelenterazine moiety in native aequorin can be replaced by a simple process. First, aequorin is luminesced by the addition of Ca^{2+}, and then the apoaequorin produced is regenerated with an analogue of coelenterazine in the presence of EDTA, 2-mercaptoethanol (or DTT), and molecular oxygen. The products are called semi-synthetic aequorins and are identified with an italic prefix (see Table 1.2). Semi-synthetic aequorins can be prepared from both native aequorin and recombinant aequorin, using various synthetic analogues of coelenterazine. A large number of coelenterazine analogues were synthesized, and about 50 kinds of semi-synthetic aequorins have been prepared and tested (Shimomura et al. 1988, 1989, 1990, 1993). Some semi-synthetic aequorins are significantly different from the native type of aequorin in various properties, including spectral characteristics (Shimomura 1995a) and sensitivity to Ca^{2+}, the rate of luminescence reaction, and the rise time of luminescence (Table 1.2). The relationship between Ca^{2+} concentration and the initial light intensity of various semi-synthetic aequorins is shown in Fig. 1.2.

As is apparent from the data of Table 1.2 and Fig. 1.2, the tolerance of the central cavity of the aequorin molecule in accommodating the coelenterazine moiety is surprisingly wide. The apparent limitations in the substitution of the coelenterazine moiety, and the changes in the Ca^{2+} sensitivity caused by the substitution, are as follows.

1. The group R^1 must be aromatic. A replacement of the original *p*-hydroxyphenyl group with a group of larger size tends to decrease Ca^{2+} sensitivity.
2. The group R^2 must be lager than the ethyl group. The replacement of the original phenyl group with a smaller non-aromatic group increases Ca^{2+} sensitivity.
3. The group R^3 must be an OH group, and no substitution is allowed on the phenyl group bearing this OH.

Table 1.2 Selected semi-synthetic aequorins derived from recombinant aequorin (Shimomura et al. 1993).

No. (Prefix)	*Structural modification of coelenterazine*[a)]	*Lumines-cence max (nm)*	*Relative luminescence capacity*[b)]	*Relative intensity at 10^{-6} or 10^{-7} M Ca^{2+}* [c)]	*Half-total time(s)*[d)]
1	None	466	1.00	1.00	M
2 (*h*)	R^1: C_6H_5	466	0.75	16	M
3 (*f*)	R^1: $C_6H_4F(p)$	472	0.80	20	M
4 (*f2*)	R^1: $C_6H_3F_2(m,p)$	470	0.80	30	M
6 (*cl*)	R^1: $C_6H_4Cl(p)$	464	0.92	0.6	5
9 (*n*)	R^1: β-naphthyl	468	0.25	0.15	5
9′ (*n*/J)[e)]	R^1: β-naphthyl	467	0.30	0.07	5
12 (*cp*)	R^2: cyclopentyl	442	0.63	28	F
13 (*ch*)	R^2: cyclohexyl	453	1.00	15	F
17 (*fb*)	R^1: $C_6H_4F(p)$, R^2: *n*-butyl	460	0.20	1100	2
19 (*hcp*)	R^1: C_6H_5, R^2: cyclopentyl	445	0.65	500	F
21 (*hch*)	R^1: C_6H_5, R^2: cyclohexyl	450	0.52	80	F
22 (*fch*)	R^1: C_6H_4F, R^2: cyclohexyl	462	0.43	73	M
23 (m5)	R^4: methyl	440	0.37	2	M
24 (*e*)	R^5: CH_2CH_2	405, 472	0.50	6	F
26 (*ef*)	R^1: $C_6H_4F(p)$, R^5: CH_2CH_2	405, 470	0.35	40	F
27 (*ech*)	R^2: cyclohexyl, R^5: CH_2CH_2	402, 440	0.40	8	F
Fluorescein-labeled[f)]		528	1.00	2	M

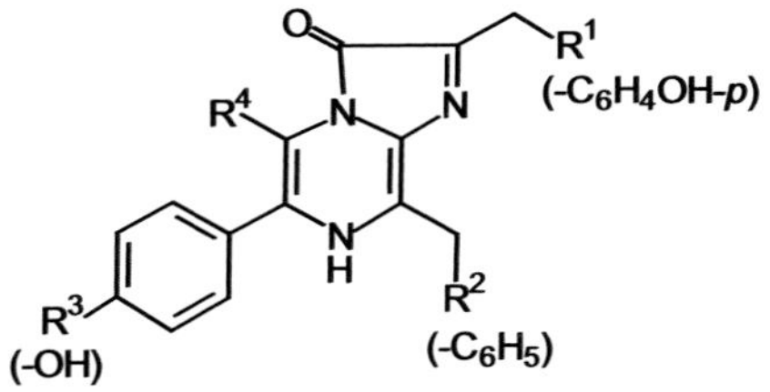

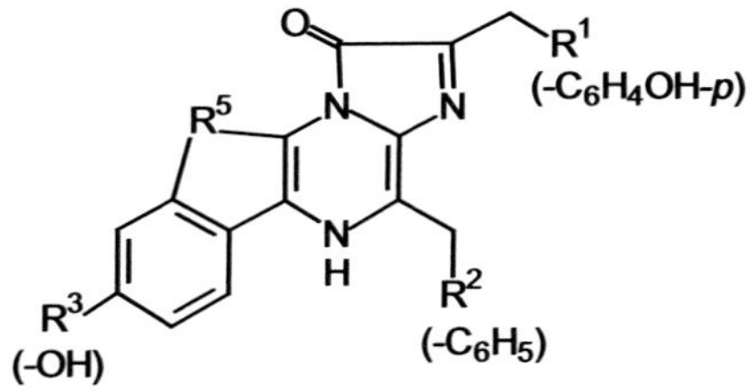

a) Only the changes from the coelenterazine structure are shown in this column. Those unchanged are shown in parentheses in the above structures.
b) The ratio in luminescence capacity: semi-synthetic aequorin/unmodified aequorin.
c) The ratio in luminescence intensity: semi-synthetic aequorin/unmodified aequorin, in 10^{-7} M Ca^{2+} for a value of 1 and larger and in 10^{-6} M Ca^{2+} for a value of less than 1.
d) The time required to emit 50% of the total light in 10 mM calcium acetate: F, 0.15–0.3 s; M, 0.4–0.8 s. The half-rise time of luminescence: F, 2–4 ms, all others, 6–20 ms.
e) Prepared from aequorin isoform J.
f) Fluorescein was chemically bound to apoaequorin, followed by regeneration using unmodified coelenterazine.

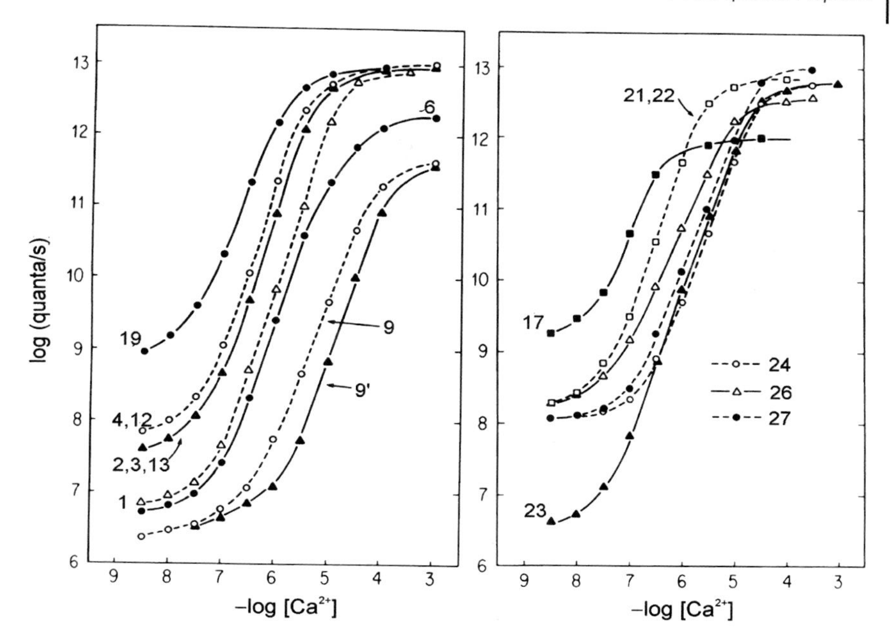

Fig. 1.2 Relationship between Ca^{2+} concentration and the initial light intensity of various recombinant semi-synthetic aequorins and *n*-aequorin J (a semi-synthetic natural aequorin made from an isoform, aequorin J). The curve number corresponds to the photoprotein number used in Table 1.2. A photoprotein (3 μg) was added to 3 mL of Ca^{2+} buffer with various pCa values (pH 7.0) containing 1 mM total EGTA, 100 mM KCl, 1 mM free Mg^{2+}, and 1 mM MOPS, at 23–24 °C. The data are taken from Shimomura et al. (1993).

1.4.8.1 *e*-Aequorins

e-Aequorins, containing a ligand of *e*-coelenterazine, show properties significantly different from other aequorins (Shimomura et al. 1988, 1989, 1990, 1993). In the structure of *e*-coelenterazines, the 5 position of the imidazopyrazinone structure is bound with the α position of the 6-(*p*-hydroxyphenyl) group through an ethylene linkage, thus restraining the two ring systems into the same plane. The luminescence reactions of *e*-aequorins are fast, with a half-rise time of 2–4 ms and a half-total time of 0.15–0.3 s, like *ch*-aequorins with an 8-cyclohexylmethyl substituent. The luminescence spectra are bimodal, with peaks at 400–405 nm and 440–475 nm. The ratio of the two peaks is variable not only with the type of aequorin but also with the measurement conditions, such as the concentration of Ca^{2+} and pH. *e*-Coelenterazines scarcely luminesce in the presence of apoaequorin, Ca^{2+}, and 2-mercaptoethanol in air (Shimomura 1995a).

References

Anctil, M., Shimomura, O. Mechanism of photoinactivation and re-activation in the bioluminescence system of the ctenophore *Mnemiopsis*. *Biochem. J.* **1984**, *221*, 269–272.

Ashley, C. C., Campbell, A. K., Eds. Detection and Measurement of Free Ca^{2+} in Cells. Elsevier/North-Holland Biomedical Press, Amsterdam, **1979**.

Blinks, J. R., Harrer, G. C. Multiple forms of the calcium-sensitive bioluminescent protein aequorin. *Fed. Proc.* **1975**, *34*, 474.

Blinks, J. R., Prendergast, F. G., Allen, D. G. Photoproteins as biological calcium indicators. *Pharmacol. Rev.* **1976**, *28*, 1–93.

Blinks, J. R., Mattingly, P. H., Jewell, B. R., van Leeuwen, M., Harrer, G. C., Allen, D. G. Practical aspects of the use of Aequorea as a calcium indicator: Assay, preparation, microinjection, and interpretation of signals. *Method. Enzymol.* **1978**, *57*, 292–328.

Campbell, A. K. Extraction, partial purification and properties of obelin, the calcium-activated luminescent protein from the hydroid *Obelia geniculata*. *Biochem. J.* **1974**, *143*, 411–418.

Campbell, A. K., Hallett, M. B., Daw, R. A., Ryall, M. E. T., Hatr, R. C., Herring, P. J. Application of the photoprotein obelin to the measurement of free Ca^{2+} in cells. In *Bioluminescence and Chemiluminescence, Basic Chemistry and Analytical Application* (DeLuca, M. A., McElroy, W. D., Eds.), pp. 601–607. Academic Press: New York, **1981**.

Charbonneau, H., Walsh, K. A., McCann, R. O., Prendergast, F. G., Cormier, M. J., Vanaman, T. C. Amino acid sequence of the calcium-dependent photoprotein aequorin. *Biochemistry* **1985**, *24*, 6762–6771.

Dunstan, S. L., Sala-Newby, G. B., Fajardo, A. B., Taylor, K. M., Campbell, A. K. Cloning and expression of the bioluminescent photoprotein pholasin from the bivalve mollusc *Pholas dactylus*. *J. Biol. Chem.* **2000**, *275*, 9403–9409.

Fujii, T., Ahn, J.-Y., Kuse, M., Mori, H., Matsuda, T., Isobe, M. A novel photoprotein from oceanic squid (*Symplectoteuthis oualaniensis*) with sequence similarity to mammalian carbon-nitrogen hydrolase domains. *Biochem. Biophys. Res. Commun.* **2002**, *293*, 874–879.

Hastings, J. W., Davenport, D. The luminescence of the millipede, Luminodesmus sequoiae. *Biol. Bull.* **1957**, *113*, 120–128.

Hastings, J. W., Gibson, Q. H. Intermediates in the bioluminescent oxidation of reduced flavin mononucleotide. *J. Biol. Chem.* **1963**, *238*, 2537–2554.

Hastings, J. W., Morin, J. G. Calcium activated bioluminescent protein from ctenophores (*Mnemiopsis*) and colonial hydroids (*Obelia*). *Biol. Bull.* **1968**, *135*, 422.

Hastings, J. W., Nealson, K. H. Bacterial bioluminescence. *Ann. Rev. Microbiol.* **1977**, *31*, 549–595.

Hastings, J. W., Mitchel, G., Mattingly, P. H., Blinks, J. R., van Leeuwen, M. Response of aequorin bioluminescence to rapid changes in calcium concentration. *Nature* **1969**, *222*, 1047–1050.

Head, J. F., Inouye, S., Teranishi, K., Shimomura, O. The crystal structure of the photoprotein aequorin at 2.3A resolution. *Nature* **2000**, *405*, 372–376.

Henry, J. P., Monny, C. Proteiprotein interaction in the *Pholas dactylus* system of bioluminescence. *Biochemistry* **1977**, *16*, 2517–2525.

Henry, J. P., Isambert, M. F., Michelson, A. M. Studies in bioluminescence. III. The *Pholas dactylus* system. *Biochim. Biophys. Acta* **1970**, *205*, 437–450.

Henry, J. P., Isambert, M. F., Michelson, A. M. Studies in bioluminescence. IX. Mechanism of the *Pholas dactylus* system. *Biochimie* **1973**, *55*, 83–93.

Henry, J. P., and Michelson, A. M. Studies in bioluminescence. IV. Properties of luciferin from *Pholas dactylus*. *Biochim. Biophys. Acta* **1970**, *205*, 451–458.

Henry, J. P., Monny, C., Michelson, A. M. Characterization and properties of *Pholas* luciferase as a metalloglycoprotein. *Biochemistry* **1973b**, *14*, 3458–3466.

Illarionov, B. A., Bondar, V. S., Illarionova, V. A., Vysotski, E. S. Sequence of the cDNA encoding the Ca^{2+}-activated photoprotein obelin from the hydroid polyp *Obelia longissima*. *Gene* **1995**, *153*, 273–274.

Illarionov, B. A., Frank, L. A., Illarionova, V. A., Bondar, V. S., Vysotski, E. S., Blinks, J. R. Recombinant obelin: cloning and expression of cDNA, purification, and characterization as a calcium indicator. *Method. Enzymol.* **2000**, *305*, 223–249.

Inouye, S., Tsuji, F. I. Cloning and sequence analysis of cDNA for the Ca^{2+}-activated photoprotein, clytin. *FEBS Lett.* **1993**, *315*, 343–346.

Inouye, S., Noguchi, M., Sakaki, Y., Takagi, Y., Miyata, T., Iwanaga, S., Miyata, T., Tsuji, F. I. Cloning and sequence analysis of cDNA for the luminescent protein aequorin. *Proc. Natl. Acad. Sci. USA* **1985**, *82*, 3154–3158.

Inouye, S., Sakaki, Y., Goto, T., Tsuji, F. I. Expression of apoaequorin complementary DNA in *Escherichia coli*. *Biochemistry* **1986**, *25*, 8425–8429.

Isobe, M., Fujii, T., Kuse, M., Miyamoto, K., Koga, K. ^{19}F-Dehydrocoelenterazine as probe to investigate the active site of symplectin. *Tetrahedron* **2002**, *58*, 2117–2126.

Johnson, F. H., Shimomura, O. Preparation and use of aequorin for rapid microdetermination of Ca^{2+} in biological systems. *Nature* **1972**, *237*, 287–288.

Johnson, F. H., Shimomura, O. Introduction to the bioluminescence of medusae, with special reference to the photoprotein aequorin. *Method. Enzymol.* **1978**, *57*, 271–291.

La, S. Y., Shimomura, O. Fluorescence polarization study of the Ca^{2+}-sensitive photoprotein aequorin. *FEBS Lett.* **1982**, *143*, 49–51.

Levine, L. D., Ward, W. W. Isolation and characterization of a photoprotein, "phialidin", and a spectrally unique green-fluorescent protein from the bioluminescent jellyfish *Phialidium gregarium*. *Comp. Biochem. Physiol.* **1982**, *72B*, 77–85.

Loomis, H. F., Davenport, D. A luminescent new xystodesmid milliped from California. *J. Wash. Acad. Sci.* **1951**, *41*, 270–272.

Loschen, G., Chance, B. Rapid kinetic studies of the light emitting protein aequorin. *Nature New Biology* **1971**, *233*, 273–274.

Markova, S. V., Vysotski, E. S., Blinks, J. R., Burakova, L. P., Wang, B.-C., Lee, J. Obelin from the bioluminescent marine hydroid *Obelia geniculata*: cloning, expression, and comparison of some properties with those of other Ca^{2+}-regulated photoproteins. *Biochemistry* **2002**, *41*, 2227–2236.

Michelson, A. M. Purification and properties of *Pholas dactylus* luciferin and luciferase. *Method. Enzymol.* **1978**, *57*, 385–406.

Morin, J. G., Hastings, J. W. Biochemistry of the bioluminescence of colonial hydroids and other coelenterates. *J. Cell. Physiol.* **1971**, *77*, 305–311.

Morise, H., Shimomura, O., Johnson, F. H., Winant, J. Intermolecular Energy Transfer in the bioluminescent system of *Aequorea*. *Biochemistry* **1974**, *13*, 2656–2662.

Müller, T., Campbell, A. K. The chromophore of pholasin: a highly luminescent protein. *J. Biolumin. Chemilumin.* **1990**, *5*, 25–30.

Musicki, B., Kishi, Y., Shimomura, O. Structure of the functional part of photoprotein aequorin. *Chem. Commun.* **1986**, 1566–1568.

Nicolas, M. T., Bassot, J. M., Shimomura, O. Polynoidin: a membrane photoprotein isolated from the bioluminescent system of scale-worms. *Photochem. Photobiol.* **1982**, *35*, 201–207.

Prasher, D., McCann, R. O., Cormier, M. J. Cloning and expression of the cDNA coding for aequorin, a bioluminescent calcium-binding protein. *Biochem. Biophys. Res. Commun.* **1985**, *126*, 1259–1268.

Prasher, D. C., McCann, R. O., Longiaru, M., Cormier, M. J. Sequence comparisons of complememtary DNAs encoding aequorin isotypes. *Biochemistry* **1987**, *26*, 1326–1332.

Reichl, S., Arnhold, J., Knight, J., Schiller, J., Arnold, K. Reactions of pholasin with peroxidases and hypochlorous acid. *Free Radical Biology & Medicine* **2000**, *28*, 1555–1563.

Ridgway, E. B., Snow, A. E. Effects of EGTA on aequorin luminescence. *Biophys. J.* **1983**, *41*, 244a.

Roberts, P. A., Knight, J., Campbell, A. K. Pholasin – A bioluminescent indicator for detecting activation of single neutrophils. *Anal. Biochem.* **1987**, *160*, 139–148.

Shimomura, O. A new type of ATP-activated bioluminescent system in the millipede *Luminodesmus sequoiae. FEBS Lett.* **1981**, *128*, 242–244.

Shimomura, O. Porphyrin chromophore in *Luminodesmus* photoprotein. *Comp. Biochem. Physiol.* **1984**, *79B*, 565–567.

Shimomura, O. Bioluminescence in the sea: photoprotein systems. *Symp. Soc. Exp. Biol.* **1985**, *39*, 351–372.

Shimomura, O. Bioluminescence of the brittle star *Ophiopsila californica. Photochem. Photobiol.* **1986a**, *44*, 71–674.

Shimomura, O. Isolation and properties of various molecular forms of aequorin. *Biochem. J.* **1986b**, *234*, 271–277.

Shimomura, O. Cause of spectral variation in the luminescence of semisynthetic aequorins. *Biochem. J.* **1995a**, *306*, 537–543.

Shimomura, O. Luminescence of aequorin is triggered by the binding of two calcium ions. *Biochem. Biophys. Res. Commun.* **1995b**, *211*, 359–363.

Shimomura, O. A short story of aequorin. *Biol. Bull.* **1995c**, *189*, 1–5.

Shimomura, O., Johnson, F. H. Partial purification and properties of the *Chaetopterus* luminescence system. In *Bioluminescence in Progress* (Johnson, F. H., Haneda, A., Eds.), pp. 495–521. Princeton University Press: Princeton, NJ, **1966**.

Shimomura, O., Johnson, F. H. Extraction, purification and properties of the bioluminescence system of the euphausiid shrimp *Meganyctiphanes norvegica. Biochemistry* **1967**, *6*, 2293–2306.

Shimomura, O., Johnson, F. H. *Chaetopterus* photoprotein: crystallization and cofactor requirements for bioluminescence. *Science* **1968**, *159*, 1239–1240.

Shimomura, O., Johnson, F. H. Properties of the bioluminescent protein aequorin. *Biochemistry* **1969**, *8*, 3991–3997.

Shimomura, O., Johnson, F. H. Calcium binding, quantum yield, and emitting molecule in aequorin bioluminescence. *Nature* **1970**, *227*, 1356–1357.

Shimomura, O., Johnson, F. H. Further data on the specificity of aequorin luminescence to calcium. *Biochem. Biophys. Res. Commun.* **1973**, *53*, 90–494.

Shimomura, O., Johnson, F. H. Regeneration of the photoprotein aequorin. *Nature* **1975a**, *256*, 236–238.

Shimomura, O., Johnson, F. H. Specificity of aequorin bioluminescence to calcium. In *Analytical Application of Bioluminescence and Chemiluminescence* (Chappelle, E. W., Picciolo, G. L., Eds.), NASA SP-388 ed., pp. 89–94. National Aeronautics and Space Administration: Washington, DC, **1975b**.

Shimomura, O., Johnson, F. H. Calcium-triggered luminescence of the photoprotein aequorin. *Symp. Soc. Exp. Biol.* **1976**, *30*, 41–54.

Shimomura, O., Johnson, F. H. Peroxidized coelenterazine, the active group in the photoprotein aequorin. *Proc. Natl. Acad. Sci. USA* **1978**, *75*, 2611–2615.

Shimomura, O., Johnson, F. H. Chemistry of the calcium-sensitive photoprotein aequorin. In *Detection and Measurement of Free Calcium Ions in Cells* (Ashley, C. C., Campbell, A. K., Eds.), pp. 73–83. Elsevier/North-Holland: Amsterdam, **1979a**.

Shimomura, O., Johnson, F. H. Comparison of the amounts of key components in the bioluminescence system of various coelenterates. *Comp. Biochem. Physiol.* **1979b**, *64B*, 105–107.

Shimomura, O., Shimomura, A. Resistivity to denaturation of the apoprotein of aequorin and reconstitution of the luminescent photoprotein from the partially denatured apoprotein. *Biochem. J.* **1981**, *199*, 825–828.

SHIMOMURA, O., SHIMOMURA, A. EDTA-binding and acylation of the Ca^{2+}-sensitive photoprotein aequorin. *FEBS Lett.* **1982**, *138*, 201–204.

SHIMOMURA, O., SHIMOMURA, A. Effect of calcium chelators on the Ca^{2+}-dependent luminescence of aequorin. *Biochem. J.* **1984**, *221*, 907–910.

SHIMOMURA, O., SHIMOMURA, A. Halistaurin, phialidin and modified forms of aequorin as Ca^{2+} indicator in biological systems. *Biochem. J.* **1985**, *228*, 745–749.

SHIMOMURA, O., JOHNSON, F. H., SAIGA, Y. Extraction, purification and properties of aequorin, a bioluminescent protein from the luminous hydromedusan, Aequorea. *J. Cell. Comp. Physiol.* **1962**, *59*, 223–239.

SHIMOMURA, O., JOHNSON, F. H., SAIGA, Y. Extraction and properties of halistaurin, a bioluminescent protein from the hydromedusan *Halistaura*. *J. Cell. Comp. Physiol.* **1963**, *62*, 9–16.

SHIMOMURA, O., JOHNSON, F. H., MORISE, H. Mechanism of the luminescent intramolecular reaction of aequorin. *Biochemistry* **1974**, *13*, 3278–3286.

SHIMOMURA, O., MUSICKI, B., KISHI, Y. Semi-synthetic aequorin: an improved tool for the measurement of calcium ion concentration. *Biochem. J.* **1988**, *251*, 405–410.

SHIMOMURA, O., MUSICKI, B., KISHI, Y. Semi-synthetic aequorins with improved sensitivity to Ca^{2+} ions. *Biochem. J.* **1989**, *261*, 913–920.

SHIMOMURA, O., INOUYE, S., MUSICKI, B., KISHI, Y. Recombinant aequorin and recombinant semi-synthetic aequorins. *Biochem. J.* **1990**, *270*, 309–312.

SHIMOMURA, O., MUSICKI, B., KISHI, Y., INOUYE, S. Light-emitting properties of recombinant semi-synthetic aequorins and recombinant fluorescein-conjugated aequorin for measuring cellular calcium. *Cell Calcium* **1993**, *14*, 373–378.

STEPHENSON, D. G., SUTHERLAND, P. J. Studies on the luminescent response of the Ca^{2+}-activated photoprotein, obelin. *Biochim. Biophys. Acta* **1981**, *678*, 65–75.

TAKAHASHI, H., ISOBE, M. Photoprotein of luminous squid, *Symplectoteuthis oualaniensis* and reconstruction of the luminous system. *Chem. Lett.* **1994**, *5*, 843–846.

TSUJI, F. I., LEISMAN, G. B. K^{+}/Na^{+}-triggered bioluminescence in the oceanic squid *Symplectoteuthis oualaniensis*. *Proc. Natl. Acad. Sci. USA* **1981**, *78*, 6719–6723.

WARD, W. W., SELIGER, H. H. Properties of mnemiopsin and berovin, calcium-activated photoproteins. *Biochemistry* **1974**, *13*, 1500–1510.

WARD, W. W., SELIGER, H. H. Action spectrum and quantum yield for the photoinactvation of mnemiopsin, a bioluminescent photoprotein from the ctenophore *Mnemiopsis* sp. *Photochem. Photobiol.* **1976**, *23*, 351–363.

WOOD, K. V., DE WET, J. R., DEWJI, N., DELUCA, M. Synthesis of active firefly luciferase by in vitro translation of RNA obtained from adult lanterns. *Biochem. Biophys. Res. Commun.* **1984**, *124*, 592–596.

2
Luminous Marine Organisms

Steven H.D. Haddock

2.1
Introduction

In the ocean, bioluminescence is found across nearly all taxa, from bacteria to fish (Fig. 2.1 and review by Herring 1987). In many senses, it is easier to name the few *non*-luminous phyla than it is to detail each luminous group. However, even in light of recent breakthroughs in molecular biology, we are hardly closer than E. Newton Harvey (1952) was to being able to explain why luminescence is distributed among phyla the way it is. For example, among the phytoplankton, many dinoflagellates are luminescent, while the silica-encapsulated diatoms are not. Among protists, in contrast, the calcium carbonate-shelled foraminifera are not luminous, while their siliceous cousins the radiolarians are. In some groups, such as ctenophores and siphonophores, nearly all species may be luminous, while other groups may have only a single luminous representative. Luminescence seems to be more prevalent in taxa that live in the water column and in the deep sea–environments well suited to communication with light. Both habitats are fairly transparent to blue-green wavelengths, and the deep sea is in a constant state of dim illumination or darkness during day and night.

Within this diverse assemblage, it is difficult to estimate the number of times that bioluminescence has independently evolved. Hastings (1983) estimates that this has occurred at least 30 times. This number would be heavily dependent on the definition of "independent". For example, once bacteria become luminous, is it considered an independent event for fish to use bacterial light in a specialized organ? If ostracods develop the ability to make light, is it an independent event for fish to use their substrate as the basis for their luminescence system? Several authors have argued that the evolution of luminescence has proceeded by co-opting chemistries that originated for other purposes, specifically as mechanisms for reducing oxidative stress (Case et al. 1994; Labas et al. 2001; Rees et al. 1998). Because marine luciferins are readily available in the food chain, it would appear that, after the initial evolutionary events, the diversification of bioluminescent organisms could be relatively easy and recent (Haddock et al. 2001). An organism

Photoproteins in Bioanalysis. Edited by Sylvia Daunert and Sapna K. Deo

ISBN: 3-527-31016-9

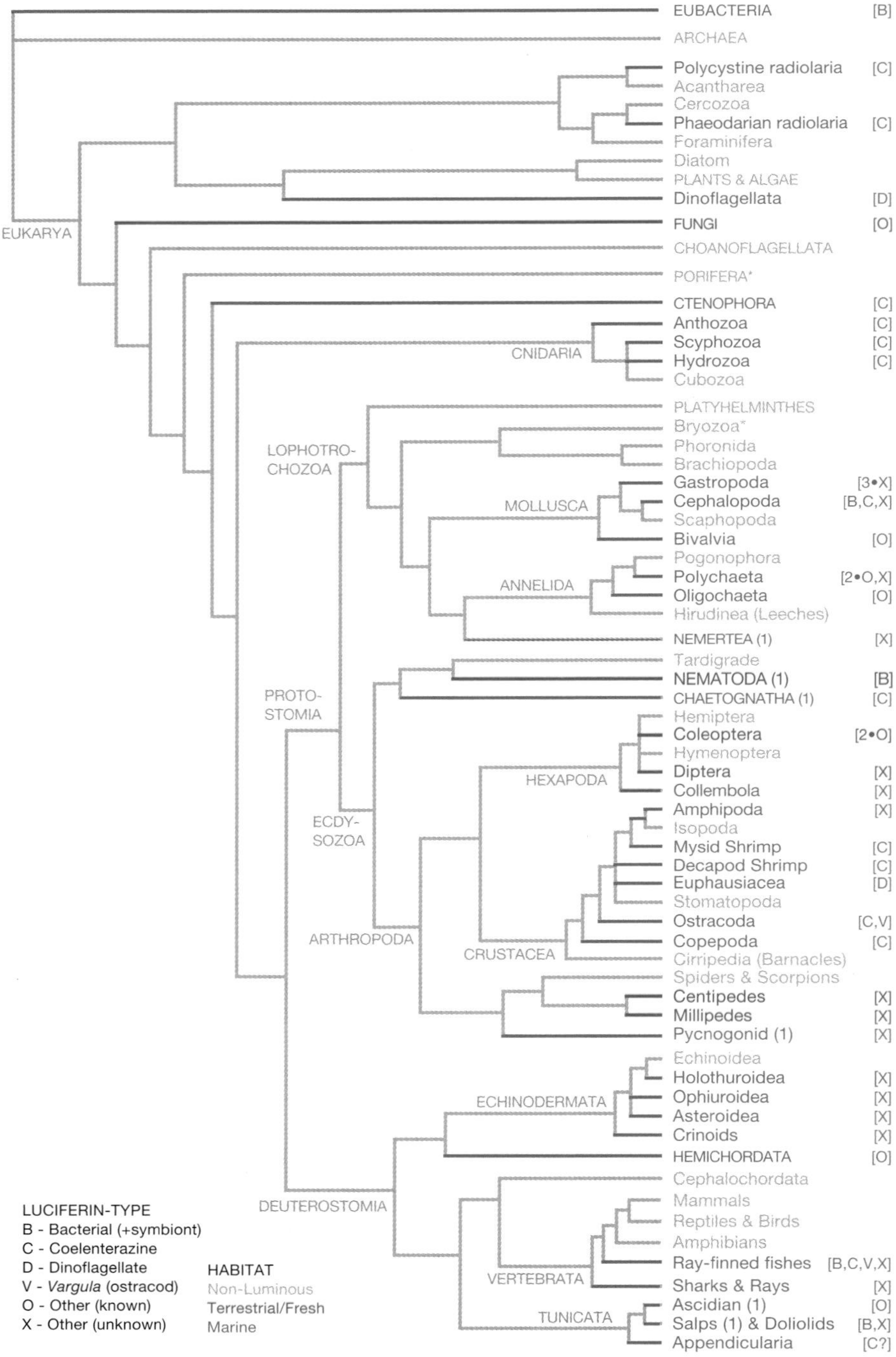

EUBACTERIA [B]
ARCHAEA
EUKARYA
Polycystine radiolaria [C]
Acantharea
Cercozoa
Phaeodarian radiolaria [C]
Foraminifera
Diatom
PLANTS & ALGAE
Dinoflagellata [D]
FUNGI [O]
CHOANOFLAGELLATA
PORIFERA*
CTENOPHORA [C]
CNIDARIA
Anthozoa [C]
Scyphozoa [C]
Hydrozoa [C]
Cubozoa
LOPHOTRO-CHOZOA
PLATYHELMINTHES
Bryozoa*
Phoronida
Brachiopoda
MOLLUSCA
Gastropoda [3•X]
Cephalopoda [B,C,X]
Scaphopoda
Bivalvia [O]
ANNELIDA
Pogonophora
Polychaeta [2•O,X]
Oligochaeta [O]
Hirudinea (Leeches)
NEMERTEA (1) [X]
PROTO-STOMIA
Tardigrade
NEMATODA (1) [B]
CHAETOGNATHA (1) [C]
ECDY-SOZOA
HEXAPODA
Hemiptera
Coleoptera [2•O]
Hymenoptera
Diptera [X]
Collembola [X]
ARTHROPODA
CRUSTACEA
Amphipoda [X]
Isopoda
Mysid Shrimp [C]
Decapod Shrimp [C]
Euphausiacea [D]
Stomatopoda
Ostracoda [C,V]
Copepoda [C]
Cirripedia (Barnacles)
Spiders & Scorpions
Centipedes [X]
Millipedes [X]
Pycnogonid (1) [X]
DEUTEROSTOMIA
ECHINODERMATA
Echinoidea
Holothuroidea [X]
Ophiuroidea [X]
Asteroidea [X]
Crinoids [X]
HEMICHORDATA [O]
Cephalochordata
VERTEBRATA
Mammals
Reptiles & Birds
Amphibians
Ray-finned fishes [B,C,V,X]
Sharks & Rays [X]
TUNICATA
Ascidian (1) [O]
Salps (1) & Doliolids [B,X]
Appendicularia [C?]

LUCIFERIN-TYPE
B - Bacterial (+symbiont)
C - Coelenterazine
D - Dinoflagellate
V - *Vargula* (ostracod)
O - Other (known)
X - Other (unknown)

HABITAT
Non-Luminous
Terrestrial/Fresh
Marine

Fig. 2.1 Phylogenetic tree of bioluminescence.

Fig. 2.1 Phylogenetic tree of bioluminescence. Mapping the distribution of bioluminescent organisms onto a phylogenetic tree emphasizes the number of times that luminescence has arisen. The data here suggest that this has happened at least 40 times. Groups are highlighted if one or more member is luminous; if there is only one, this is noted in parentheses. The luciferin employed within the group is indicated in square brackets to the right of the group name, as indicated. The principal backbone of the tree is based the molecular phylogeny of Philippe (2005) and the protistan tree of Cavalier-Smith (2004). More detailed relationships of the terminal groups were compiled from numerous individual phylogenies, mostly molecular. The reports of bioluminescence are based on Herring (1987), supplemented with more recent discoveries as noted in the text.

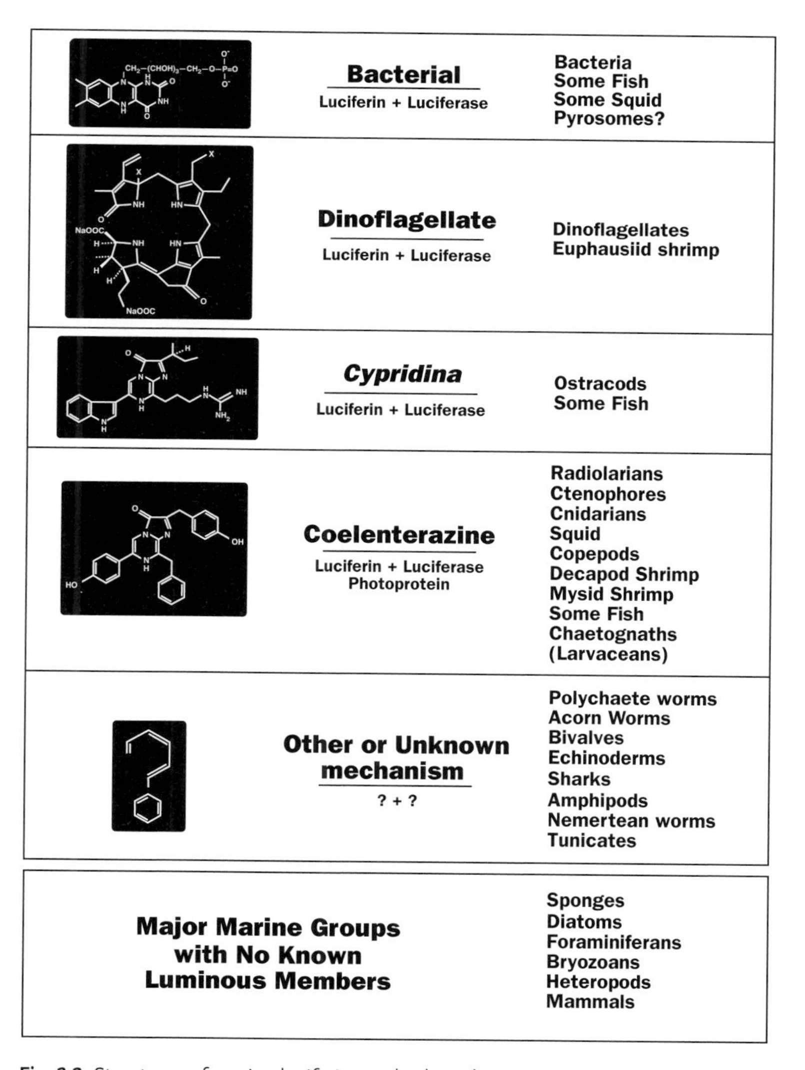

Fig. 2.2 Structures of marine luciferins and selected organisms that employ them.

needs only to evolve the catalyst – a luciferase or photoprotein – to become luminous. Clearly this is not far from the truth because of the large number of organisms synthesizing proteins to act on a few marine luciferins (Fig. 2.2). The number of distinct origins becomes apparent when one views an evolutionary tree compiled from various molecular phylogenies, onto which the known examples of bioluminescence have been mapped (Fig. 2.1). Given the recent discoveries of luminous taxa, and assuming that each different luminescent chemistry within a taxonomic group represents an independent event, we can infer that luminescence has originated at least 40 times.

2.1.1
Non-luminous Taxa

There are two difficulties in assembling a list of luminous taxa that make it necessary to provide continual updates. First is the presence of "false-positive" reports. In observing bioluminescence, it is easy to be misled by iridescence and bright pigmentation or by the presence of microscopic contaminating organisms. In some cases, such as that of the brightly colored copepod *Sapphirina*, it is easy to provide an alternative explanation for false reports of luminescence. Questionable observations, however, are like malicious gossip: the suggestion is difficult to "disprove", even with a large amount of evidence to the contrary (e.g., Haddock and Case 1995). In this chapter, sponges, bryozoans, and the salp *Cyclosalpa* are conservatively considered non-luminous despite early reports, because there have been no supporting observations in the last 60 years. Someday, they too may be convincingly demonstrated to produce light.

The opposite situation occurs when luminescence has not been observed simply because intact specimens are difficult to collect. This has been the case for a chaetognath, as well as for certain deep-sea doliolids (salp relatives) and pteropods (unpublished observations). These "false negatives" are easier to clarify than false positives, once the appropriate material is obtained.

The list of major marine taxa without a single known luminescent representative includes diatoms, foraminifera, sponges, heteropods, and mammals. Sponges have been variously reported as luminous, but these observations have never been confirmed in detail, and given their impressive filtration rates, it would be difficult to exclude the possibility of contamination by accumulated plankton. Despite the counter-example of the heteropods (a diverse planktonic group), many other mollusks are highly luminous, including squids and even bivalves (*Pholas*). Arrow worms (Chaetognatha) were thought to be one of the most abundant planktonic groups with no luminous members until the recent discovery of bioluminescence in a deep-sea species (Haddock and Case 1994). Marine mammals are not known to be intrinsically luminous, leaving the title of largest bioluminescent organism to the siphonophores (see below) and squid. It has been suggested that the rare "megamouth" shark *Megachasma* is bioluminescent, but this assertion has been challenged on lack of evidence (Herring 1985b). Some other sharks are in fact luminous.

2.1.2
Luminous Taxa

Although there are numerous examples of the independent evolution of bioluminescence, four types of luciferin are responsible for light production in the majority of marine phyla (Fig. 2.2). These light-emitting compounds are conserved, while the catalyzing proteins (either luciferases or photoproteins) are diverse and encoded in the genome of each luminous organism. One explanation, perhaps the most parsimonious, for the few types of luciferins in many phyla is that they are passed along through dietary links. Exogenous origins have been demonstrated for the existence of ostracod luciferin in the midshipman fish *Porichthys* (Warner and Case 1980), for the presence of coelenterazine in hydromedusae (Haddock et al. 2001), and for the use of dinoflagellate luciferin by krill (Dunlap et al. 1980). In nature, as in molecular labs, evolution appears to have targeted genes able to act on a few types of readily available luciferins in the food chain, leading to very diverse catalysts acting on a few substrates.

The survey of bioluminescent organisms that follows is organized by shared luciferins, not by an implied taxonomic hierarchy. The chemistries of some of the luminous systems mentioned here are detailed in Chapter 1 of this volume.

2.2
Taxonomic Distribution of Bioluminescence

2.2.1
Bacterial Luminescence

Luminous **bacteria** (Fig. 2.3A) such as *Vibrio fischeri* produce a low, steady glow for hours in the presence of oxygen. In the best-studied species, transcription of light-producing genes is regulated by the presence of an "autoinducer" in the environment (Nealson and Hastings 1979). This autoinducer is produced by the bacteria and accumulates as the population density increases, so that a dilute culture will not produce appreciable light. The genetic "cassette" responsible for light production is now well understood (Meighen 1991), and these pathways have been modified and manipulated to form the basis for a suite of biosensors that produce light in proportion to the concentration of target molecules in the environment (e.g., Ramanathan et al. 1997).

The production of continual light at high concentrations makes luminous bacteria well suited for symbiotic relationships with other organisms. It is sometimes assumed that these mutualisms are responsible for most marine luminescence, but they are actually rather uncommon. Examples are known mainly from the light organs of **anglerfish** (Fig. 2.6A) (Munk 1999), **flashlight fish** (Haygood and Distel 1993), and **ponyfish** (Haneda and Tsuji 1976; Wada et al. 1999), as well as from certain species of luminous **squid** (Ruby and McFall-Ngai 1992). The best-studied symbiosis is probably that of the bobtailed squid *Euprymna*

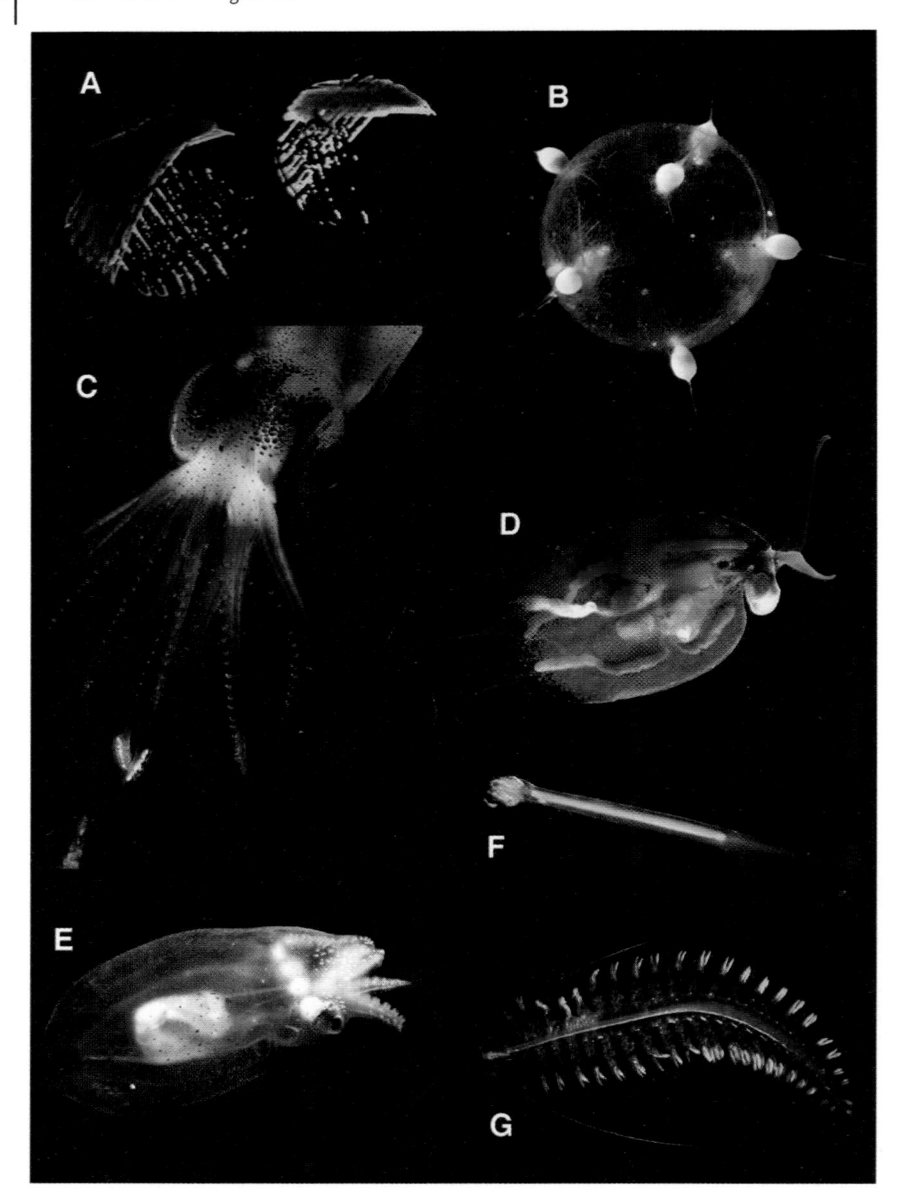

Fig. 2.3 Luminous organisms (various).
(A) Bacteria streaked on a plate;
(B) phaeodarian radiolarian *Tuscarantha luciae*;
(C) squid with light organs at tentacle tips;
(D) pelagic nudibranch *Phylliroe*;
(E) pelagic octopus *Japetella*;
(F) chaetognath *Caecosagitta macrocephala*;
(G) yellow luminous polychaete *Tomopteris*.

scolopes and its bacterial symbiont *Vibrio*. A special light organ hosts a bacterial culture that is regularly vented into the environment. The bacterial inoculation is tightly regulated and controlled by the host, and this relationship has been used as a model for pathogenic microbial systems (Ruby and McFall-Ngai 1992).

Pyrosomes (pelagic colonial urochordates) are also suspected of producing light through bacterial symbioses, although this has not been clearly demonstrated. The most dramatic example of light produced by bacteria is found in the form of rare milky seas, where the surface of the water glows from horizon to horizon. These events occur most frequently in the Indian Ocean, and knowledge of their occurrence has been derived mainly from the anecdotal reports of mariners (Herring and Watson 1993). Recently, a satellite-based observation revealed the massive scale of one such event, which covered an area the size of the state of Connecticut and involved more than 10^{22} bacteria (Miller et al. 2005).

2.2.2
Dinoflagellate Luciferin

A sailor or beachgoer who sees sparkling lights in the water at night is most likely witnessing the bioluminescence of **dinoflagellates**. These protists occupy a somewhat equivocal position between unicellular algae and protozoans. Some species are autotrophic, deriving energy from photosynthesis; others are heterotrophic, consuming diatoms and other plankton, and some have qualities of both. Some types are even parasitic and may cause severe health hazards, but these are not known to be luminescent. Under the right conditions – typically a calm period following a period of high productivity – dinoflagellate populations may bloom rapidly, leading to a “red tide”. The red coloration is caused by the absorbance of dinoflagellate pigments, and if the right species are involved, at night the red tide can turn blue in an intense display of luminescence. Places where these blooms occur frequently include the famous “bioluminescent bays” of Puerto Rico and Jamaica.

The tetrapyrrole luciferin of dinoflagellates is interesting because its structure is similar to the structure of chlorophyll (Fig. 2.2), differing mainly in the metal ions present (Dunlap et al. 1981; Nakamura et al. 1989). The two compounds may in fact be interconverted on a day–night cycle, as the cell alternates between photosynthesizing and luminescing. Dinoflagellates use particular luciferases, which have been shown through cloning to have three repeated catalytic domains (Bae and Hastings 1994; Liu et al. 2004). They also contain luciferin-binding proteins (LBPs), which regulate light production by limiting the availability of the luciferin and may be involved in generating their circadian cycle of light emission (Morse et al. 1989).

This same luciferin is employed by certain **euphausiid shrimp** (krill), which are thought to obtain their light-producing compounds from dinoflagellates in their diet (Dunlap et al. 1980). Krill, like many planktonic invertebrates, express their bioluminescence through numerous ventral photophores and likely use their light for counterillumination (see Section 2.3).

2.2.3
Cypridina (*Vargula*) Luciferin

Ostracods (Fig. 2.5D) are small crustaceans enclosed in a clam-like chitinous carapace. The luciferin of *Cypridina* (also called *Vargula*) was one of the first to be crystallized and chemically characterized (Kishi et al. 1966; Shimomura et al. 1957). Also known as seed shrimp or firefleas, ostracods are the source of one of the most famous anecdotes about bioluminescence. During World War II, the Japanese army collected massive quantities of ostracods and dried them on tarps on the shore. These stockpiles were to be used as a chemical illumination system. A small handful produces a steady glow when moistened and crushed in the palm. Even 60 years later, these dried crustaceans, stored in jars labeled "Japan 1944", will produce light when macerated. Ostracods dispense luciferin and luciferase into the seawater from "nozzles" near their mouth, forming discrete puffs of light. In some types, these puffs form a specific sequence of glowing or flashing dots that identify males to females of the same species (Morin 1986). This display, akin to that of terrestrial fireflies, is one of the most sophisticated uses of bioluminescence known to occur in the sea. The luciferase of *Cypridina* has been cloned and characterized (Thompson et al. 1989). Other types of ostracods use the luciferin coelenterazine (see below), emphasizing the diverse origins of luminescence even within a specific taxonomic group.

The only other organism known to use ostracod luciferin is the **midshipman fish** *Porichthys notatus*. The name "midshipman" refers to the conspicuous rows of photophores along its belly, which resemble the buttons of a naval uniform. These have been shown to act in a manner consistent with a counterillumination function (see below), growing brighter as downwelling light increases (Harper and Case 1999). *Porichthys* is one of the most thoroughly investigated examples of a dietary source of luciferin. This species is distributed along the west coast of the United States, but luminous ostracods are found only in the southern portion of the range. As a result, *Porichthys* collected to the north are not bioluminescent, but they can be made to glow if they are fed ostracods (Warner and Case 1980). Thus, this fish is dependent on its prey not only for nutrition but also for, in some sense, its chemical defense system.

2.2.4
Coelenterazine

Coelenterate-type luciferin is by far the most widely distributed marine luciferin, found in at least seven different phyla, from protists to vertebrates. Like the *Cypridina* luciferin discussed above, coelenterazine is a nitrogen-based imidazole compound. While it has been detected in numerous species (Shimomura et al. 1980; Thomson et al. 1997), its presence does not necessarily imply that it is responsible for light production, as it has been found in many non-luminous species as well (Shimomura 1987). It appears to be well represented in the food chain, and, as will be discussed, this may partially explain its ubiquity.

As its name implies, this luciferin was first discovered in a coelenterate – specifically, the **hydromedusa** *Aequorea*. It is found in other hydrozoans (Fig. 2.4D), including *Obelia* and siphonophores (Fig. 2.4E), where it is generally used in conjunction with calcium-activated photoproteins. Although *Aequorea* and some other hydromedusae have green fluorescent proteins, most hydrozoans do not. In most marine organisms, luminescence occurs primarily in the violet to green range (max: 440–505 nm). **Siphonophores** are unusual planktonic elongate hydrozoans, of which most are luminous. Only *Physalia* (the Portuguese man-of-war), some benthic species, *Lychnagalma*, and perhaps *Physophora* are not known to be luminous. Because certain siphonophores can reach lengths of many decameters (Fig. 2.4E), they are arguably the longest organisms on earth. In the genus *Erenna*, the bioluminescence appears to function in a pendant lure, attracting fish to batteries of stinging cells (Haddock et al. 2005). Such functions are not typical for jellies, which are thought to use bioluminescence as a defensive, distractive display. This lure is also unique because its light is modified by surrounding fluorescent material to produce an orange-red coloration.

Coelenterazine is responsible for light production in the other **cnidarian** groups as well. The **anthozoan** sea pansy *Renilla* was another early subject of groundbreaking science, and the chemistry of its luciferin–luciferase system was one of the first to be examined (Hori and Cormier 1965). It has been used as a model system in a large range of neurological and pharmacological studies (e.g., Anctil 1989). In addition to a luciferase that catalyzes coelenterazine oxidation, *Renilla* has a luciferin-binding protein, analogous to that of the dinoflagellates. This is bound to the luciferin and regulates its accessibility to the luciferase in a calcium-dependent manner, thus controlling light emission somewhat like a photoprotein. The amino acid sequence of the LBP (Kumar et al. 1990) and the luciferase (Lorenz et al. 1991) of *Renilla* are both known. *Renilla* also contains a green fluorescent protein that has been fully characterized but is published only in a patent. Other luminous octocorals include the sea pens *Halipterus* (Fig. 2.4C), *Stylatula*, *Umbellula*, and *Ptilosarcus*.

Another major group of anthozoans is the hexacorallia (anemones and corals), and a few of these are bioluminescent as well. *Parazoanthus* grows parasitically to encrust gorgonians, giving the appearance that the gorgonian itself is producing light. Most anemones and corals, however, while frequently laden with fluorescent proteins, are not bioluminescent.

The **scyphomedusae**, or "true jellyfish", were among the first marine organisms in which luminescence was noticed, dating back to when Pliny reported poking one with a stick in the first century AD. Forskål acknowledged the "night lights" of *Pelagia* when he gave it the specific name *noctiluca* in 1775, just as Linnaeus had done with the glowworm *Lampyris noctiluca*. (Given this criterion, most scyphozoa might be named *noctiluca*, as bioluminescence is common in the group.) Some of the most long-lasting and spectacular luminescence displays are produced by coronate medusae, including *Periphylla* and *Atolla* (Fig. 2.4B) (Herring 1990; Herring and Widder 2004). Given sufficient stimulus, these species will begin to produce a frenzied display that can persist for 30 seconds or more. The luciferase

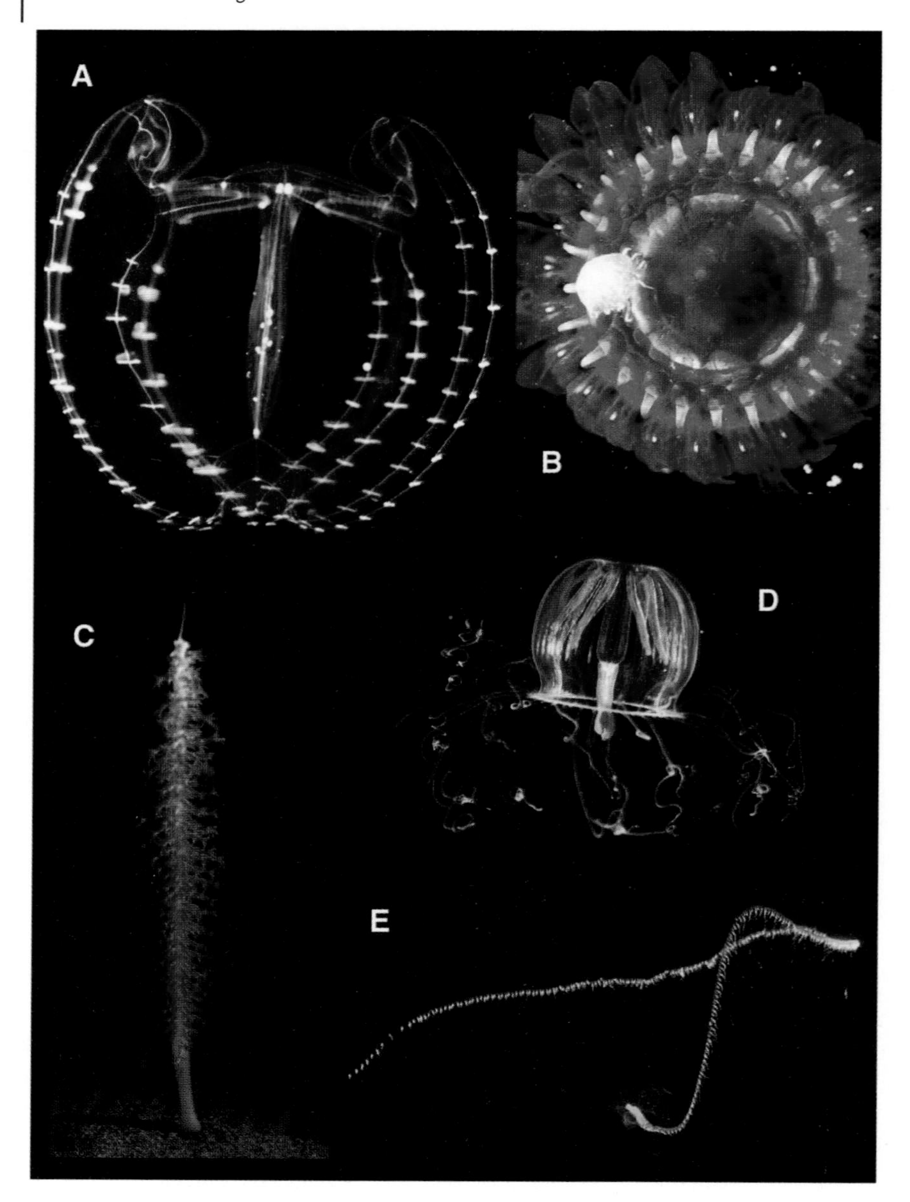

Fig. 2.4 Luminous organisms (jellies).
(A) Ctenophore *Deiopea*;
(B) scyphomedusa *Atolla* (with isopod);
(C) anthozoan sea pen *Halipterus*;
(D) hydromedusa *Aglantha*;
(E) siphonophore *Praya*.
(Colors are from the camera strobe, not from luminescence).

of *Periphylla* has been described chemically (Shimomura and Flood 1998), and efforts are underway to clone the gene.

Despite the prevalence of luminescence among cnidarians, two groups are not known to have any luminous members. These are the Stauromedusae (stalked jellyfishes that attach to sea grasses) and Cubomedusae (box jellies). As with many phyla, it is not clear why these particular taxa would not be luminous; one is an example of specialization for a sessile lifestyle, and the other is a highly mobile pelagic predator.

Another phylum that is almost entirely luminous is the **Ctenophora** or "comb jellies" (Fig. 2.4A). These mysterious jellies propel themselves through the water using tiny ciliated paddles. They have been historically grouped with cnidarians as "coelenterates", although their similarities are essentially due to convergence, and molecular work suggests that are not closely related to each other (Podar et al. 2001). Ctenophores, like hydrozoans, produce light using calcium-regulated photoproteins with coelenterazine as the luciferin. More then 90% of the genera in this small phylum are luminous (Haddock and Case 1995). The species *Beroe forskalii* can be stimulated into a frenzied display like that of the coronate scyphomedusae, while others such as *Euplokamis, Mertensia*, and *Bathyctena* can shoot out glowing bioluminescent particles when disturbed.

Although coelenterazine was named for the cnidarians and ctenophores where it was first discovered, there is evidence that these groups cannot, in fact, produce their own luciferin (Haddock et al. 2001). They must instead obtain it through their diet, with crustaceans being the most likely ultimate source. As disturbing as this might seem to our concept of the early evolution of cnidarian bioluminescence, it does help to explain the tremendous diversity of coelenterazine-based luminescence in the sea.

Coelenterazine is widely distributed in **crustaceans** (Fig. 2.5A, D, E), and they are probably the group that synthesizes the molecule and distributes it through the food chain. Evidence of the synthesis of coelenterazine comes from a study of the **decapod** shrimp *Systellaspis* (Thomson et al. 1995), in which the presence of coelenterazine was assayed during the development of eggs. A related shrimp, *Oplophorus*, can disgorge great amounts of luminous fluid, making it an excellent candidate for chemistry studies, and its luciferase has been cloned (Inouye et al. 2000). External displays are seen in the deep-sea lophogastrid (formerly mysid) *Gnathophausia*. Interestingly, this species appears to have a dietary requirement for coelenterazine (Frank et al. 1984), underscoring the risks of overgeneralization in the case of bioluminescent abilities. Most crustaceans have been found to use a luciferase–luciferase reaction and have evolved individual luciferases to use in conjunction with coelenterazine. Shrimp also may use their luminescence for counterillumination when it is expressed through ventral photophores, and some of the most elegant studies of this mode of camouflage were performed on the decapod *Sergestes* (Latz and Case 1982; Warner et al. 1979).

Under most conditions, **copepods** are the most numerous zooplankton in the sea, and they also use coelenterazine as their light-emitting molecule. Because they are the "bugs" of the sea, and are consumed by a diverse array of other creatures,

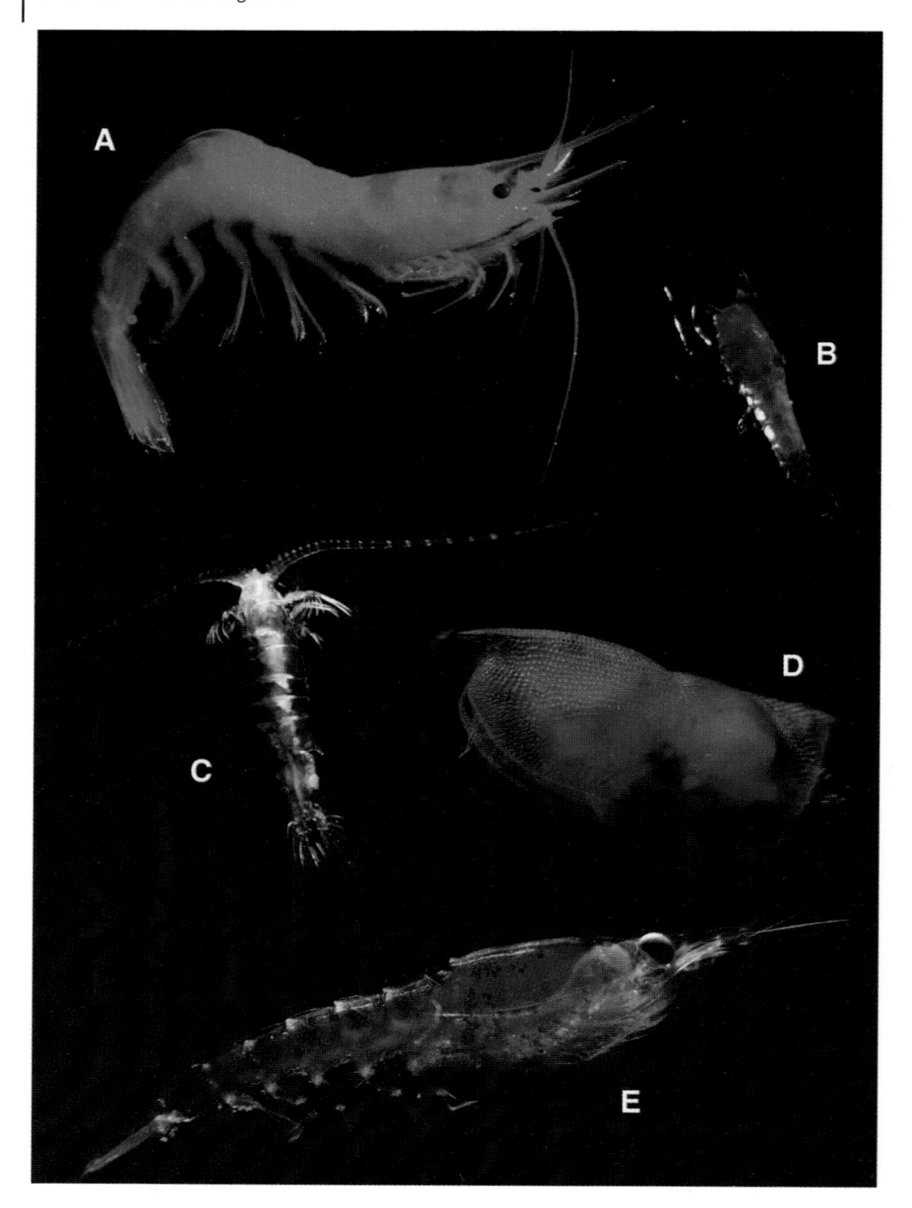

Fig. 2.5 Luminous organisms (crustaceans).
(A) Decapod shrimp *Acanthephyra*;
(B) amphipod *Scina*;
(C) copepod *Gaussia*;
(D) ostracod *Conchoecia*;
(E) krill *Thysanoessa*.

they would also be an excellent source of luciferin in the food supply. Notable copepods are the giant (1 cm) black-pigmented *Gaussia princeps* (Fig. 2.5C) and the abundant cold-water genus *Metridia*. The luciferases of *Gaussia, Pleuromamma*, and *Metridia* have been cloned by separate groups (Markova et al. 2004; Szent-Gyorgyi et al. 2003). The common genus *Calanus* is not luminescent. Copepods appear to use bioluminescence for defensive purposes, as several species shoot out luminous fluid from special nozzles as part of their escape responses.

Although most **ostracods** have their own special form of luciferin described above, the genus *Conchoecia* (Fig. 2.5D) uses coelenterazine to make light (Oba et al. 2004). Therefore, even within this fairly small group, there are at least two examples of the independent evolution of bioluminescence. The luciferin of **amphipods** (Fig. 2.5B) is not yet known, but because these crustaceans often have symbiotic relationships with jellies, it would not be surprising to find that they also use coelenterazine. Occasional specimens of the amphipod *Cyphocaris* may produce orange-colored luminescence, which is very rare in the sea.

Nearly all species of arrow worm, in the phylum **Chaetognatha**, are *not* luminous. This is surprising because their transparent morphology and planktonic life are characteristic of many luminous species. However, they are also suggested to be planktonic cousins of the non-luminous nematodes, a speciose group in which only occasional terrestrial species are known to be infected by luminous bacteria. One deep-sea species of chaetognath (Fig. 2.3F) was discovered to have bioluminescent granules containing coelenterazine and luciferase (Haddock and Case 1994). These particles are shed into the water as part of the escape response, leaving behind a sparkling cloud.

Radiolarians are amoeba-like protists that occur from shallow tropical waters to depths greater than 2500 m. There are two main types: a shallow colonial form called Polycystines, which insert glass spicules into their gelatinous matrix, and the deep-living Phaeodarians (Fig. 2.3B), which use silica to form glass capsules. Molecular evidence shows that these two groups are deeply divergent, and they are each more closely related to other non-luminous protists than to each other (Polet et al. 2004). Because both groups appear to use calcium-regulated photoproteins involving coelenterazine, this may be another example of multiple convergent evolution (Campbell et al. 1979). However most of the work has been done on Polycystine species, and therefore more in-depth investigation is required. It is tempting to speculate that calcareous protists, such as foraminiferans, may be non-luminous because they are unable to sequester calcium. The function of bioluminescence in these planktonic species is unknown.

Among marine organisms, some of the most complex bioluminescent displays are produced by **cephalopods**, especially **squid** (Fig. 2.3C). These mollusks may have three or more different types of photophores at different locations across the body of an individual. For example, many have small photophores, sometimes of different colors, along their ventral surface for counterillumination; large photophores near their eyes for warning flashes (and counterillumination); and additional light organs at the tips of their tentacles, perhaps for defense or prey attraction. The vampire squid, *Vampyroteuthis*, which is distinct enough that it

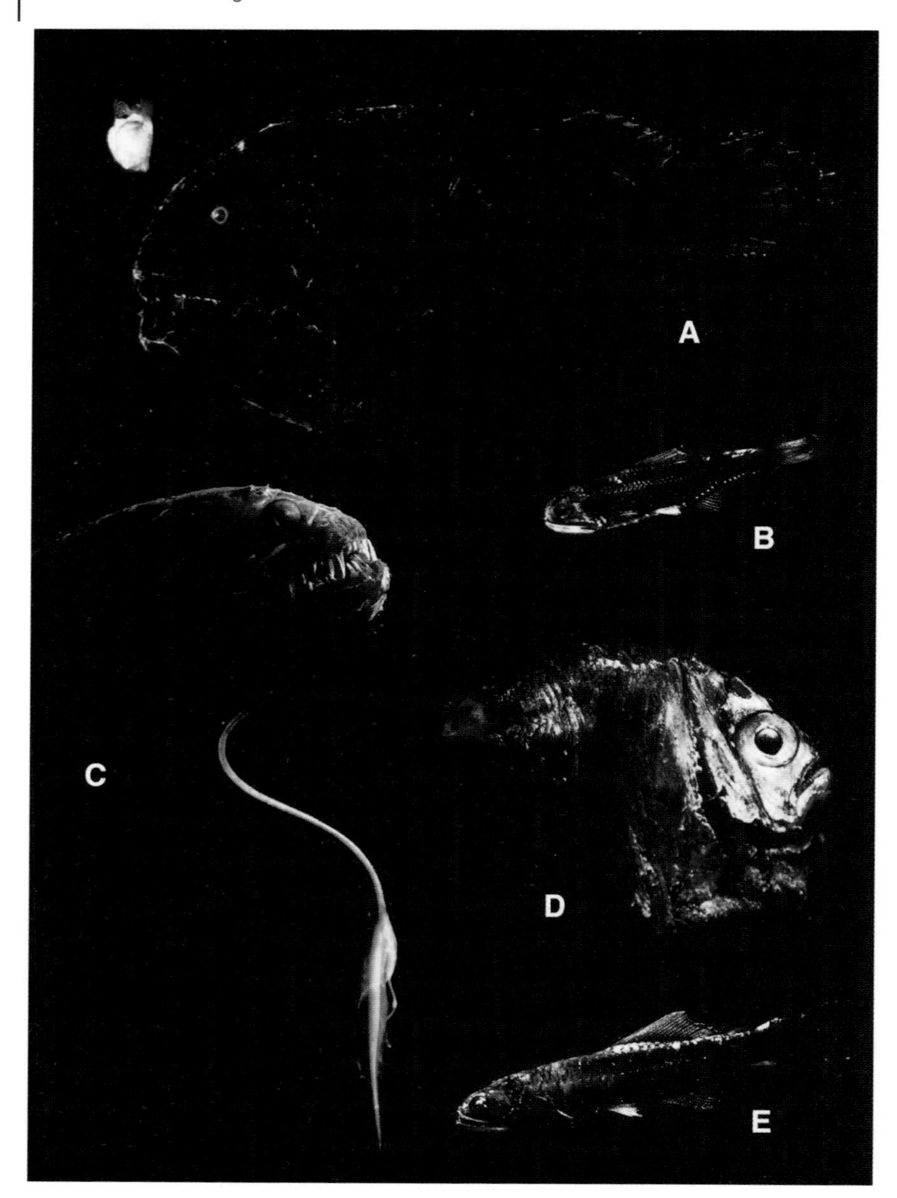

Fig. 2.6 Luminous organisms (fish).
(A) Anglerfish *Chaenophryne*;
(B) myctophid lanternfish;
(C) dragonfish *Idiacanthus*;
(D) hatchetfish;
(E) sternchaser myctophid.

has been classified between the squids and octopods, exudes coelenterazine-based granules from its tentacle tips (Robison et al. 2003), in addition to bearing several bright lights across its body. The squid *Symplectoteuthis* has a distinct photoprotein that is triggered by monovalent ions, and it uses dehydrocoelenterazine as the light-emitting substrate (Takahashi and Isobe 1994). Other squids may have bacterial symbionts as described above. Females of the midwater octopod *Japetella* (Fig. 2.3E) develop a unique ring of luminescence around their mouth, leading to the suggestion that this signal may be used for intraspecific communication (Herring 2000). The chemistry of this particular reaction is unknown. Another dramatic display of luminescence is found in the cirrate octopod *Stauroteuthis*. This deep-living species has web-like membranes connecting its arms, and along these arms, the suckers themselves glow (Johnsen et al. 1999).

The final major group known to use coelenterazine is the bony fishes (Fig. 2.6). In addition to the bacterial and *Cypridina*-type luciferin described above, fish are also known to have high levels of coelenterazine in their photophores and eggs, along with related luciferase activity (Mallefet and Shimomura 1995; Rees et al. 1990). Fish rival squid in the variety and extent of their adaptations to produce light. Most of the species living in the midwater have ventral photophores, which may include a complex array of reflective surfaces, lenses, and filters to modify the light signal used in counterillumination (Herring 1977). Hatchet fish (Fig. 2.6D), for example, are flattened into a vertical plane, and their narrow silhouette is well covered by extensive light organs. Several kinds of fish have glowing (nonbacterial) barbels dangling from their chin or at the end of modified fin rays, apparently for use as lures. One of the most spectacular luminous fish is the black dragonfish *Idiacanthus* (Fig. 2.6C). In addition to an ornate barbel and luminous fin rays, this stomiid fish has its entire upper and lower surfaces lined with extremely bright photophores. Lanternfish, or myctophids (Fig. 2.6A), have species-specific patterns of spot-like photophores along the sides of their bodies. Along with their blue-green photophores, a few of the loosejaw fishes such as *Malacosteus* and *Aristostomias* use filters to produce red bioluminescence – the longest wavelength luminescence known from the sea (O'Day and Fernandez 1974). The luciferases of bony fishes in general have not yet been characterized.

A few types of small elasmobranch sharks are also luminous, mostly in the form of numerous tiny photophores along their ventral surface. In an interesting twist, it has been suggested that they use counterillumination for offensive, rather than defensive, purposes. The cookie-cutter shark *Isitius*, in particular, preys by gouging disc-shaped bites from large pelagic fish such as tuna. Its uniform illumination pattern is broken by a dark patch under the throat, and this is thought to appear like a small fish, luring the predators close enough for a quick counterattack (Widder 1998).

2.2.5
Other Luciferins: Known and Unknown

Although the four types of luciferin listed above can account for a large portion of the phylogenetic diversity of marine bioluminescence, there are several other taxa whose light-producing chemistries are either unique or as yet unknown. Some of these organisms are described below.

The **bivalve** *Pholas*, a rock-boring clam, was one of the first organisms to have its luminescence chemistry investigated in detail (Dubois 1887). Its chemistry is a variation on the ones presented thus far: the luciferin is bound within a photoprotein, pholasin, which reacts when exposed to a luciferase or any reactive oxygen species. The photoprotein itself has been cloned, and its sequence is unique among those known (Dunstan et al. 2000).

Another **mollusk** that does not fit into the categories listed above is the gastropod **nudibranch** *Phylliroe* (Fig. 2.3D). Gastropods are a diverse group that includes many benthic snails, as well as some pelagic species such as pteropods and heteropods. For the most part, these pelagic forms are not luminous, but *Phylliroe* and its relatives are. Again, this might be related to their diet, as they are known to prey upon gelatinous plankton such as hydromedusae and siphonophores. This makes coelenterazine the most likely luciferin to be involved, although the chemistry has not yet been investigated. Other interesting exceptions among the gastropods are the freshwater pulmonate limpet *Latia*, which is one of a handful of freshwater species known to luminesce, and the terrestrial snail *Dyakia*. The luciferin and luciferase of *Latia* have been examined in detail (Shimomura and Johnson 1968a, 1968b).

As with chaetognaths, there is another small phylum of marine worms that have a single known luminous member. These are the Nemertea, or **ribbon worms**. Not much has been added to the initial reports of light from the benthic species *Emplectonema* (Kanda 1939). With the large numbers of deep-sea planktonic nemertean species that are being discovered (Roe and Norenburg 2001), it is hoped that there will be additional luminous nemerteans added to the list of taxa.

The phylum Hemichordata includes the **acorn worms**, which have luminous larvae and adults. Recently, the chemistry of the luciferin of the enteropneust *Ptychodera flava* was described (Kanakubo and Isobe 2005), and it is unlike other known luminescence systems. When considering the functions of luminescence, it is important to view the entire life cycle, because many benthic organisms have planktonic larvae for which luminescence may serve very different roles.

Luminescence occurs in an extremely diverse subset of the segmented **polychaete worms**. These annelids have many forms of luminescent expression, including rapid flashing, luminous "scales", and long-duration glowing exudates. The odontosyllids are famous for their spawning displays, where swarms of females are surrounded by circling males, both emitting light. Polychaetes also have some of the shortest wavelength displays (< 440 nm max) from the terebellid *Polycirrus* (Huber et al. 1989), as well as the long-wavelength, gold-colored light of the planktonic *Tomopteris* (Fig. 2.3G). The basic chemistry of some polychaetes has been investigated but not fully solved. Photoproteins – four to six times larger than coelenterate photoproteins and involving iron and peroxide rather than calcium – have been isolated from *Chaetopterus* and from the polynoid *Harmothoe* (described in Chapter 1). No photoproteins or luciferases have been cloned, and the luciferins of the group are unknown. Terrestrial annelids, specifically **earthworms,** can also produce a luminous slime, and glowing species have been found in Europe, North America, Australia, and New Zealand (Rota et al. 2003).

The urochordates are a small but morphologically diverse marine group that includes benthic tunicates, and planktonic **salps**, **doliolids**, and **larvaceans**. They generally filter feed by pumping large volumes of water through fine-mesh "baskets". Although most tunicates and salps are not luminescent, there are, as usual, exceptions. The colonial pyrosome, mentioned above, is thought to use bacteria to produce light. Other planktonic salps are not confirmed to be luminescent, despite the descriptor of a "luminous organ" commonly applied to portions of *Cyclosalpa* that appear bluish in white-light illumination. Doliolids are closely related to salps, and the most common types are not luminescent, although some brightly luminous deep-sea species have recently been discovered (personal observation). There is a single benthic **tunicate**, *Clavelina*, that was recently found to be luminous (Aoki et al. 1989).

Larvaceans are tadpole-shaped planktonic urochordates that build mucous "houses" to filter food from the water. During the formation of the house, they embed bioluminescent particles into the matrix (Galt 1978; Galt and Sykes 1983). These particles sparkle when the houses are disturbed. They may contain coelenterazine, although this is not fully demonstrated (Galt, personal communication). Because larvaceans may produce and discard many houses through the course of the day, they generate far more luminescence than their numbers would indicate. In the deep sea, giant larvaceans build houses with outer filters reaching a meter in diameter.

Another exclusively marine group that may seem an unlikely place in which to find bioluminescence, because of their predominantly benthic life, is the phylum **Echinodermata**. Although sea urchins are not known to be luminous, the other groups all have luminescent members. Even the familiar **sea stars** have luminous species, although they are not the types one would find in a tide pool. **Crinoids** are well represented in the fossil record, and their branched shape looks more like a coral or algae than a relative of the sea stars. Several sea cucumbers are luminous, including the benthic *Pannychia* (Herring 1985a) and the strange, deep pelagic (or at least bentho-pelagic) *Enypiastes* (Robison 1992).

Luminescence of the **brittle stars** (ophiuroids) is undoubtedly the most thoroughly examined system among the echinoderms. They have luminescent organs down the length of their arms, and these are thought to deter predators (Grober 1988). When disturbed, they can shed their arms and move quickly away or wave them about with a rapidly scintillating display. The pharmacology and physiological control of these displays has been well studied (Brehm et al. 1973; De Bremaeker et al. 2000; Dewael and Mallefet 2002). Surprisingly, however, little is known of the chemistry. In *Ophiopsila*, there is a photoprotein of sorts, which may be activated with hydrogen peroxide (Shimomura 1986). It is difficult to examine calcium dependence because of the high levels of calcium present in the skeletal elements of the animals.

2.3
Functions

There is no single role that can explain the ubiquity of bioluminescence in the marine environment. Producing a flash of light is energetically expensive, but it is also an extremely effective means for a small organism to communicate over long distances in the ocean. Some of the more widely proposed functions are discussed below.

2.3.1
Startle or Distract

There are two mechanisms that have been offered to explain the function of dinoflagellate bioluminescence (and both are similarly applicable to other organisms). The first is that the sudden flash of light startles a predator or triggers a rapid rejection response before the cell is ingested (Buskey et al. 1983, 1987). This kind of flash could also apply to jellies, which may be trying to minimize the damage caused by an encounter, or advertising their toxicity. Disruptive flashes are seen in the myctophid fish that have "sternchaser" organs (Fig. 2.6E) – extremely bright photophores near their tail that would be effective if used in combination with a burst of swimming. Many displays are triggered in conjunction with escape responses, and when these take the form of a glowing cloud or burst, it seems reasonable to assume that they would convey an advantage by distracting a predator. Clouds of luminous material are ejected from squids, a chaetognath, various shrimp, and searsiid fishes.

2.3.2
Burglar Alarm

The second proposed function for dinoflagellate bioluminescence, which can also apply to other abundant plankton, is that of the widely known "burglar alarm" hypothesis. In this scenario, grazing crustaceans, for example, will leave a luminous wake (the alarm) that attracts predators to them. This would then have an indirect effect on the fitness of the flashing dinoflagellate. This hypothesis has garnered some experimental support (Mensinger and Case 1992) using both fish and squid as secondary predators.

A variation on this method may explain the dark pigmentation that often lines the gut of otherwise transparent organisms. When an organism eats luminous prey, it puts itself at risk of being illuminated from within, revealing itself in the darkness. If a ctenophore, for example, can regenerate its tissue, then it would be worth sacrificing a glowing bit if that puts the predator at greater risk.

2.3.3 Counterillumination

Many organisms, most notably shrimps, fishes, and squids, have photophores concentrated on their undersurfaces that project light which can be seen from below. The intensity of this light typically varies to match the average ambient light field that is coming down from above. This luminescence, called counter-illumination, serves to camouflage their silhouette either by disruption or by blending of the shadow against the background light field. This is one of the most clearly demonstrated functions of luminescence, by both experimental and inferential evidence (Latz and Case 1982; Mensinger and Case 1992; Warner et al. 1979). Organisms that are thought to counterilluminate will control the intensity, angular distribution, and color of the light they are producing. Some squid even modulate their color to match whether they are against a backdrop of sunlight or moonlight (Young and Mencher 1980).

2.3.4 Mating Displays

There are several examples of species in which the males and females have different photophore patterns, especially among fish and squid. However, the use of bioluminescence in mating displays has scant data to support it. Some examples are noted in the taxonomic treatments above, but the best-supported case is in the species-specific patterns shown by ostracods (Morin 1986). The possibility of using bioluminescence for sexual displays has been examined by Herring (2000).

2.3.5 Prey Attraction

Probably the most intuitive role for bioluminescence is also the rarest. This is the use of bioluminescence to lure prey. Such a function has been suggested for anglerfish, with their luminous organ, and for some squids. Unfortunately, witnessing a capture resulting from a glowing lure is nearly impossible in the deep sea. Recently, another example of glowing lures was discovered in a deep-sea, fish-eating siphonophore (Haddock et al. 2005).

Clearly, bioluminescence is an important component of marine ecology. The wide range of luminescent organisms provides innumerable prospects for further experimentation and discovery.

References

Anctil, M. Modulation of a rhythmic activity by serotonin via cyclic AMP in the coelenterate *Renilla köllikeri*. *J. Comp. Physiol. A* **1989**, *159*, 491–500.

Aoki, M., Hashimoto, K., Watanabe, H. The intrinsic origin of bioluminescence in the ascidian, *Clavelina miniata*. *Biol. Bull.* **1989**, *176*, 57–62.

Bae, Y. M., Hastings, J. W. Cloning, sequencing and expression of dinoflagellate luciferase DNA from a marine alga. *Biochim. Biophys. Acta* **1994**, *1219*, 449–456.

Brehm, P. H., Morin, J. G., Reynolds, G. T. Bioluminescent characteristics of the ophiuroid, *Ophiopsila californica*. *Biol. Bull.* **1973**, *145*, 426.

Buskey, E., Mills, L., Swift, E. The effects of dinoflagellate bioluminescence on the swimming behavior of a marine copepod. *Limnol. Oceanogr.* **1983**, *28*, 575–579.

Buskey, E. J., Mann, C. G., Swift, E. Photophobic responses of calanoid copepods: possible adaptive value. *J. Plankton Res.* **1987**, *9*, 857–870.

Campbell, A. K., Lea, T. J., Ashley, C. C. Coelenterate photoproteins. In *Detection and measurement of free Ca^{2+} in cells* (Ashley, C. C., Campbell, A. K., Eds.), pp. 13–72, *1979*.

Case, J. F., Haddock, S. H. D., Harper, R. D. The ecology of bioluminescence. In *Bioluminescence and chemiluminescence: fundamentals and applied aspects* (Campbell, A. K., Kricka, L. J., Stanley, P. E., Eds.), pp. 115–122. John Wiley and Sons, New Yor, *1994*.

Cavalier-Smith, T. Only six kingdoms of life. *Proc. R. Soc. Lond. B Biol. Sci.* **2004**, *271*, 1251–1262.

De Bremaeker, N., Baguet, F., Mallefet, J. Effects of catecholamines and purines on luminescence in the brittlestar *Amphipholis squamata* (Echinodermata). *J. Exp. Biol.* **2000**, *203*, 2015–2023.

Dewael, Y., Mallefet, J. Luminescence in ophiuroids (Echinodermata) does not share a common nervous control in all species. *J. Exp. Biol.* **2002**, *205*, 799–806.

Dubois, R. Note sur la fonction photogénique chez le *Pholas dactylus*. *C.r. Séanc. Soc. Biol. Fr.* **1887**, *39*, 564–566.

Dunlap, J., Shimomura, O., Hastings, J. W. Crossreactivity between the light-emitting systems of distantly related organisms: Novel type of light-emitting compound. *Proc. Nat. Acad. Sci. USA* **1980**, *77*, 1394–1397.

Dunlap, J. C., Hastings, J. W., Shimomura, O. Dinoflagellate luciferin is structurally related to chlorophyll. *FEBS* **1981**, *135*, 273–276.

Dunstan, S. L., Sala-Newby, G. B., Fajardo, A. B., Taylor, K. M., Campbell, A. K. Cloning and expression of the bioluminescent photoprotein pholasin from the bivalve mollusc *Pholas dactylus*. *J. Biol. Chem.* **2000**, *275*, 9403–9409.

Frank, T. M., Widder, E. A., Latz, M. I., Case, J. F. Dietary maintenance of bioluminescence in a deep-sea mysid. *J. Exp. Biol.* **1984**, *109*, 385–389.

Galt, C. P. Bioluminescence: dual mechanism in a planktonic tunicate produces brilliant surface display. *Science* **1978**, *200*, 70–72.

Galt, C. P., Sykes, P. F. Sites of bioluminescence in the appendicularians *Oikopleura dioca* and *O. labradoriensis* (Urochordata: Larvacea). *Biol. Bull.* **1983**, *77*, 155–159.

Grober, M. S. Brittle-star bioluminescence functions as an aposematic signal to deter crustacean predators. *Animal Behavior* **1988**, *36*, 493–501.

Haddock, S. H. D., Case, J. F. A bioluminescent chaetognath. *Nature* **1994**, *367*, 225–226.

Haddock, S. H. D., Case, J. F. Not all ctenophores are bioluminescent: *Pleurobrachia*. *Biol. Bull.* **1995**, *189*, 356–362.

Haddock, S. H. D., Dunn, C. W., Pugh, P. R., Schnitzler, C. E. Bioluminescent and red-fluorescent lures in a deep-sea siphonophore. *Science* **2005**, *309*, 263.

Haddock, S. H. D., Rivers, T. J., Robison, B. H. Can coelenterates make coelenterazine? Dietary requirement for luciferin in cnidarian bioluminescence. *Proc. Nat. Acad. Sci.* **2001**, *98*, 11148–11151.

Haneda, Y., Tsuji, F. I. The luminescent system of pony fishes. *J. Morphol.* **1976**, *150*, 539–552.

Harper, R. D., Case, J. F. Disruptive counterillumination and its anti-predatory value in the plainfish midshipman *Porichthys notatus*. *Mar. Biol.* **1999**, *134*, 529–540.

Harvey, E. N. *Bioluminescence*. Academic Press, New York, *1952*.

Hastings, J. W. Biological diversity, chemical mechanisms, and the evolutionary origins of bioluminescent systems. *J. Mol. Evol.* **1983**, *19*, 309–321.

Haygood, M. G., Distel, D. L. Bioluminescent symbionts of flashlight fishes and deep-sea anglerfishes form unique lineages related to the genus *Vibrio*. *Nature* **1993**, *363*, 110–111.

Herring, P. J. Luminescence in cephalopods and fish. *Symp. Zool. Soc. Lond.* **1977**, *38*, 127–159.

Herring, P. J. Bioluminescent echinoderms: unity of function in diversity of expression? In *Echinoderm research* (Emson, R. H., Smith, A. B., Campbell, A. C., Eds.), pp. 1–17. Balkema, Rotterdam, **1985a**.

Herring, P. J. Tenuous evidence for the luminous mouthed shark. *Nature* **1985b**, *318*, 238.

Herring, P. J. Systematic distribution of bioluminescence in living organisms. *J. Biolum. Chemilum.* **1987**, *1*, 147–163.

Herring, P. J. Bioluminescent responses of the deep-sea scyphozoan *Atolla wyvillei*. *Mar. Biol.* **1990**, *106*, 413–417.

Herring, P. J. Species abundance, sexual encounter, and bioluminescent signalling in the deep sea. *Phil. Trans. R. Soc. Lond. B* **2000**, *355*, 1273–1276.

Herring, P. J., Watson, M. Milky Seas: a bioluminescent puzzle. *The Marine Observer* **1993**, *63*, 22–30.

Herring, P. J., Widder, E. A. Bioluminescence of deep-sea coronate medusae (Cnidaria: Scyphozoa). *Mar. Biol.* **2004**, *146*, 39–51.

Hori, K., Cormier, M. J. Studies on the bioluminescence of *Renilla reniformis*. V. Absorption and fluorescence characteristics of chromatographically pure luciferin. *Biochim. Biophys. Acta* **1965**, *102*, 386–396.

Huber, M. E., Arneson, A. C., Widder, E. A. Extremely blue bioluminescence in the polychaete *Polycirrus perplexus* (Terebellidae). *Bull. Mar. Sci.* **1989**, *44*, 1236–1239.

Inouye, S., Watanabe, K., Nakamura, H., Shimomura, O. Secretional luciferase of the luminous shrimp *Oplophorus gracilirostris*: cDNA cloning of a novel imidazopyrazinone luciferase *FEBS* **2000**, *481*, 19–25.

Johnsen, S., Balser, E. J., Widder, E. A. Light-emitting suckers in an octopus. *Nature* **1999**, *398*, 113–114.

Kanakubo, A., Isobe, M. Isolation of brominated quinones showing chemiluminescence activity from luminous acorn worm, *Ptychodera flava*. *Bioorganic, Medicinal Chemistry* **2005**, *13*, 2741–2747.

Kanda, S. The luminescence of a nemertean, *Emplectonema kandai*, Kato. *Biol. Bull.* **1939**, *77*, 166–173.

Kishi, Y., Goto, T., Hirata, Y., Shimomura, O., Johnson, F. H. *Cypridina* bioluminescence I: structure of *Cypridina* luciferin. *Tetrahedron Lett.* **1966**, *7*, 3427–3436.

Kumar, S., Harrylock, M., Walsh, K. A., Cormier, M. J., Charbonneau, H. Amino acid sequence of the Ca^{2+}-triggered luciferin binding protein of *Renilla reniformis*. *FEBS* **1990**, *268*, 287–290.

Labas, Y. A., Matz, M. V., Zakhartchenko, V. A. On the origin of bioluminescent systems. In *Bioluminescence and Chemiluminescence* (Case, J. F., Herring, P. J., Robison, B. H., Haddock, S. H. D., Kricka, L. J., Stanley, P. E., Eds.), pp. 91–94. World Scientific, Asilomar, CA, *2001*.

Latz, M. I., Case, J. F. Light organ and eyestalk compensation to body tilt in the luminescent midwater shrimp, *Sergestes similis*. *J. Exp. Biol.* **1982**, *98*, 83–104.

Liu, L., Wilson, T., Hastings, J. W. Molecular evolution of dinoflagellate luciferases, enzymes with three catalytic domains in a single polypeptide. *Proc. Nat. Acad. Sci.* **2004**, *101*, 16555–16560.

Lorenz, W. W., Mccann, R. O., Longiaru, M., Cormier, M. J. Isolation and expression of a cDNA encoding *Renilla reniformis* luciferase. *Proc. Nat. Acad. Sci.* **1991**, *88*, 4438–4442.

Mallefet, J., Shimomura, O. Presence of coelenterazine in mesopelagic fishes from the Strait of Messina. *Mar. Biol.* **1995**, *124*, 381–385.

Markova, S., Golz, S., Frank, L. A., Kalthof, B., Vysotski, E. S. Cloning and expression of cDNA for a luciferase from the marine copepod *Metridia longa*. *Journal of Biological Chemistry* **2004**, *279*, 3212–3217.

Meighen, E. A. Molecular biology of bacterial bioluminescence. *Microbiological Reviews* **1991**, *55*, 123–142.

Mensinger, A. F., Case, J. F. Dinoflagellate luminescence increases susceptibility of zooplankton to teleost predation. *Mar. Biol.* **1992**, *112*, 207–210.

Miller, S. D., Haddock, S. H. D., Elvidge, C. D., Lee, T. H. Detection of a bioluminescent milky sea from space. *Proc. Nat. Acad. Sci.* **2005**, *102*, 14181–14184.

Morin, J. G. "Firefleas" of the sea: Luminescence signaling in marine ostracode crustaceans. *Florida Entomologist* **1986**, *69*, 105–121.

Morse, D., Pappenheimer, A. M. J., Hastings, J. W. Role of a luciferin-binding protein in the circadian bioluminescent reaction of *Gonyaulax polyedra*. *Journal of Biological Chemistry* **1989**, *264*, 11822–11826.

Munk, O. The escal photophore of ceratioids (Pisces; Ceratioidei) a review of structure and function. *Acta Zool.* **1999**, *80*, 265–284.

Nakamura, H., Kishi, Y., Shimomura, O., Morse, D., Hastings, J. W. Structure of dinoflagellate luciferin and its enzymatic and nonenzymatic air-oxidation products. *J. Am. Chem. Soc.* **1989**, *111*, 7607–7611.

Nealson, K. H., Hastings, J. W. Bacterial Bioluminescence: Its control and ecological significance. *Microbiological Reviews* **1979**, *43*, 4496–4518.

O'day, W. T., Fernandez, H. R. *Aristostomias scintillans* (Malacosteidae): a deep-sea fish with visual pigments apparently adapted to its own bioluminescence. *Vision Res.* **1974**, *14*, 545–550.

Oba, Y., Tsuduki, H., Kato, S., Ojika, M., Inouye, S. Identification of the luciferin-luciferase system and quantification of coelenterazine by mass spectrometry in the deep-sea luminous ostracod *Conchoecia pseudodiscophora*. *Chembiochem* **2004**, *5*, 1495–1499.

Philippe, H., Lartillot, N., Brinkmann, H. Multigene analyses of bilaterian animals corroborate the monophyly of Ecdysozoa, Lophotrochozoa, and Protostomia. *Mol. Biol. Evol.* **2005**, *22*, 1246–1253.

Podar, M., Haddock, S. H. D., Sogin, M. L., Harbison, G. R. A molecular phylogenetic framework for the phylum Ctenophora using 18S rRNA genes. *Mol. Phylo. Evol.* **2001**, *21*, 218–230.

Polet, S., Berney, C., Fahrni, J., Pawlowski, J. Small-subunit ribosomal RNA gene sequences of Phaeodarea challenge the monophyly of Haeckel's Radiolaria. *Protist* **2004**, *155*, 53–63.

Ramanathan, S., Shi, W., Rosen, B. P., Daunert, S. Sensing antimonite and arsenite at the subattomole level with genetically engineered bioluminescent bacteria. *Anal. Chem.* **1997**, *69*, 3380–3384.

Rees, J. F., De Wergifosse, B., Noiset, O., Dubuisson, M., Janssens, B., Thompson, E. M. The origins of marine bioluminescence: Turning oxygen defence mechanisms into deep-sea communication tools. *J. Exp. Biol.* **1998**, *201*, 1211–1221.

Rees, J. F., Thompson, E. M., Baguet, F., Tsuji, F. I. Detection of coelenterazine and related luciferase activity in the tissues of the luminous fish, Vinciguerria attenuata. *Comp. Biochem. Physiol. B* **1990**, *96A*, 425–430.

Robison, B. H. Bioluminescence in the benthopelagic holothurian Enypniastes eximia. *J. Mar. Biol. Assoc. U. K.* **1992**, *72*, 463–472.

Robison, B. H., Reisenbichler, K. R., Hunt, J. C., Haddock, S. H. Light production by the arm tips of the deep-sea cephalopod *Vampyroteuthis infernalis*. *Biol. Bull.* **2003**, *205*, 102–109.

ROE, P., NORENBURG, J. L. Morphology and taxonomic distribution of a newly discovered feature, postero-lateral glands, in pelagic nemerteans. *Hydrobiol.* **2001**, *456*, 133–144.

ROTA, E., ZALESSKAJA, N. T., RODIONOVA, N. A., PETUSHKOV, V. N. Redescription of *Fridericia heliota* (Annelida, Clitellata: Enchytraeidae), a luminous worm from the Siberian taiga, with a review of bioluminescence in the Oligochaeta. *J. Zool., Lond.* **2003**, *260*, 291–299.

RUBY, E. G., MCFALL-NGAI, M. J. A squid that glows in the night: development of an animal-bacterial mutualism. *J. Bacteriol.* **1992**, *174*, 4865–4870.

SHIMOMURA, O. Bioluminescence of the brittle star *Ophiopsila californica*. *Photochem. Photobiol.* **1986**, *44*, 671–674.

SHIMOMURA, O. Presence of coelenterazine in non-bioluminescent marine organisms. *Comp. Biochem. Physiol.* **1987**, *86B*, 361–363.

SHIMOMURA, O., FLOOD, P. Luciferase of the scyphozoan medusa *Periphylla periphylla*. *Biol. Bull.* **1998**, *194*, 244–252.

SHIMOMURA, O., GOTO, T., HIRATA, Y. Crystalline *Cypridina* luciferin. *Bull. Chem. Soc. Japan* **1957**, *30*, 929–933.

SHIMOMURA, O., INOUE, S., JOHNSON, F. H., HANEDA, Y. Widespread occurrence of coelenterazine in marine bioluminescence. *Comp. Biochem. Physiol.* **1980**, *65B*, 435–437.

SHIMOMURA, O., JOHNSON, F. H. Purification and properties of the luciferase and of a protein cofactor in the bioluminescence system of *Latia neritoides*. *Biochemistry* **1968a**, *7*, 2574–2580.

SHIMOMURA, O., JOHNSON, F. H. The structure of *Latia* luciferin. *Biochemistry* **1968b**, *7*, 1734–1738.

SZENT-GYORGYI, C., BALLOU, B. T., DAGNAL, E., BRYAN, B. Cloning and characterization of new bioluminescent proteins. *Proceedings of SPIE* **2003**, *3600*, 4–11.

TAKAHASHI, H., ISOBE, M. Photoprotein of luminous squid, *Symplectoteuthis oualaniensis* and reconstruction of the luminous system. *Chem. Lett.* **1994**, *May*, 843–846.

THOMPSON, E. M., NAGATA, S., TSUJI, F. I. Cloning and expression of cDNA for the luciferase from the marine ostracod Vargula hilgendorfii. *Proc. Nat. Acad. Sci.* **1989**, *86*, 6567–6571.

THOMSON, C. M., HERRING, P. J., CAMPBELL, A. K. Evidence for *de novo* biosynthesis of coelenterazine in the bioluminescent midwater shrimp, *Systellaspis debilis*. *J. Mar. Biol. Assoc. U. K.* **1995**, *75*, 165–171.

THOMSON, C. M., HERRING, P. J., CAMPBELL, A. K. The widespread occurrence and tissue distribution of the imidazolopyrazine luciferins. *J. Biolum. Chemilum.* **1997**, *12*, 87–91.

WADA, M., AZUMA, N., MIZUNO, N., KUROKURA, H. Transfer of symbiotic luminous bacteria from parental *Leiognathus nuchalis* to their offspring. *Mar. Biol.* **1999**, *135*, 683–687.

WARNER, J. A., CASE, J. F. The zoogeography and dietary induction of bioluminescence in the midshipman fish, *Porichthys notatus*. *bb* **1980**, *159*, 231–246.

WARNER, J. A., LATZ, M. I., CASE, J. F. Cryptic bioluminescence in a midwater shrimp. *Science* **1979**, *203*, 1109–1110.

WIDDER, E. A. A predatory use of counterillumination by the squaloid shark, *Isistius brasiliensis*. *Environ. Biol. Fishes* **1998**, *53*, 267–273.

YOUNG, R. E., MENCHER, F. M. Bioluminescence in mesopelagic squids: Diel color change during counterillumination. *Science* **1980**, *208*, 1286–1288.

3 Beetle Luciferases: Colorful Lights on Biological Processes and Diseases

Vadim R. Viviani and Yoshihiro Ohmiya

3.1 Introduction

Bioluminescence, the emission of cold light by living organisms for communicative purposes, has been extensively used to convey information of living processes and for bioanalytical purposes. Among the bioluminescent systems, the luciferin–luciferase system of fireflies is one of the most studied and exploited for bioanalytical purposes. It has been used to detect and quantify ATP in living cells and for microbiological purposes. More recently, the luciferase gene has been extensively used as a luminescent biomarker to study gene expression and to follow up the course of infections and cancer development in animal models, helping to devise new therapy strategies. New luciferases producing different bioluminescent colors, including red, are improving these methods and promise to extend the range of bioanalytical applications in the fields of biophotonics, biotechnology, medicine, and environment.

Bioluminescence results from the conversion of potential energy of chemical bonds in light. In such exergonic processes, molecules generally known as luciferins are oxidized, through the intermediacy of peroxides, forming electronically excited-state products that decay and thereby emit light [1, 2]. Luciferases are the key enzymes catalyzing such reactions. Bioluminescence in nature is used for communicative purposes such as for sexual attraction [2]. For humans, bioluminescence is currently used as one of the most versatile and sensitive tools to convey information about specific living processes in basic research and for bioanalytical purposes. The luciferase genes of several bioluminescent organisms have been cloned, sequenced, and expressed in different cells [2]. Among them, the firefly luciferases are one of the most used systems [3, 4], being originally used as analytical tools to measure ATP and later as one of the most versatile reporter genes [5]. Many beetle luciferases producing a range of bioluminescent colors have been cloned and are expanding the range of applications as reporter genes [5]. Currently, under the keyword "luciferase" more than 120 new papers per month and 2000 papers per year are added to the Pubmed database, most of them about

Photoproteins in Bioanalysis. Edited by Sylvia Daunert and Sapna K. Deo

ISBN: 3-527-31016-9

applications. It would be impossible and outside the scope of this chapter to detail all the applications using the firefly luciferin–luciferase system. Therefore, we will give an overview of the main lines of applications of these fantastic enzymes and the prospects of using new luciferases in biotechnology and biomedicine.

3.2 Beetle Luciferases

Bioluminescent beetles are found in three main families: Lampyridae (fireflies), Phengodidae (railroad worms) (Fig. 3.1), and Elateridae (click beetles). They emit a wide variety of bioluminescence colors using the same luciferin (Fig. 3.1): fireflies emit in the green-yellow region, click beetles in the green-orange region, and railroad worms in the green-red region of the spectrum [4]. The bioluminescent system of beetles involves the ATP-activated oxidation of a benzothiazolic luciferin [6] (Fig. 3.2). In the first step, luciferase catalyzes the activation of D-luciferin by adenylation at expenses of ATP. During the second step luciferase catalyzes the oxidation of luciferyl adenylate, followed by departure of AMP and concomitant formation of a cyclic peroxide intermediate, named dioxetanone, whose cleavage results in excited singlet oxyluciferin and carbon dioxide. The decay of excited oxyluciferin generates a photon, usually in the green-yellow range of the spectrum, with a reported efficiency of 88%. Three basic mechanisms have been proposed to explain bioluminescence color modulation by beetle luciferases: (1) nonspecific effects including the solvent effect and orientation polarizability [7, 8]; (2) specific interactions between active-site residues and oxyluciferin functional groups, mainly acid–base and electrostatic interactions [9]; and (3) the degree of rotational freedom of oxyluciferin thiazinic rings under the influence of active-site geometry [10]. The two former hypotheses are supported by experimental evidence.

Many beetle luciferases, mainly from fireflies, have been cloned and sequenced. They are polypeptides consisting of 543–550 amino acid residues with an average molecular weight of 60 kDa [11–24] and a C-terminal peroxisomal targeting peptide, Ser–Lys–Leu, which drives them into peroxisomes [3, 4]. They are homologous to the family of AMP ligases that include fatty acid CoA synthetases, aromatic acid CoA synthetases, coumarate CoA synthetases, and peptidyl synthetases [3, 4, 25].

The three-dimensional structure of firefly luciferase was solved in the absence of substrates [26]. It shows a main N-terminal domain bound by a hinge to a smaller C-terminal domain. The active site was proposed to be located in a cavity in the N-terminal domain facing the C-terminal domain, and two models were constructed [27, 28]. The C-terminal domain was found to play an essential role for efficient yellow-green light emission [29]. Beetle luciferases are classified in two functional groups: (1) the pH-sensitive firefly luciferases and (2) the pH-insensitive railroad worm luciferases. Studies based on modeling, photoaffinity, and site-directed mutagenesis identified important residues for the catalytic function [30–33] and for bioluminescence color determination in firefly luciferases

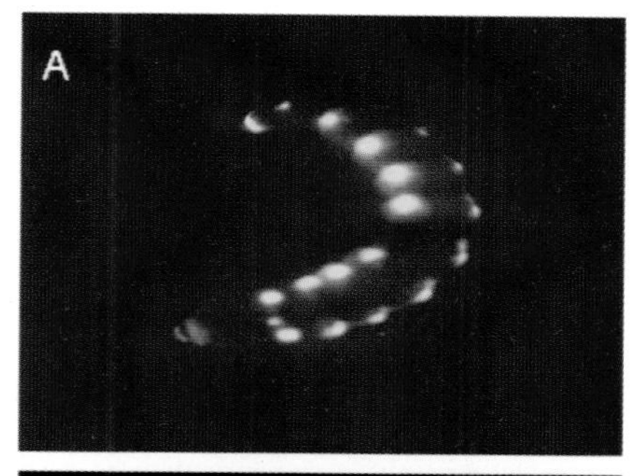

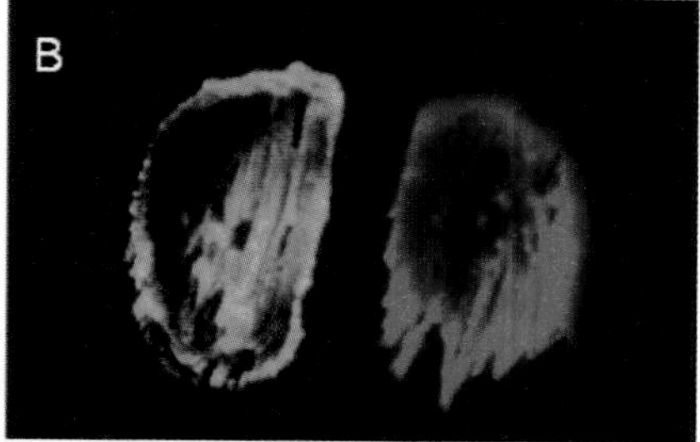

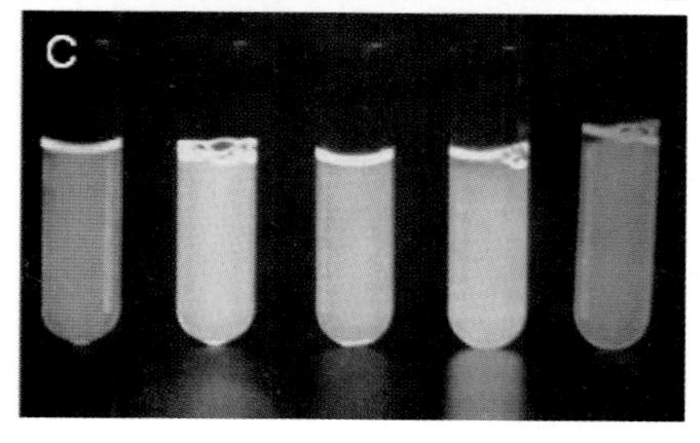

Fig. 3.1 Beetle bioluminescence:
(A) *Phrixotrix* railroadworm;
(B) *E. coli*-expressing *Phrixotrix* railroad worm green- (left) and red-emitting (right) luciferases;
(C) *in vitro* assay using different beetle luciferases.

[34–37], click beetles, and railroad worm luciferases [38–42]. They fall into two main classes: those directly involved in the active site and those that indirectly affect the active-site conformation. The pH-sensitive firefly luciferases were found to be especially prone to be affected by mutations outside the active site, whereas the pH-insensitive ones were more resistant [43].

3.3
Bioanalytical Assays of ATP

Since the purification of the *Photinus pyralis* firefly luciferase and synthesis of its luciferin, about 50 years ago, the luciferin–luciferase system of fireflies has been extensively used in applications involving the detection and quantification of ATP (Table 3.1, Fig. 3.3). It is the most sensitive method available, superior to fluorescence and colorimetric methods, detecting down to 0.01 fmol of ATP and showing the widest linear range (1 pM–1 μM) [44]. Methods of ATP extraction depend on the sample to be analyzed, and several methods have been devised [45]. Kits for analysis of ATP in biological samples are available.

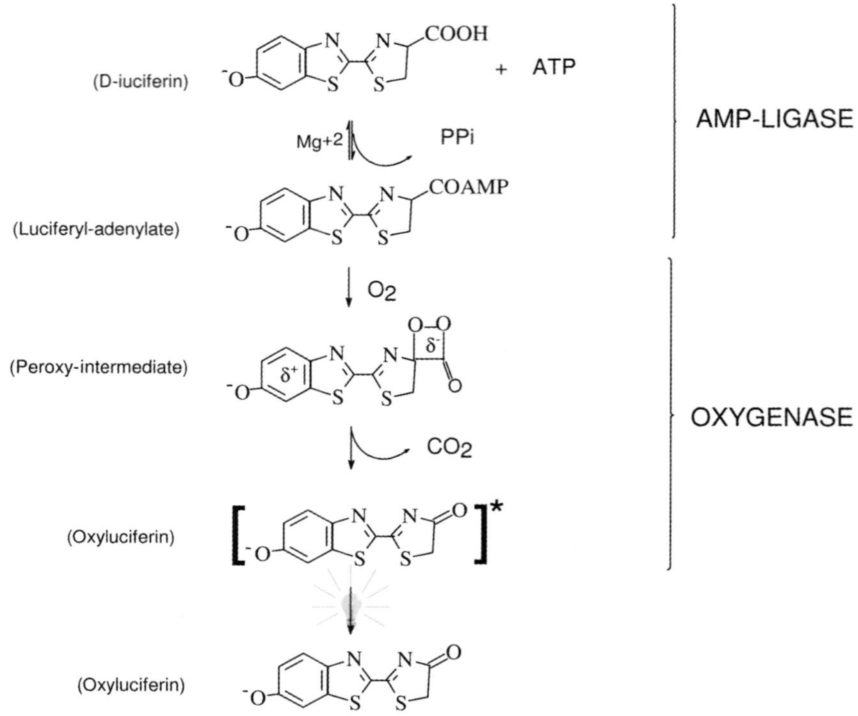

(D-iuciferin) + ATP

Mg+2 PPi

(Luciferyl-adenylate)

O_2

(Peroxy-intermediate)

CO_2

(Oxyluciferin)

(Oxyluciferin)

AMP-LIGASE

OXYGENASE

Fig. 3.2 Bioluminescent reaction catalyzed by beetle luciferases.

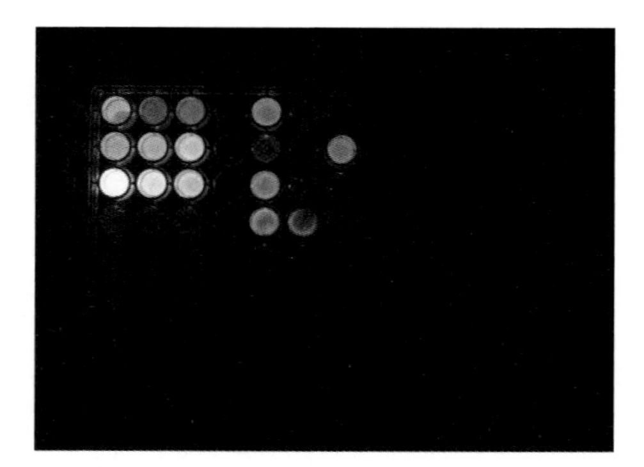

Fig. 3.3 Bioluminescence of beetle luciferin–luciferase system in the presence of different concentrations of ATP.

Table 3.1 Analytical applications of beetle luciferases.

Major area	*Specific application*		*Examples*
ATP assays	ATP assays		
	Hygiene		Food, beverage, textile, hospital
	Microbiology		Bacteriuria
	Biomass		Seawater, water treatment plants
	Cell viability		Platelets, erythrocytes, macrophages, spermatozoids
	Enzyme assays	Formation	Creatine kinase, oxidative phosphorylation, pyruvate kinase, adenylate kinase, phosphodiesterase
		Degradation	Hexokinase, ATPases, apyrase cGMP phosphodiesterase
Reporter gene	Transfection		
	Promoter activity assays		Bacteria, yeast, plants, insects, mammals
	Biophotonic imaging/gene delivering		
	• Pathogenic virus		HIV, HSV, HTMLV
	• Pathogenic Mycobacteria		*M. tuberculosis*
	• Pathogenic bacteria		*Salmonella*, *Staphylococcus*
	• Pathogenic yeast		*Candida albicans*
	• Tumor progression		Glioma, human cervical carcinoma
	• Cell trafficking		Lymphocytes
	• Transgenic expression		Mice
	Cytotoxicity		
	Biosensors • Environmental disruptors		Hg, As, naphthalenes, phenols, agrochemicals

3.3.1 Biomass Estimation and Microbiological Contamination

The luciferin–luciferase system of fireflies was extensively used in applications indirectly involving ATP measurement for biomass estimation [46, 47] (Table 3.1). Examples are their use for microbiological contamination evaluation in several industries, including food, beverages, milk, and textiles. This system is being used in water treatment plants and in hospitals for hygiene tests. Some companies commercially provide specific kits developed for assaying the contamination of these products [48, 49]. Cell activation from platelets, nerve endings, muscle, and plasma also can be measured based on ATP [44]. The luciferin–luciferase system has been successfully used to monitor the continuous production of ATP by mitochondria and ATP release by nerve endings and platelets. Tests for analyzing contamination of biological fluids and infections, among them bacteriuria, have been developed [50].

3.3.2
Cytotoxicity and Cell Viability Tests

The concentration of ATP in normal living cells ranges from 1 mM to 10 mM, thus saturating for firefly luciferase. Based on the ATP content of healthy cells, and its depletion in dying ones, assays for cell viability (erythrocytes, platelets, white cells, spermatozoids, and cultured cells, etc.) have been developed [44]. The same principle has been applied for tests to evaluate the cytotoxicity of several compounds.

3.3.3
Enzymatic Assays

The luciferin–luciferase system also has been used for assays of enzymes directly or indirectly involved in ATP generation and degradation [44] (Table 3.1). It has been used to study the direct generation of ATP in metabolic processes such as oxidative phosphorylation and photosynthesis, using mitochondria and chloroplasts, or to investigate the mechanisms of ATPases and ion pumps. An assay kit for myocardial infarction involving the measurement of creatine kinase MB isoenzyme, released from damaged myocardial cells, has been produced commercially [51].

3.4
Luciferases as Reporter Genes

After cloning of the cDNA for firefly luciferase, a new range of applications involving the use of the luciferase gene as a reporter gene appeared. Since then, the firefly luciferase gene has been extensively used to investigate transformation and transfection efficiencies in different cells, analysis of promoter activities, and location of gene expression, etc. [52–54]. Viral vectors such as adenovirus, HSV, vaccinia virus, etc., were labeled by inserting the firefly luciferase gene and were used to monitor their gene expression and dissemination in cells and tissues [55–58]. Regulation of gene expression of human cytomegalovirus (HCMV) and human immunodeficiency virus (HIV) in individual intact HeLa cells has been imaged using constructs with firefly luciferase gene downstream of viral promoters [54]. The firefly luciferase gene has been expressed in several mammalian cell lines such as monkey kidney cells, Chinese hamster ovary cells, HeLa cells, and rat pituitary tumor cells [51–58]. A fusion protein of firefly luciferase–aequorin was used to monitor ATP and Ca^{+2} intracellular changes in HeLa cells [59].

3.4.1
Dual and Multiple Reporter Assays

For practical purposes, two promoter activities, for target and control genes, must be analyzed, because expression of the reporter enzyme depends on conditions in

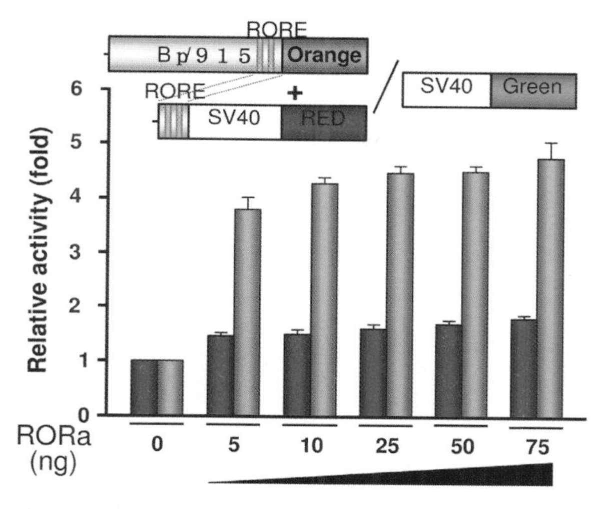

Fig. 3.4 Multicolor reporter system showing the simultaneous monitoring of mRORa4 dose-dependent induction of RORE-mediated (red bars) and mBmal1 promoter fragment-driven (orange bars) transcriptions [60].

the cell as estimated by the expression of the control gene. The firefly luciferase is used to estimate a transcriptional activity, whereas the *Renilla* luciferase is used as an internal control for normalization of the firefly luciferase activity to minimize experimental variability caused by differences in cell viability or transfection efficiency, the so-called dual-assay system (e.g., the Dual Luciferase Reporter Assay System; Promega, Madison, WI, USA; http://www.promega.com/tbs/tm040/tm.html/). However, multiple transcriptions, which simultaneously progress in the cells, cannot be monitored in this system, since only one-gene transcriptional activity can be measured. Recently, in order to measure two-gene transcriptional activities simultaneously, a tricolor reporter *in vitro* assay system was developed in which the expression of three genes can be monitored simultaneously by splitting the emissions from green-, orange-, and red-emitting luciferases with optical filters (Fig. 3.4) [60]. This system allows us to simply and rapidly monitor three gene expressions simultaneously (two are test reporters and one is an internal control) by using a single luminescent substrate in one tube (e.g., the MultiReporter Assay System -Tripluc-, TOYOBO, Osaka, Japan; http://www.toyobo.co.jp/seihin/xr/lifescience.html).

3.5 Biophotonic Imaging in Animals: A Living Light on Diseases

One of the most exciting and promising applications of beetle luciferase genes is the real-time imaging of normal and pathogenic biological processes in living organisms (Fig. 3.5).

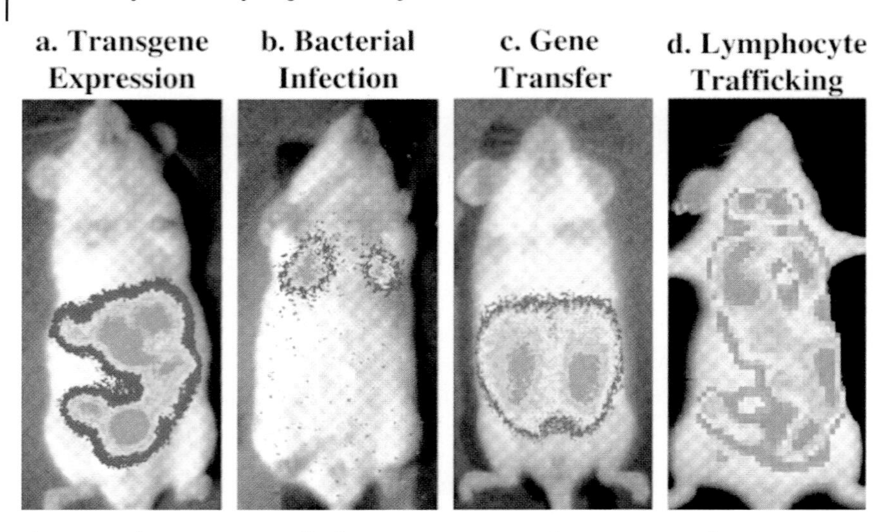

Fig. 3.5 Biophotonic imaging of bioluminescent mice according to Contag and Bachmann [61].

3.5.1
Pathogen Infection in Living Models

The firefly luciferase gene has been transiently and stably expressed in mammalian organisms for a variety of purposes [52–54, 61–64]. It has been used to follow up bacterial infections in model animals using *Streptococcus, Salmonella,* and *Staphylococcus aureus* strains and mycobacterial infections such as *Mycobacterium tuberculosis* [64–67]. Bioluminescent *Candida albicans* strains also have been engineered with the firefly luciferase gene to follow the course of infections in living mice models. Similar studies have been done using viral infections such as herpes (HSV) [57] and adenovirus [56]. Transgenic mice containing viral promoter fusions such as HTMLV have been developed and tested to study the range of tissues and cells that are capable of supporting viral expression. HIV infection can be followed by real-time imaging in mammalian cell cultures and in live animals. Assays for HIV using a plasmid that contains the firefly luciferase gene under the control of a viral promoter have been developed [63, 68]. Transgenic mice containing the LTR promoter of HIV fused to the firefly luciferase gene were produced as a useful model to study *in vivo* regulation of viral gene expression [69].

3.5.2
Drug Screening

Thus, based on the above technology, the firefly luciferase gene has been used to evaluate antimicrobial activity in pathogenic strains of *Mycobacterium tuberculosis* [66] and *Staphylococcus aureus* and antiviral activity in HIV-infected mice [69, 70], helping to speed up drug screening procedures.

3.5.3 Tumor Proliferation and Regression Studies

The firefly luciferase gene is also being used for noninvasive assessment of tumor cell proliferation in animal models [71], which is helping in the development of new antineoplastic therapies [72]. Such procedures have been employed to study progression and regression of human cervical carcinoma cells in living mice. Tumors have been imaged by the selective expression of bioluminescence by using virus-carried luciferase genes linked to tissue- or tumor-specific promoters [73]. Furthermore, the tropism, migration, and infiltration of certain mammalian cells such as leucocytes into tumors can be used to bioluminescently label these cells to reveal the location of tumors and metastases [73]. Besides being noninvasive, this technology allows for a sensitive detection and location of metastases during the initial stages of tumorigenesis, allowing us to devise new therapeutic strategies before the late manifestation of cancer. Perhaps in the future these bioluminescent technologies could be used for diagnostic and therapy evaluation purposes in human beings themselves.

3.5.4 Gene Delivery and Gene Therapy

Similarly, the firefly luciferase gene has been used to monitor delivery, location, and pattern of transgene expression in gene therapy assays [52]. Through the use of viral and retroviral vectors, genes encoding therapeutic proteins as well as the luciferase markers have been introduced into mammalian cells to evaluate the treatment of diseases such as Parkinson's disease, cystic fibrosis, and cancer.

3.5.5 Luciferase as Biomarkers for Cell Trafficking Studies

Transgenic mice expressing the firefly luciferase gene and fusion protein with GFP have been engineered as a source of bioluminescently labeled cells to track the fate of specific cells during transplants and to avail cell trafficking of cells such as lymphocytes.

3.5.6 Immunoassays

Biotinylated recombinant firefly luciferase was also used as a probe to detect proteins and nucleic acids in blots. A method for detecting protein A-bearing *Staphylococcus aureus* using biotinylated luciferase has been developed: this method is more sensitive and rapid than conventional colorimetric assays, detecting down to 1 pg ml^{-1} of protein A [74]. A fusion protein of protein A and the firefly luciferase gene was constructed and successfully used in sensitive dot and Western blotting assays, detecting down to 5 pg of α-fetoprotein [75]. Firefly luciferase gives results comparable to alkaline phosphatase.

3.6 Biophotonic Imaging in Plants

The firefly luciferase gene has been used as a reporter gene in transgenic plants with the aim of studying gene expression under stress conditions, infection, and circadian control studies [76, 77].

3.7 Biosensors: Sensing the Environment

Firefly luciferase-engineered bacteria, cyanobacteria, yeast, algae, plants, and multicellular animals have been successfully used as whole-cell biosensors [78–84]. Because firefly luciferase depends on ATP, its bioluminescent signal can be conveniently linked to the energetic status of the cell. In the simplest cases these cells can be directly used in general toxicity tests. More sophisticated bioassays for specific toxic agents have been constructed by fusion of a luciferase gene to a stress-inducible transducer, the luminescence being directly related to the degree of activation, detecting down to ppm amounts of these compounds. The latter are exemplified by whole-cell biosensors for heavy metals such as mercury and arsenic, phenols, agrochemicals, hormones [84], and other environmental disruptors. They can also be used as reporter genes to monitor the bioavailability of nutrients such a nitrogen and iron.

3.8 Novel Luciferases: Different Colors for Different Occasions

Until recently, only firefly luciferases had been used for bioanalytical purposes. However, they produce only yellow-green light under physiological conditions and have the drawback of being pH sensitive, reducing the efficiency of signal detection for most applications. Red luciferase mutants were produced by genetic engineering, and some of them are currently in use for applicative purposes. However, their spectrum is not very red-shifted and furthermore is very broad [36, 85–87]. Click beetle luciferases, despite producing a wider range of bioluminescence colors and being pH insensitive, have not acquired the same level of popularity as firefly luciferases. More recently, our group has cloned several beetle luciferases from railroad worms, click beetles [19, 21, 22], and fireflies [24, 43]. Among them, the *Phrixotrix hirtus* railroad worm luciferase is the only true red-emitting luciferase, displaying a very narrow spectrum [19] (Fig. 3.6). The luciferase from the click beetle *Pyrearinus termitilluminans*, besides displaying

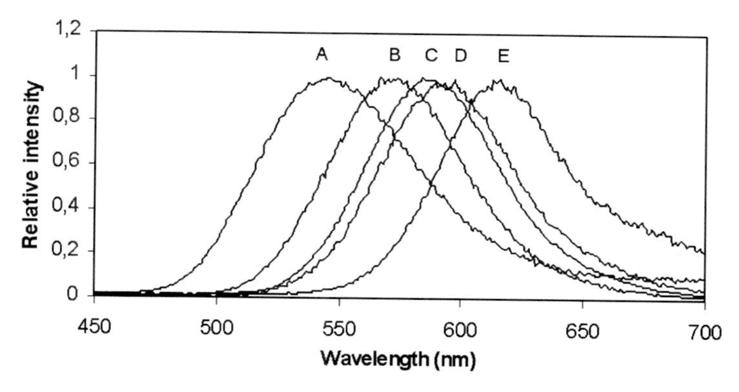

Fig. 3.6 Bioluminescence spectra of recombinant beetle luciferases cloned and engineered by our laboratories [19, 21, 42].
(A) *Pyrearinus termitilluminans*;
(B) *Phrixotrix* mutant yellow-emitting luciferase;
(C and D) *Phrixotrix* mutant orange-emitting luciferases;
(E) *Phrixotrix hirtus* red-emitting luciferase.

the most blue-shifted spectrum among beetle luciferases, is also relatively thermostable (Fig. 3.6). Furthermore, our laboratories succeeded in engineering beetle luciferases producing a wide range of colors varying from green to red (Fig. 3.5) [40–42]. Recently, Nakajima et al. succeeded in expressing red luciferase in mammalian cells [89]. For these reasons, red-emitting luciferase is assuming important new applications in multicolor reporter systems for bacterial and mammalian cells [89, 90]. Red-emitting luciferase has further potential to be used in pigmented biological samples where the luminescent signal of firefly luciferases is considerably decreased. A photodetection system using interference filters has been developed to monitor the bioluminescent signal of distinct luciferases simultaneously in multicolor reporter assays. Thermostable mutants were produced by random and site-directed mutagenesis [91–93]. Recently, *Photinus* firefly luciferase that is stable at 60 °C was developed by Promega.

Cloning of new luciferases with different properties and their engineering can expand their range of applications. For example, the engineering of these luciferases to produce more red-shifted and more blue-shifted emitting enzymes is underway. Theoretically, based solely on the chemical structure of beetle D-luciferin, it is possible to extend the bioluminescence color up to 638 nm for the red end of the spectrum and to 530 nm for the blue end. However, combinatorial chemistry can further expand these limits. For example, luciferin analogues producing blue- and red-shifted colors have been developed already [94, 95]. The use of blue-shifted luciferases with orange and red fluorescent proteins can expand the use of the BRET system to beetle luciferases. The development of new and more sensitive photodetecting systems promises to bring bioluminescence technology to routine laboratory and biomedical activities.

Acknowledgments

This work was supported by grants from Fundação de Ámparo à Pesquisa do Estado de São Paulo (FAPESP 00/05467-4, São Paulo Brazil) and from the Japan Society for the Promotion of Science (JSPS).

References

1 Wilson, T., Hastings, J. W. *Annu. Rev. Cell Dev. Biol.* **1998**, *14*, 197–230.

2 Hastings, J. W. Cell Physiology (3rd ed.), Academic Press, NY, **2001**, 1115.

3 Wood, K. V. *Photochem. Photobiol.* **1995**, *62*, 662–673.

4 Viviani, V. R. *Cell. Mol. Life Sci.* **2002**, *59*, 1833–1850.

5 Roda, A., Pasini, P., Michelini, E., Guardigli, M. *Trends in Biotechnol.* **2004**, *22*, 296–303.

6 McElroy, W. D., DeLuca, M. Bioluminescence in Action (Herring, P., Ed.), pp. 109–127, Academic Press, New York, **1978**.

7 DeLuca, M. *Biochemistry* **1969**, *8*, 160–166.

8 Sandalova, T. P., Ugarova, N. N. *Biochemistry* (Moscow) **1999**, *64*, 962–967.

9 White, E. H., Branchini, B. *J. Am Chem. Soc.* **1975**, *97*, 1243–1245.

10 McCapra, F., Gilfoyle, D. J., Young, D. W., Church, N. J., Spencer, P. Bioluminescence and Chemiluminescence: Fundamental and Applied Aspects (Campbell, A. K., Kricka, L. J., Stanley, P. E., Eds.), pp. 387–391, John Wiley and Sons, Chichester, UK, **1994**.

11 De Wet, J. R., Wood, K. V., Helinsky, D. R., DeLuca, M. *Proc. Natl. Acad. Sci. USA* **1985**, *82*, 7870–7873.

12 Tatsumi, H., Masuda, T., Kajiyama, N., Nakano, E. *J. Biolum. Chemilum.* **1989**, *3*, 75–78.

13 Tatsumi, H., Kajiyama, N., Nakano, E. *Biochim. Biophys. Acta* **1992**, *1131*, 161–165.

14 Devine, J. H., Kutuzova, G. D., Green, V. A., Ugarova, N. N., Baldwin, T. O. *Biochem. Biophys. Acta* **1993**, *1173*, 121–132.

15 Ohmiya, Y., Ohba, N., Toh, H., Tsuji, F. I. *Photochem. Photobiol.* **1995**, *62*, 309–313.

16 Sala-Newby, G. B., Thomson, C. M., Campbell, A. K. *Biochem. J.* **1996**, *313*, 761–767.

17 Li Ye, Buck, L. M., Scaeffer, H. J., Leach, F. R. *Biochim. Biophys. Acta* **1997**, *1339*, 39–52.

18 Wood, K. V., Lam, Y. A., Seliger, H. H., McElroy, W. D. *Science* **1989**, *244*, 700–702.

19 Viviani, V. R., Silva, A. C. R., Perez, G. L. O., Santelli, R. V., Bechara, E. J. H., Reinach, F. C. *Photochem. Photobiol.* **1999**, *70*, 254–260.

20 Gruber, M., Kutuzova, G. D., Wood, K. Bioluminescence and Chemiluminescence: Molecular reporting with photons, Proceedings of the 9th International Symposium (Hastings, J. W., Kricka, L. J., Stanley, P. E., Eds.), pp. 244–247, John Wiley & Sons, Chichester, UK, **1996**.

21 Viviani, V. R., Bechara, E. J. H., Ohmiya, Y. *Biochemistry* **1999**, *38*, 8271–8279.

22 Ohmiya, Y., Mina, S., Viviani, V. R., Ohba, N. *Sci. Rept. Yokosuka City Mus.* **2000**, *47*, 31–38.

23 Lee, K. S., Park, H. J., Bae, J. S., Goo, T. W., Kim, I., Sohn, H. D., Jin, B. R. *J. Biotechnol.* **2001**, *92*, 9–19.

24 Viviani, V. R., Arnoldi, F. G. C., Brochetto-Braga, Ohmiya, Y. *Comp. Biochem. Physiol.* Part B **2004**, *139*, 151–156.

25 Schroeder, S. *Nucl. Acid Res.* **1989**, *17*, 460.

26 Conti, E., Franks, N. P., Brick, P. *Structure* **1996**, *4*, 287–298.

27 Branchini, B. R., Magyar, R. A., Murtishaw, M. H., Anderson, S. M.,

ZIMMER, M. *Biochemistry* **1998**, *37*, 15311–15319.

28 SANDALOVA, T. P., UGAROVA, N. N. *Biochemistry* (Moscow) **1999**, *64*, 962–967.

29 ZAKO, T., AYABE, K., ABURATANI, T., KAMIYA, N., KITAYAMA, A., UEDA, H., NAGAMUNE, T. *Biochem. Biophys. Acta.* **2003**, *1649*, 183–189.

30 BRANCHINI, B. R., MAGYAR, R. A., MARCANTONIO, K. M., NEWBERRY, K. J., STROH, J. G., HINZ, L. K., MURTISHAW, M. H. *J. Biol. Chem.* **1997**, *272*, 19359–19364.

31 BRANCHINI, B. R., MAGYAR, R. A., MURTISHAW, M. H., ANDERSON, S. M., HELGERSON, L. C., ZIMMER, M. *Biochemistry* **1999**, *38*, 13223–13230.

32 BRANCHINI, B. R., MURTISHAW, M. H., MAGYAR, R. A., ANDERSON, S. M. *Biochemistry* **2000**, *39*, 5433–5440.

33 BRANCHINI, B. R., MAGYAR, R. A., MURTISHAW, M. H., PORTIER, N. C. *Biochemistry* **2001**, *40*, 2410–2418.

34 KAJIYAMA, N., NAKANO, E. *Protein Eng.* **1991**, *4*, 691–693.

35 UEDA, H., YAMANOUCHI, H., KITAYAMA, A., INOUE, K., HIRANO, T., SUZUKI, E., NAGAMUNE, T., OHMIYA, Y. Bioluminescence and Chemiluminescence: Molecular Reporting with Photons. Proc. 9th International Symposium (HASTINGS, J. W., KRICKA, L. J., STANLEY, P. E., Eds.), pp. 216–219, John Wiley & Sons, Chichester, U, **1996**.

36 MAMAEV, S. V., LAIKHTER, A. L., ARSLAN, T., HECHT, S. M. *JACS.* **1996**, *118*, 7243–7244.

37 OHMIYA, Y., HIRANO, T., OHASHI, M. *FEBS Lett.* **1996**, *384*, 83–86.

38 WOOD, K. V. *J. Biol. Chemilum.* **1990**, *5*, 107–114.

39 VIVIANI, V. R., OHMIYA, Y. *Photochem. Photobiol.* **2000**, *72*, 267–271.

40 VIVIANI, V. R., UCHIDA, A., SUENAGA, N., RYUFUKU, M., OHMIYA, Y. *Biochem. Biophys. Res. Comm.* **2001**, 280, 1286–1291.

41 VIVIANI, V. R., UCHIDA, A., VIVIANI, W., OHMIYA, Y. *Photochem. Photobiol.* **2002**, *76*, 538–544.

42 VIVIANI, V. R., SILVA NETO, A. J., OHMIYA, Y. *Prot. Eng. Des. Select.* **2004**, *17*, 113–117.

43 VIVIANI, V. R., OHELMEYER, T. L., ARNOLDI, F. G. C., BROCHETTO-BRAGA,M. R. *Photochem. Photobiol.* **2005**, *81*, 843–848.

44 CAMPBELL, A. K. Chemiluminescence: Principles and Applications in Biology and Medicine. VCH, Chichester, UK, **1988**.

45 STANLEY, E. *Methods in Enzymol.* **1986**, *133*, 14–22.

46 IM-HANSEN, O., KARL, D. M. Biomass and adenylate energy charge determination in microbial cell extracts and environmental samples. *Meth. Enzymol.* **1978**, *52*, 73–84.

47 CHAPPELLE, E. W., PICCIOLO, G. L., DEMING, J. W. *Meth. Enzymol.* **1978**, *52*, 65–72.

48 STANLEY, P. E. *J. Biolum. Chemilum.* **1993**, *8*, 51–63.

49 ATP 96-An International Symposium on Industrial Applications of Bioluminescence. Berkshire, UK. *J. Biolum. Chemilum.* *11*, 317–335.

50 HANNA, B. A. *Methods in Enzymol.* **1986**, *133*, 22–27.

51 LUNDIN, A. *Meth. Enzymol.* **1978**, *52*, 56–64.

52 NAYLOR, L. H. *Biochem. Pharm.* **1999**, *58*, 749–757.

53 SALA-NEWBY, G. B., KENDALL, J. M., JONES, H. E., TAYLOR, K. M., BADMINTON, M. N., LLEWELLYN, D. H., CAMPBELL, A. K. Fluorescent and luminescent probes for biological activity (WATSON, W. T., Ed.), pp. 251–271, Academic Press, London, **1999**.

54 GREER, L. F., SZALAY, A. S. *Luminescence* **2002**, *17*, 43–74.

55 BRASIER, A. R., RON, D. *Methods in Enzymol.* **1992**, *216*, 386–396.

56 MITTAL, S. K., MCDERMOTT, M. R., JOHNSON, D. C., PREVEC, L., GRAHAM, F. L. *Viral Res.* **1993**, *28*, 67–90.

57 KOVACS, F., METTENLEITER, T. C. *J. Gen. Virol.* **1991**, *72*, 2999–3008.

58 RODRÍGUEZ, J. F., RODRÍGUEZ, D., RODRÍGUEZ, J. R., MCGOWAN, E. B., ESTEBAN, M. *PNAS* **1988**, *85*, 1667–1671.

59 SALANEWBY, G. B., TAYLOR, K. M. et al. *Immunology* **1998**, *93*, 601–609.

60 NAKAJIMA, Y., KIMURA, T., SUGATA, K., ENOMOTO, T., ASAKAWA, A., KUBOTA, H.,

Ikeda, M., Ohmiya, Y. *Biotechniques* **2005**, *38*, 891–894.

61 Contag, C., Bachmann, M. H. *Annu. Rev. Biomed. Eng.* **2002**, *4*, 235–260.

62 Hooper, C. E., Ansorge, R. E., Browne, H. M., Tomkins, P. *J. Biolum. Chemilum.* **1990**, *5*, 123–130.

63 Contag, C. H., Spilman, S. D., Contag, P. R., Oshiro, M., Eames, B., Dennery, P. et al. *Photochem. Photobiol.* **1997**, *66*, 523–531.

64 Contag, C. H., Contag, P. R., Mullins, J. I., Spilman, S. D., Stevenson, D. K., Benaron, D. A. *Mol. Microbiol.* **1995**, *18*, 593–603.

65 Brovko, L. Y. et al. *J. Food Prot.* **2003**, *66*, 2160–2163.

66 Hazbon, M. H., Guarin, N., Ferro, B. E., Rodríguez, A. L., Labrada, L. A., Tovar, R., Riska, P. F., Jacobs, W. R. *J. Clin. Microbiol.* **2003**, *41*, 4865–4869.

67 Deb, D. K., Srivastava, K. K., Srivastava, R., Srivastava, B. S. *Biochem. Biophys. Res. Comm* **2000**, *279*, 457–461.

68 Schwartz, O., Virelzier, J. L., Montagnier, L., Hazan, U. *Gene* **1990**, *88*, 197–205.

69 Morrey, J. D., Bourn, S. M., Bunch, T. D., Jackson, M. K., Sidwell, R. W., Barrows, L. R., Daynes, R. A., Rosen, C. A. *J. Virol.* **1991**, *65*, 5045–5051.

70 Jekinson, S., McCoy, D. C., Kerner, S. A., Ferris, R. G., Lawrence, W. K., Clay, Condreay, J. P., Smith, C. D. *J. Biomol. Screen.* **2003**, *8*, 463–470.

71 Edinger, M., Sweeney, T. J., Tucker, A. A., Olomu, A. B., Negrin, R. S., Contag, C. H. *Neoplasia* **1999**, *1*, 303–310.

72 Sweeney, T. J., Mailander, V., Tucker, A. A., Olomu, A. B., Zhang, W., Cao, Y., Negrin, R. S., Contag, C. H. *PNAS* **1999**, *96*, 12044–12049.

73 Yu, Y. A., Timiryasova, T., Zhang, Q., Beltz, R., Szalay, A. A. *Anal. Bioanal. Chem.* **2003**.

74 Fukuda, S., Tatsumi, H., Igarashi, H., Igimi, S. *Lett. Appl. Microbiol.* **2000**, *31*, 134–138.

75 Zhang, X., Kobatake, E., Kobayashi, K., Yanagida, Y., Aizawa, M. *Anal. Biochem.* **2000**, *282*, 65–69.

76 Ow, D. W., Wood, K. V., DeLuca, M., DeWet, J. R., Helinski, D. R., Howell, S. H. *Science* **1986**, *234*, 56–859.

77 Millar, A. J., Carre, I. A. et al. *Science* **1995**, *267*, 1161–1163.

78 Hollis, R. P., Killham, K., Glover, L. A. *Appl. Environm. Microbiol.* **2000**, *66*, 1676–1679.

79 Bachmann, T. *Trends in Biotechnol.* **2003**, *21*, 247–249.

80 Souza, S. F. *Biosesnsors and Bioelectronics* **2001**, *16*, 337–353.

81 Taurianen, S., Virta, M., Chang, W., Karp, M. *Anal. Biochem.* **2000**, *272*, 191–198.

82 Lagido, C., Pettit, J., Porter, A. J. R., Paton, G. I., Glover, L. A. *FEBS Lett.* **2001**, *493*, 36–39.

83 Belkin, S. *Curr. Opin Microbiol* **2003**, *6*, 206–212.

84 Pedahzur, R., Polyak, B., Belkin, S. *J. Appl. Toxicol.* **2004**, *24*, 343–348.

85 Kajiyama, N., Nakano, E. *Protein Eng.* **1991**, *4*, 691–693.

86 Ueda, H., Yamanouchi, H., Kitayama, A., Inoue, K., Hirano, T., Suzuki, E., Nagamune, T., Ohmiya, Y. Bioluminescence and Chemiluminescence: Molecular Reporting with Photons (Hastings, J. W., Kricka, L. J., Stanley, P. E., Eds.), pp. 216–219, Wiley & Sons, Chichester, **1996**.

87 Kitayama, A., Yoshizaki, H., Ohmiya, Y., Ueda, H., Nagamune, T. *Photochem. Photobiol.* **2003**, *77*, 333–338.

88 Nakajima, Y., Kimura, T., Suzuki, C., Ohmiya, Y. Improved expression of novel red- and green-emitting luciferases of *Phrixotrix* railroad worms in mammalian cells. *Biosci. Biotechnol. Biochem.* **2004**, *68*, 948–951.

89 Kitayama, Y., Kondo, T., Nakahira, Y., Nishimura, H., Ohmiya, Y., Oyama, T. *Plant Cell. Physiol.* **2004**, *45*, 109–113.

90 Nakajima, Y., Ikeda, M., Kimura, T. Honma, S., Ohmiya, Y., Honma, K. *FEBS Lett.* **2004**, *565*, 122–126.

91 White, P. J., Squirrel, D. J., Arnaud, P., Lowe, C. R., Murray, J. A. H. *Biochem. J.* **1996**, *319*, 343–350.

92 Kajiyama, N., Nakano, E. *Biochemistry* **1992**, *32*, 13795–13799.

93 Willey, T. L., Squirrell, D. J., White, P. J. Bioluminescence and Chemiluminescence: Proceedings of the 11th International Symposium (Case, J. F., Herring, P. J., Robison, B. H., Haddock, S. H. D., Kricka, L. J., Stanley, P. E., Eds.), pp. 201–204, World Scientific, Singapore, **2001**.

94 White, E. H., Worther, H., Seliger, H. H., McElroy, W. D. *J. Am. Chem. Soc.* **1966**, *88*, 2015–2019.

95 Branchini, B. R., Hayward, M. M., Bamford, S., Brennan, P. M., Lajiness, E. J. *Photochem. Photobiol.* **1989**, *49*, 689–695.

4 Split Luciferase Systems for Detecting Protein–Protein Interactions in Mammalian Cells Based on Protein Splicing and Protein Complementation

Yoshio Umezawa

4.1 Introduction

Protein–protein interactions (PPIs) are known to play key roles in structural and functional organization of living cells. Many unsolved problems currently studied in molecular biology and biochemistry are related to PPIs. Identification of these interactions and characterization of their physiological significance constitute one of the main goals of current research in different biological fields [1–4].

The yeast two-hybrid system has been extremely useful for detecting and identifying PPIs *in vivo* [5, 6], where a library of proteins is screened for interaction with a "bait" protein. The two-hybrid method, a standard functional assay, facilitates the identification of PPIs and has been proposed as a method for the generation of protein interaction maps [7–9]. This approach allows rapid detection of protein-binding partners, including the relevant interacting domains, and immediately provides the gene that encodes the identified interacting proteins; there are, however, cases where this is not applicable. For example, in cases where a set of detectable protein interactions are those occurring in the nucleus, in proximity to the reporter gene, this method is not useful for the study of membrane proteins. Moreover, if one of the proteins is a transcriptional activator, it may itself induce transcription of the reporter gene. Conceptually, the biological information generated by two-hybrid analysis is often questioned because of the inherent artificial nature of the assay. Nonetheless, numerous previously unknown protein interactions have been identified using this yeast two-hybrid system.

To overcome the limitations of the yeast two-hybrid assay, various permutations of the two-hybrid method have been described, including the split-ubiquitin system (USPS) [10–12], the SOS recruitment system [13–15], dihydrofolate reductase complementation [16, 17], β-galactosidase complementation [18], and the G-protein fusion system [19]. These systems are well suited for assaying interactions between cytoplasm- and membrane-proximal proteins, but they can be utilized only in appropriately engineered cells and/or are prone to false-positive signals.

Photoproteins in Bioanalysis. Edited by Sylvia Daunert and Sapna K. Deo

ISBN: 3-527-31016-9

Fluorescence ratio imaging has also been used to study protein interactions in living cells [20]. The detection mechanisms are based on the fluorescence resonance energy transfer (FRET) that occurs when donor and acceptor fluorophores are in sufficient proximity (< 10 nm) and an appropriate relative orientation. The use of FRET has allowed us to detect protein interactions or conformational changes in real time in living cells, or to analyze the distances between the two proteins as a molecular ruler. However, it is limited by the requirement that the fluorescent labels on the interacting proteins be sufficiently close to permit efficient energy transfer. Also, the labeled proteins need to be introduced into the cells at relatively high concentrations. Clearly, a method that would allow a direct examination of molecular interactions with fewer size constraints, at the site where they occur within a eukaryotic cell, would be advantageous.

Recently, we developed a protein splicing-based split green fluorescent protein (GFP) system for the analysis of interactions between soluble proteins that provides an alternative method for the *in vivo* analysis of protein interactions [21]. The protein splicing is a post-translational processing event involving precise excision of an internal protein segment, the intein, from a primary translation product with concomitant ligation of the flanking sequences, the exteins (*ex*ternal pro*teins*) [22]. The N- and C-terminal halves of intein, derived from *Saccharomyces cerevisiae*, are fused to split the N- and C-terminal halves of enhanced GFP, respectively, and each is connected to the interacting protein pairs of interest. In this system, PPIs facilitated the protein splicing *in trans* and the two exteins of split EGFP were ligated to yield the EGFP fluorophore, which was detected *in vivo* in *E. coli* cells by fluorescence. This system has the advantage that neither reporter genes nor substrates for enzymes are needed, in contrast to the earlier methods. For wider applications of this method, detection of protein–protein interactions in mammalian cells is naturally targeted. However, it appears that the available detection limit of the method has yet to be improved for its use in mammalian cells.

4.2
Protein Splicing-based Split Firefly Luciferase System [23]

Here we describe a protein splicing-based split luciferase to monitor protein–protein interactions in mammalian cells, using an intein of DnaE derived from the cyanobacterium *Synechocystis* sp. strain PCC6803. The DnaE intein was selected among several known inteins because it has the natural splicing ability to ligate accompanying proteins *in trans*. This principle is shown in Fig. 4.1. The Ssp DnaE intein is a naturally split intein, composed of 123 amino acid residues in the N-terminal half and 36 amino acid residues in the C-terminal half, which possesses an ability to ligate N- and C-exteins [24]. The amino terminus of DnaE is fused to the N-terminal half of the luciferase enzyme and the carboxy terminus of DnaE to the rest of the luciferase. Each of these fusion proteins is linked to a protein of interest (protein A) and its target protein (protein B). When interaction occurs between the two proteins, the N- and C-termini of DnaE are brought in close proximity and undergo correct folding, which induces a splicing event. The N- and C-terminal fragments of the split luciferase thereby directly link to each other by a peptide bond. The matured luciferase thus formed creates its active center to emit light, which can be detected by a luminometer. The extent of the protein–protein interaction can be evaluated by measuring the magnitude of luminescence intensity originating from the formation of luciferase. By thus using characteristic features of luciferase for highly sensitive detection together with DnaE for high splicing efficiency, we demonstrated that the sensitivity of this method is much improved compared to the earlier split EGFP method, and PPIs triggered by outer-membrane signals became detectable in mammalian cells.

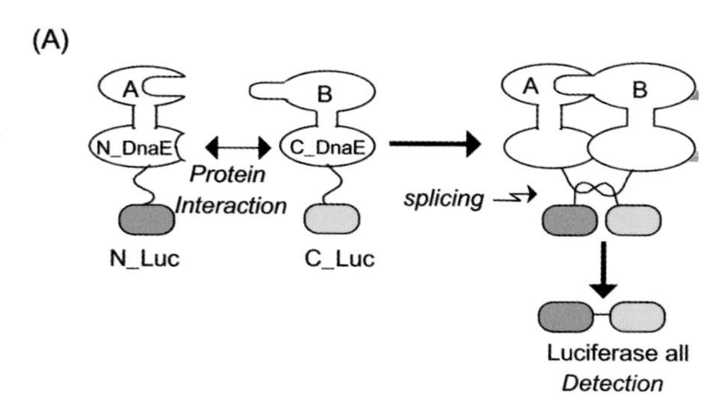

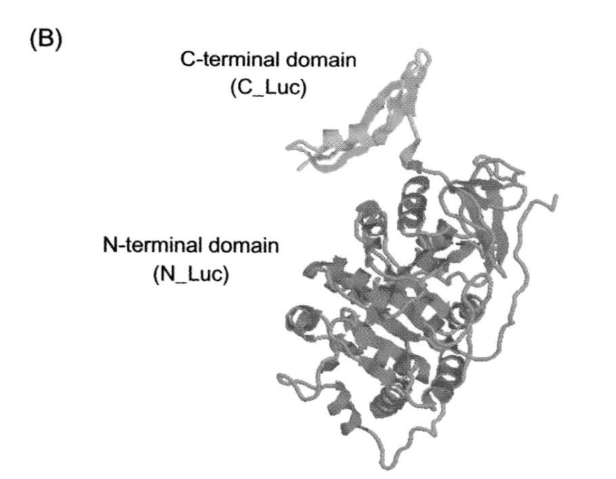

Fig. 4.1 (A) Principle for the present protein splicing-based split luciferase enzyme system. The N-terminal half of DnaE (N_DnaE) and the C-terminal half of DnaE (C_DnaE) are connected with the N-terminal half of luciferase (1–437 amino acids; green) and the C-terminal half of luciferase (438–544 amino acids; yellow), respectively. Interacted protein A and protein B are linked to opposite ends of those DnaE. Interaction between protein A and protein B accelerates the folding of N_DnaE and C_DnaE, and protein splicing results. The N- and C-terminal halves of luciferase are linked together by a normal peptide bond to recover its enzymatic activity.
(B) 3D structure of firefly luciferase. N- and C-terminal halves of the enzymes are shown as green and yellow, respectively. This figure was generated using Rasmol [23].

4.2.1
Split Luciferase Works as a Probe for Protein Interaction

To examine whether some particular PPIs in eukaryotic cells facilitate the splicing event to yield matured luciferase, phosphorylation of IRS-1 and its target SH2N domain derived from phosphatidyl inositol 3-kinase, both of which are involved in physiologically relevant insulin signaling, was chosen as a known protein interaction pair [25]. After the CHO-HIR cells were transiently co-transfected with pIRES-DSL(Y/S) and pRL-TK vectors, the cells were stimulated with 1.0×10^{-7} M insulin for 72 h, 3 h, or with the same concentration of insulin for 5 min, followed by replacing the insulin solution with FCS-free medium without insulin for 175 min. The results are shown in Table 4.1. When CHO-HIR cells were stimulated by insulin for 72 h, 3 h and 5 min, the magnitudes of the normalized luciferase activities were 1.0, 0.73, and 0.17, respectively. Upon stimulation with FCS-free medium in the absence of insulin, the magnitude of the background luminescent intensity was 0.15, which was nearly identical to that obtained with the 5-min treatment of insulin.

The firefly luciferase activity obtained with the stimulation of insulin for more than 3 h was found to be four times higher than the background signal. This increase in the kinase activity is explained by the phosphorylated Y941 peptide, generated by the kinase reaction of the insulin receptor and selectively bound to SH2N, which facilitated folding and splicing *in trans* of DnaE inteins; as a result, the two external regions of the N- and C-terminals of luciferase were ligated to recover its activity. To confirm this, the effects of a point mutation on the splicing efficiency were examined by comparing the Y941 peptide and its mutant. The phosphorylation site of the 941 tyrosine residue in the Y941 peptide was replaced by an Ala residue that was not subjected to phosphorylation by the insulin receptor. The results are shown in Table 4.2. Upon expression of the Y941 mutant in CHO-HIR cells, the magnitude of the normalized luminescence intensity was the same as the background level, demonstrating that phosphorylation of Y941 peptide is indispensable for producing the luciferase enzyme by the splicing reaction.

Table 4.1 Time-dependent increases in protein-splicing events upon addition of insulin.

Time	*RLU*	
	Insulin (100 nM)	*Insulin (0 M)*
5 min	0.17 ± 0.10	–
3 h	0.73 ± 0.26	0.15 ± 0.04
72 h	1.00 ± 0.28	–

CHO-HIR cells were cultured in 6-well plates and were transfected with 2 μg of the plasmid pIRESDSL(Y/S) and 0.02 μg of the control plasmid pRL-TK. After incubation for 45 h, the cells were stimulated with 100 nM of insulin for the indicated time at 37 °C. RLU; relative luminescence intensity per unit protein expression.

Table 4.2 Effect of a point mutation in Y941 peptide on protein splicing.

	RLU	
	Insulin (100 nM)	***Insulin (0 M)***
Y941 peptide	0.50 ± 0.95	0.11 ± 0.04
Y941 A mutant	0.16 ± 0.05	0.15 ± 0.05

CHO-HIR cells were cultured in 6-well plates and were transfected with 2 μg of pIRES-DSL(Y/S) or its mutant together with 0.02 μg of pRL-TK. After incubation for 45 h, the cells were stimulated with or without 100 nM of insulin for 3 h and measured their luminescence.

It has been shown that the kinase activity of the insulin receptor increases within a minute after insulin binds to its receptor [26]; therefore, upon stimulation with insulin for 5 min, the Y941 peptide was expected to be phosphorylated by the insulin receptor. However, the magnitude of the luminescence intensity upon stimulation with insulin for 5 min was almost the same as the background signal. This result probably occurred because the folding and splicing reaction of DnaE inteins did not proceed far enough to produce firefly luciferase and thus its luminescence could not be detected, even though the phosphorylated Y941 peptide bound to the SH2N domain. In contrast, upon treatment with insulin for 3 h, the magnitude of luminescence intensity increased greatly relative to the background signal. These results demonstrate that a reaction time of 3 h needed for the protein splicing to occur was enough to discriminate from the background signal the luminescence activity generated by protein interactions.

The background luminescence intensity of 0.15 seems relatively high in comparison to the one obtained in the case of untransfected CHO-HIR cells. It is well known that a naturally split Ssp DnaE intein has the *trans*-splicing ability to ligate its exteins [24, 27, 28]. This implies that the splicing reaction of the two intein halves in the absence of insulin stimulation was partly, but not completely, induced by the association of the intein fragments. The partial association of DnaE inteins may trigger the splicing event that causes the relatively high background signals. The fact that the background signal in the case of the Y941 peptide was almost the same as the one for the Y941A mutant (Table 4.2) suggests that the nonspecific phosphorylation of the Y941 peptide negligibly contributed to the background signal.

To show the feasibility of quantitative detection of the insulin-induced protein interaction in CHO-HIR cells, luminescence from firefly luciferase was evaluated as to its insulin concentration dependence. After the CHO-HIR cells were transiently co-transfected with pIRES-DSL(Y/S) and pRL-TK vectors, the cells were stimulated with various concentrations of insulin for 30 min, and thereafter each cell lysate was subjected to the luminescence measurements. The results are shown in Fig. 4.2. The magnitude of the luminescence intensity increased with an increasing concentration of insulin from 1.0×10^{-10} M to 1.0×10^{-6} M. At insulin concentration levels lower than 1.0×10^{-11} M, no change in the luminescence was observed.

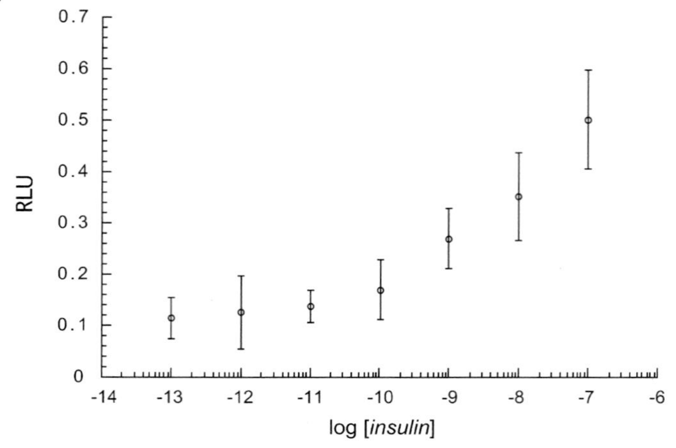

Fig. 4.2 Insulin concentration dependence on relative luminescence unit (RLU). CHO-HIR cells were cultured in 6-well plates and were transfected with 2 μg of the plasmid pIRES-DSL(Y/S) and 0.02 ng of the control plasmid pRL-TK. After incubation for 45 h, the cells were stimulated with insulin for 3 h at 37 °C. The concentration of insulin ranged from 1.0×10^{-13} M to 1.0×10^{-7} M [23].

The relationship between the logarithmic insulin concentration and the magnitude of luminescence intensities was not sigmoidal. It has been shown that typical kinase assays of kinase activity of insulin receptor with [γ-^{32}P]ATP and synthetic substrates such as random copolymers of poly(Glu : Tyr)(4 : 1) consist of insulin-dependent ^{32}P incorporation into the substrates [29–31]. This discrepancy may be due to the difference in the detection methods for the phosphorylated substrates. In the present method, the detection of the phosphorylated Y941 was performed using the splicing reaction of split luciferase, of which complicated kinetics in the living cells might have caused this non-sigmoidal relationship between the insulin concentration and the observed luciferase activity. The exact reason for this, however, remains to be worked out.

Recently, this splicing-based split luciferase system for detecting PPIs was applied for imaging a particular PPI in living mice [32]. The reconstituted firefly luciferase can emit a broadband near to infrared, which is tissue transparent. A PPI of two strongly interacting proteins, My-D and Id, was imaged in cells implanted into a particular portion of living mice. M-pD and Id are members of the helix-loop-helix family of nuclear proteins.

4.3 Split Renilla Luciferase Complementation System [33]

For spatial and quantitative kinetic analysis of PPIs in living mammalian cells, we have developed a split *Renilla* luciferase complementation method. The split *Renilla* luciferase complementation method relies on the spontaneous emission of luminescence upon PPI-induced complementation of the split *Renilla* luciferase, with the cell membrane-permeable substrate coelenterazine. Consequently, unlike conventional complement enzymes that lead to stable diffusive fluorescent products [34–38], this split *Renilla* luciferase complementation readout is capable of locating the PPIs with emission of bioluminescence only at the sites and time of their occurrence in living cells.

To date, *Renilla* luciferase is one of the major reporter proteins that have been used for optical imaging studies in living cells and rodents. This enzyme, which was cloned and sequenced by Lorentz [39], is a monomeric photoprotein with a molecular weight of 36 kDa, half the molecular weight of firefly luciferase, and is easily expressed within mammalian cells. The crystal structure of *Renilla* luciferase is not known, but it is known that its N-terminal and several cysteine residues are important for its luminescence activity [40, 41]. In contrast to firefly luciferase [32], the enzymatic reaction of *Renilla* luciferase does not require ATP. To add to these advantages of *Renilla* luciferase, its substrate coelenterazine is known to penetrate through the mammalian cell membrane and rapidly diffuse throughout the cytosol [42, 43]. *Renilla* luciferase catalyzes the oxidation by O_2 of coelenterazine to its excited-state product (oxycoelenterazine monoanion) to emit luminescence with a broadband (ca. 400–630 nm) covering a tissue-transparent near-infrared region (Fig. 4.3).

The principle of the present split *Renilla* luciferase complementation strategy is shown in Fig. 4.3. To monitor the interaction between two proteins A and B, the N-terminal half of the split *Renilla* luciferase is fused to protein A, and protein B is fused to the C-terminal half. Interaction between protein A and protein B and the consequent juxtapositioning of the split *Renilla* luciferase simultaneously lead to formation of the complement *Renilla* luciferase, thereby spontaneously emitting bioluminescence with its membrane-permeable substrate, coelenterazine, *in situ* in living mammalian cells.

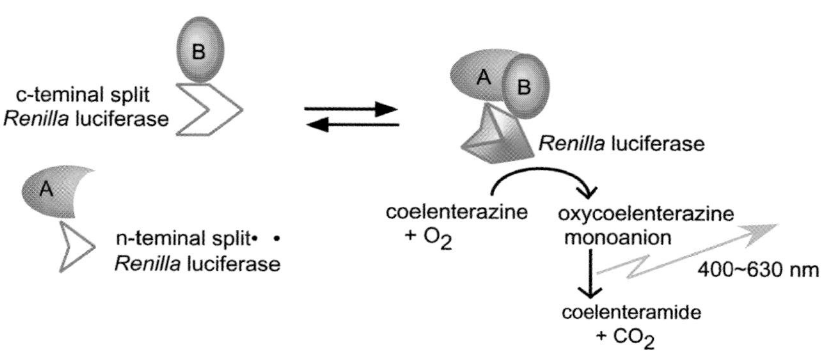

Fig. 4.3 A schematic diagram of the complementation strategy based on split *Renilla* luciferase. The split *Renilla* luciferase complementation method for locating PPIs relies on the spontaneous emission of luminescence upon PPI-induced complementation of the split *Renilla* luciferase, with a cell membrane-permeable substrate, coelenterazine, *in situ* in living mammalian cells [23].

We now will validate this split *Renilla* luciferase complementation strategy for its use in visualizing a known PPI between the Y941 peptide and the N-terminal SH2 domain (SH2n) upon protein phosphorylation in living Chinese hamster ovary cells overexpressing with human insulin receptors (CHO-HIR) [23, 44, 45] (see above). The 941 tyrosine residue in the Y941 peptide derived from insulin receptor substrate-1 (IRS-1) is phosphorylated by the insulin receptor upon insulin stimulation and is dephosphorylated by a protein tyrosine phosphatase [45, 46]. The SH2n from the p85 subunit ($p85_{330-429}$) of phosphatidylinositol 3-kinase binds with the phosphorylated 941 tyrosine residue within the IRS-1 [47–49] (see above). The two proteins, the N-terminal half of the split *Renilla* luciferase to connect to the Y941 and SH2n to connect to the C-terminal half, were expressed in the cells. The interaction between the Y941 and the SH2n upon insulin stimulation leads to the complementation of *Renilla* luciferase.

4.3.1 Time Course of the Interaction Between Y941 and SH2n

The time course of interaction between Y941 and SH2n observed with sRL91 is shown in Fig. 4.4a. The luminescence intensities increased within 5 min after insulin stimulation and gradually decreased afterward. This time dependence of the interaction is due to tyrosine phosphorylation and dephosphorylation, which is in good agreement with the immunoblot analysis of anti-phosphotyrosine antibody shown in Fig. 4.4b. The results indicate that the luminescence activity of sRL91 directly reflects the ongoing PPI in living cells.

4.3.2 Location of the Interaction Between Y941 and SH2n

Spontaneous emission of luminescence by a particular PPI-induced complementation of the split *Renilla* luciferase together with its membrane-permeable substrate, coelenterazine, allows noninvasive imaging of the sites and time of its occurrence in living cells. The PPI between Y941 and SH2n in the CHO-HIR cells expressing sRL91 was thereby imaged with and without insulin stimulation as shown in Fig. 4.5a. Upon 100-nM insulin stimulation, luminescence emitted by complement *Renilla* luciferase increased only near the plasma membrane, whereas such bright contrast was not observed in the absence of insulin. This indicates that with insulin stimulation, the interaction between Y941 and SH2n occurred near the plasma membrane in the cytosol. The CHO-HIR cells expressing full-length *Renilla* luciferase (hRL124C/A) emitted luminescence uniformly throughout the cells with its substrate coelenterazine (Fig. 4.5b), which precludes the possibility that the result in Fig. 4.5a is due to an excessive accumulation of coelenterazine in and right below the plasma membrane.

The split *Renilla* luciferase complementation method was thus developed for spatial and kinetic analysis of PPIs in living mammalian cells. This split *Renilla* luciferase complementation readout was shown to work for locating a PPI between

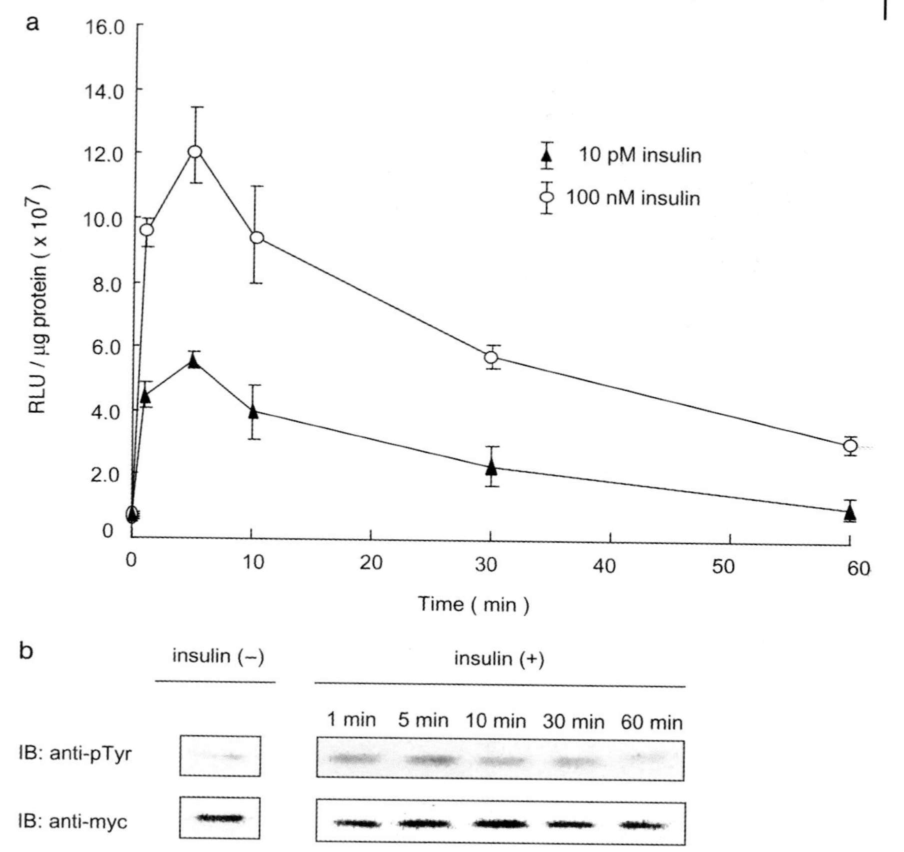

Fig. 4.4 (a) Time course of the luminescence upon sRL91 complementation. The cells expressing sRL91 were incubated in the culture medium supplemented with 100 nM or 10 pM insulin for 1 min, 5 min, 10 min, 30 min, and 60 min at 37 °C. The luminescence of these cells was immediately assessed. (b) Immunoblot analysis of tyrosine phosphorylation and dephosphorylation on Y941 in sRL91. The cells expressing sRL91 were incubated in the culture medium supplemented with 100 nM insulin for 1 min, 5 min, 10 min, 30 min, and 60 min at 37 °C. The whole cell lysates were immunoprecipitated with anti-myc antibody. The immunoblot analysis was made using anti-phosphotyrosine antibody and anti-myc antibody [33].

the tyrosine-phosphorylated peptide (Y941) of IRS-1 and the SH2 domain of PI3K in insulin signal transduction in living CHO-HIR cells. It was thereby found that the insulin-stimulated interaction occurred near the plasma membrane in the cytosol. Unlike diffusive products involved in other complement enzyme systems [5–9], the present bioluminescence readout by PPI-induced formation of the complement *Renilla* luciferase promises to locate PPIs at the sites and time of their occurrence, thereby allowing spatial and kinetic analysis of PPIs *in vivo* in living cells and organisms.

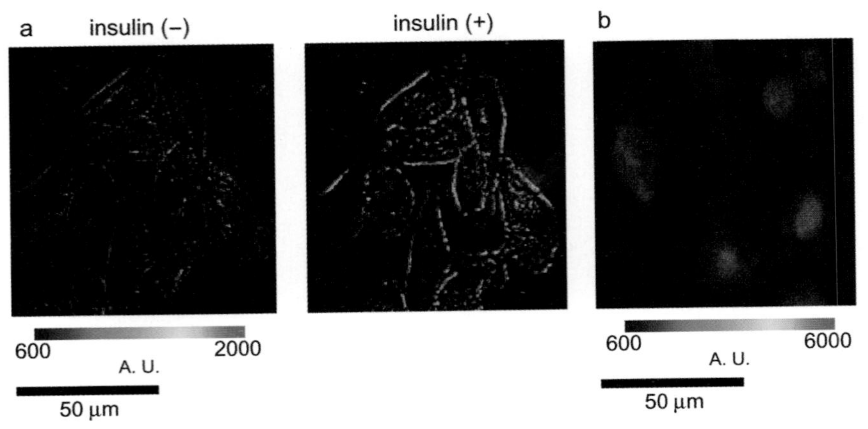

Fig. 4.5 Spatial analysis of the interaction between Y941 and SH2n in living cells. These luminescence microscopic images of CHO-HIR cells were taken with a CCD camera for 300 s (a) and 60 s (b) as exposure times, respectively, in PBS supplemented with a 20% coelenterazine substrate buffer. These images were acquired with the cells expressing sRL91 (a) in the absence (left) or presence (right) of 100 nM insulin, and with the cells expressing only full-length *Renilla* luciferase (hRL124C/A) (b). The luminescence intensity is represented with a color scale [33].

References

1 Ozawa, T., Umezawa, Y. *Curr. Opin. Chem. Biol.* **2001**, *5*, 578–583.

2 Mendelsohn, A. R., Brent, R. *Science* **1999**, *284*, 1948–1950.

3 Colas, P., Brent, R. *Trends Biotechnol.* **1998**, *16*, 355–363.

4 Lin, H., Cornish, V. W. *Angew. Chem. Int. Ed.* **2001**, *40*, 871–875.

5 Chien, C. T., Bartel, P. L., Sternglanz, R., Fields, S. *Proc. Natl. Acad. Sci. USA* **1991**, *88*, 9578–9582.

6 Fields, S., Song, O. *Nature* **1989**, *340*, 245–246.

7 Flores, A., Briand, J. F., Gadal, O., Andrau, J. C., Rubbi, L., Mullem, V., Boschiero, C., Goussot, M., Marck, C., Carles, C., Thuriaux, P., Sentenac, A., Werner, M. *Proc. Natl. Acad. Sci. USA* **1999**, *96*, 7815–7820.

8 Ito, T., Tashiro, K., Muta, S., Ozawa, R., Chiba, T., Nishizawa, M., Yamamoto, K., Kuhara, S., Sakaki, Y. *Proc. Natl. Acad. Sci. USA* **1999**, *97*, 1143–1147.

9 Walhout, A. J. M., Sordella, R., Lu, X., Hartley, J. L., Temple, G. F., Brasch, M. A., Thierry-Mieg, N., Vidal, M. *Science* **2000**, *287*, 116–122.

10 Dunnwald, M., Varshavsky, A., Johnsson, N. *Mol. Biol. Cell* **1999**, *10*, 329–344.

11 Johnsson, N., Varshavsky, A. *Proc. Natl. Acad. Sci. USA* **1994**, *91*, 10340–10344.

12 Stagljar, I., Korostensky, C., Johnsson, N., Heesen, S. *Proc. Natl. Acad. Sci. USA* **1998**, *95*, 5187–5192.

13 Aronheim, A., Zandi, E., Hennemann, H., Elledge, S. J., Karin, M. *Mol. Cell. Biol.* **1997**, *17*, 3094–3102.

14 Aronheim, A. *Nuc. Acids Res.* **1997**, *25*, 3373–3374.

15 Broder, Y. C., Katz, S., Aronheim, A. *Curr. Biol.* **1998**, *8*, 1121–1124.

16 Remy, I., Michnick, S. W. *Proc. Natl. Acad. Sci. USA* **1999**, *96*, 5394–5399.

17 Pelletier, J. N., Arndt, K. M., Plückthun, A., Michnick, S. W. *Nature Biotech.* **1999**, *17*, 683–690.

18 Rossi, F., Charlton, C. A., Blau, H. M. *Proc. Natl. Acad. Sci. USA* **1997**, *94*, 8405–8410.

19 Ehrhard, K. N., Jacoby, J. J., Fu, X.-Y., Jahn, R., Dohlman, H. G. *Nature Biotech.* **2000**, *18*, 1075–1079.
20 Pollok, B. A., Heim, R. *Trends Cell Biol.* **1999**, *9*, 57–60.
21 Ozawa, T., Nogami, S., Sato, M., Ohya, Y., Umezawa, Y. *Anal. Chem.* **2000**, *72*, 5151–5157.
22 Gimble, F. S. *Chem. Biol.* **1998**, *5*, R251–R256.
23 Ozawa, T., Kaihara, A., Sato, M., Tachihara, K., Umezawa, Y. *Anal. Chem.* **2001**, *73*, 2516–2521.
24 Wu, H., Hu, Z., Liu, X.-Q. *Proc. Natl. Acad. Sci. USA* **1998**, *95*, 9226–9231.
25 White, M. F. *Diabetologia* **1997**, *40*, S2–S17.
26 Solow, B. T., Harada, S., Goldstein, B. J., Smith, J. A., White, M. F., Jarett, L. *Mol. Endo.* **1999**, *13*, 1784–1798.
27 Scott, C. P., Abel-Santos, E., Wall, M., Wahnon, D. C., Benkovic, S. J. *Proc. Natl. Acad. Sci. USA* **1999**, *96*, 13638–13643.
28 Evans, T. C., Martin, D., Kolly, R., Panne, D., Sun, L., Ghosh, I., Chen, L., Benner, J., Liu, X.-Q., Xu, M.-Q. *J. Biol. Chem.* **2000**, *275*, 9091–9094.
29 Zhang, B., Szalkowski, D., Diaz, E., Hayes, N., Smith, R., Berger, J. *J. Biol. Chem.* **1994**, *269*, 25735–25741.
30 Zhang, B., Salituro, G., Szalkowski, D., Li, Z., Zhang, Y., Royo, I., Vilella, D., Diez, M. T., Pelaez, F., Ruby, C., Kendall, R. L., Mao, X., Griffin, P., Calaycay, J., Zierath, J. R., Heck, J. V., Smith, R. G., Moller, D. E. *Science* **1999**, *284*, 974–977.
31 Ozawa, T., Sato, M., Sugawara, M., Umezawa, Y. *Anal. Chem.* **1998**, *70*, 2345–2352.
32 Paulmurugan, Y., Umezawa, Y., Gambhir, S. *Natl. Acad. Sci. USA* **2002**, *99*, 15608–15613.
33 Kaihara, A., Kawai, Y., Sato, M., Ozawa, T., Umezawa, Y. *Anal. Chem.* **2003**, *75*, 4176–4181.
34 Blakely, B. T., Rossi, F. M., Tillotson, B., Palmer, M., Estelles, A., Blau, H. M. *Nat. Biotechnol.* **2000**, *18*, 218–222.
35 Rossi, F., Charlton, C. A., Blau, H. M. *Proc. Natl. Acad. Sci. USA* **1997**, *94*, 8405–8410.
36 Remy, I., Michnick, S. W. *Proc. Natl. Acad. Sci. USA* **1999**, *96*, 5394–5399.
37 Galarneau, A., Primeau, M., Trudeau, L. E., Michnick, S. W. *Nat. Biotechnol.* **2002**, *20*, 619–622.
38 Wehrman, T., Kleaveland, B., Her, J. H., Balint, R. F., Blau, H. M. *Proc. Natl. Acad. Sci. USA* **2002**, *99*, 3469–3474.
39 Lorenz, W. W., Mccann, R. O., Longiaru, M., Cormier, M. J. *Proc. Natl. Acad. Sci. USA* **1991**, *88*, 4438–4442.
40 Liu, J., O'Kane, D. J., Escher, A. *Gene* **1997**, *203*, 141–148.
41 Liu, J., Escher, A. *Gene* **1999**, *237*, 153–159.
42 Contag, C. H., Bachmann, M. H. *Annu. Rev. Biomed. Eng.* **2002**, *4*, 235–260.
43 Greer, L. F., Szalay, A. A. *Luminescence* **2002**, *17*, 43–74.
44 Sato, M., Ozawa, T., Inukai, K., Asano, T., Umezawa, Y. *Nat. Biotechnol.* **2002**, *20*, 287–294.
45 Sato, M., Ozawa, T., Yoshida, T., Umezawa, Y. *Anal. Chem.* **1999**, *71*, 3948–3954.
46 Golstein, B. J., Bittener-Kowalczyk, A., White, M. F., Harbeck, M. *J. Biol. Chem.* **2000**, *1275*, 4283–4289.
47 Pratipanawatr, W., Pratipanawatr, T., Cusi, K., Berria, R., Adams, J. M., Jenkinson, C. P., Maezono, K., DeFronzo, R. A., Mandarino, L. J. *Diabetes* **2001**, *50*, 2572–2578.
48 Yoshimura, R., Araki, E., Ura, S., Todaka, M., Tsuruzoe, K., Furukawa, N., Motoshima, H., Yoshizato, K., Kaneko, K., Matsuda, K., Kishikawa, H., Shichiri, M. *Diabetes* **1997**, *46*, 929–936.
49 Yonezawa, K., Ueda, H., Hara, K., Nishida, K., Ando, A., Chavanieu, A., Matsuda, H., Shii, K., Yokono, K., Fukui, Y., Calas, B., Grigorescu, F., Dhand, R., Gout, I., Otsu, M., Waterfield, M. D., Kasuga, M. *J. Biol. Chem.* **1992**, *36*, 25958–25966.

5
Photoproteins in Nucleic Acid Analysis

Theodore K. Christopoulos, Penelope C. Ioannou, and Monique Verhaegen

5.1
Hybridization Assays

The highly specific and strong interaction between two complementary nucleic acid strands forms the basis for the development of hybridization assays. Nucleic acid hybridization has become a fundamental analytical technique for the detection and quantification of specific DNA or RNA sequences and is used extensively in research and diagnostics in laboratories. Major areas of application of hybridization assays include the detection of nucleic acid sequences that are related to neoplastic disease; the detection and/or determination of various pathogens in clinical, environmental, and food samples; the detection of mutations associated with disease; the analysis of chromosomal rearrangements associated with neoplasias; the detection of genetically modified organisms; and DNA fingerprinting.

DNA or RNA probes labeled with radioisotopes (^{32}P, ^{35}S or ^{3}H) in combination with autoradiography dominated the field of hybridization assays for more than two decades. The classical methodology for nucleic acid analysis includes electrophoretic separation, transfer to a suitable membrane (Southern or Northern transfer), and hybridization with radioactive probes. However, the health hazards and problems associated with the stability, use, and disposal of radioisotopes and the long exposure times (hours to days) required for detection by autoradiography have placed limitations on the routine use of hybridization assays. In recent years, nucleic acid analysis by hybridization has undergone a transition from radioactive labels to non-radioactive alternatives, which was driven by the need to improve detectability and facilitate automation and high-throughput analysis while avoiding the aforementioned limitations.

Bioluminometric hybridization assays [1–4] that use photoproteins as reporters offer higher detectability and wider dynamic range than spectrophotometric and fluorometric methods. This is due to the fact that in bioluminescence the excited species is formed during the course of a chemical reaction (an oxidation reaction). Consequently, bioluminometric measurements do not require excitation light,

Photoproteins in Bioanalysis. Edited by Sylvia Daunert and Sapna K. Deo

ISBN: 3-527-31016-9

thereby avoiding the problems arising from the scattering of excitation radiation, fluorescence from other components of the sample, and photobleaching.

The photoprotein aequorin is an excellent reporter molecule because it can be detected down to the 10^{-18} mole level (1 atto mol) by the simple addition of excess Ca^{2+}. Furthermore, the reaction is completed within 3 s, a significant advantage over enzyme reporters (such as alkaline phosphatase) that require long incubation times.

The applications of hybridization assays that use a photoprotein as a reporter focus on the determination of nucleic acid sequences amplified by the polymerase chain reaction (PCR) or other exponential amplification techniques [5, 6]. PCR is a powerful technique for the *in vitro* exponential amplification of specific DNA sequences to levels that are several orders of magnitude higher than those in the starting material. PCR entails denaturation of the sample DNA, hybridization (annealing) of two oligodeoxynucleotide primers that flank the region of interest, and polymerization using a thermostable DNA polymerase. After repetitive cycles of denaturation, primer annealing, and enzymic extension, the DNA segment defined by the two primers is selectively amplified. Exponential amplification of specific RNA sequences can be achieved by first generating a complementary copy of DNA with reverse transcriptase.

Figure 5.1 presents general hybridization assay configurations for detection/quantification of nucleic acid sequences in a high-throughput format using photoproteins as reporters. In the first approach ("immobilized target" assay), the target DNA or RNA is immobilized on the appropriate solid surface (e.g., a microtiter well or beads), the one strand is removed by treating with NaOH, and the immobilized strand is hybridized with a specific probe that is linked to the photoprotein. In the second method ("immobilized probe" configuration), the target DNA is denatured and hybridized with a probe that is immobilized on the solid surface. The hybridized target is then linked with the photoprotein reporter. In the third configuration ("sandwich-type" assay), the target DNA is denatured and allowed to hybridize with two probes. One probe is immobilized on the solid surface and the other is linked with the photoprotein reporter. The

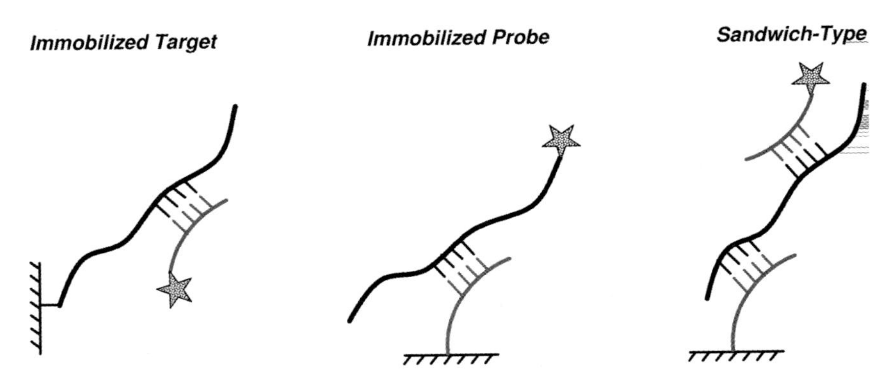

Fig. 5.1 Configurations of hybridization assays that use bioluminescent proteins as reporters.

third configuration offers higher specificity because two probes must hybridize in order to get a signal. The detectability of the hybridization assay is determined mainly by two factors: the detectability of the reporter molecule and the nonspecific binding of the detection reagents to the solid phase.

The immobilization of the target in the "immobilized target" assay is usually based on the strong and specific interaction between biotin and (strept)avidin. Polystyrene microtiter wells and polystyrene beads (including magnetic beads) are easily coated with (strept)avidin. The target DNA may be labeled through the polymerase chain reaction by using a primer biotinylated at the 5′ end or by incorporating biotin-modified dNTPs. Alternatively, the target DNA may be labeled with a hapten (e.g., digoxigenin) through PCR and captured by an anti-hapten antibody that is immobilized on the solid phase. Capture of the target DNA by a streptavidin-coated surface, however, offers the advantage that the binding withstands the NaOH treatment used for the removal of one strand. When an antibody is used for capture, the target is first heat-denatured and then added to the well.

The immobilization of the probe on the solid surface may also be accomplished by using the biotin/streptavidin or the hapten/anti-hapten antibody interaction. Oligonucleotide probes can be labeled with biotin or a hapten either during synthesis or by using the enzyme terminal deoxynucleotidyl transferase, a DNA polymerase that adds dNTPs and modified dNTPs to the 3′ end of any DNA molecule without the need of a template. Alternatively, the probe can be attached (chemically) to bovine serum albumin (BSA) and the conjugate used for coating of the polystyrene surface by physical adsorption. The probe also may be immobilized by covalent attachment to the solid surface.

The linking of a photoprotein to the hybrids is carried out either by direct covalent attachment of the photoprotein to the probe or by noncovalent bridging (with the probe or the target sequence) through biotin–streptavidin or hapten–antibody interaction as described above.

The detectabilities achieved with the above bioluminometric hybridization assays are in the low pmol range of target DNA (concentration of target DNA in the well), whereas fluorescent labels offer detectabilities in the nmol range. As a consequence, fewer PCR cycles are required for bioluminometric detection of the amplification product, and the possibility of sample contamination from amplified DNA is much lower.

The detectability of the aequorin-based bioluminometric hybridization assays can be enhanced by introducing (enzymically) multiple aequorin labels per DNA hybrid [7]. Heat-denatured DNA target is hybridized in microtiter wells with an immobilized capture probe and a digoxigenin-labeled detection probe. The hybrids react with anti-digoxigenin antibody conjugated to horseradish peroxidase. Peroxidase catalyzes the oxidation of a digoxigenin–tyramine conjugate by H_2O_2, resulting in the attachment of multiple digoxigenin moieties to the solid phase through the tyramine group, whereas the digoxigenin remains exposed. Aequorin-labeled anti-digoxigenin antibody is then allowed to bind to the immobilized haptens. The bound aequorin is determined by adding excess Ca^{2+}. The enzyme amplification improves the detectability about 8–10 times compared to the assay that does not involve a peroxidase amplification step.

5.2 Quantitative Polymerase Chain Reaction

The exponential increase of the amplification product during PCR poses serious difficulties in the application of PCR as a quantitative method for determination of the starting quantity of target DNA. Quantification requires the establishment of a reproducible relationship between the analytical signal obtained from the amplification product and the number of target DNA molecules in the sample prior to amplification. The amount (P) of product accumulated after n cycles of exponential amplification is given by the equation

$$P = T\,(1 + E)^n,$$

where T is the initial amount of target DNA and E is the average efficiency of the reaction for each cycle. The theoretical value of E is 1, i.e., the product doubles in each cycle. In reality, however, E has a smaller value depending on the reaction conditions and the nature of the sample. Variations in factors such as the concentration of polymerase, primers, dNTPs, Mg^{2+}, and cycling parameters (temperatures and times) affect the efficiency of amplification. In addition, the incorporation of primers into undesirable products leads to a decrease in the PCR yield. Furthermore, as the PCR enters a plateau phase at a high number of cycles (depending on the amount of starting template), there is a decrease in PCR efficiency, as a result of substrate saturation of the DNA polymerase and competition between strand reannealing and primer binding, as the concentration of amplified DNA increases.

A consequence of the exponential amplification is that small reaction-to-reaction variations in the efficiency have a profound effect on the quantity of the PCR product. For instance, a 5% decrease in E, from 1 to 0.95, results in a 50% decrease of the product (for $n = 25$).

One approach to quantitative PCR (QPCR) entails the co-amplification, in the same tube, of the target DNA sequence with a competitive synthetic DNA internal standard (DNA competitor) that closely resembles the target DNA [6, 8]. The internal standard (IS) uses the same primers as the target DNA and contains an insertion or deletion large enough to allow separation from the target by gel electrophoresis. Each sample is titrated with the IS. This is accomplished by adding increasing and known amounts of the IS to aliquots of the sample containing a constant amount of target DNA followed by PCR and electrophoresis. The equivalence point is determined either by inspection of the gel (same band intensities for target and IS) or, more accurately, by densitometric analysis, which should take into account the effect of the length of the DNA on the intensity of the bands. The use of an IS allows for compensation of the fluctuations of the amplification efficiency. Also, any PCR inhibitors present in the sample will equally affect the amplification of target DNA and IS so that the ratio of the amounts of their PCR products gives the ratio of the initial amounts of the two sequences in the sample. The main disadvantage of competitive PCR is the low throughput, which is a result of the multiple amplification reactions required for titration of each sample, followed by gel electrophoresis and densitometry.

Another approach to QPCR is based on the continuous monitoring of product accumulation by a homogeneous fluorometric hybridization assay (real-time PCR) with detectability at the nanomolar range. The cycle at which the fluorescence signal attains a certain preset threshold value is inversely related to the starting amount of target DNA. The amplification products are measured at the beginning of the exponential phase [6, 9]. Quantification is based on external calibration curves constructed by serial dilutions of the target. Real-time PCR methods do not employ internal standards because a DNA competitor would suppress the yield of amplified target to levels that may be undetectable [10].

The high detectability (in the low pmol concentrations) of bioluminometric hybridization assays that employ photoproteins as reporters allows accurate and precise determination of the PCR products of target and competitor DNA, despite the suppression of amplification caused by their competition for the same primer set. This enables quantification of the target sequence without the need for titration of each sample with various quantities of competitor. Instead, the target DNA is co-amplified with a constant amount of competitor. Furthermore, bioluminometric hybridization assays of the PCR products are performed in microtiter wells, thus ensuring simplicity and high throughput. A typical hybridization assay configuration used in bioluminometric quantitative PCR [11, 12] is shown in Fig. 5.2. The microtiter wells are coated with a BSA–$(dT)_{30}$ conjugate. The oligonucleotide probes consist of a segment complementary to

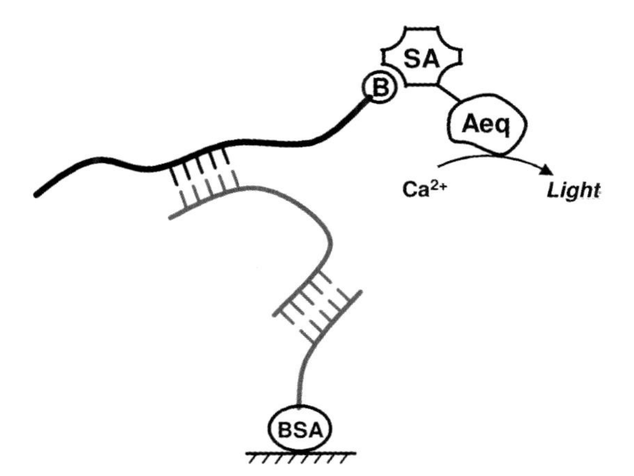

Fig. 5.2 Assay configuration for bioluminometric quantitative competitive polymerase chain reaction.
BSA: bovine serum albumin; B: biotin; SA: streptavidin; Aeq: aequorin.

the target or internal standard and a poly(dA) tail. The probes hybridize to both the biotinylated denatured PCR product and the immobilized $(dT)_{30}$ strands. A streptavidin–aequorin conjugate is used for determination of the captured hybrids. The ratio of the signals for target and internal standard is a linear function of the number of DNA molecules in the sample. The overall procedure including PCR and hybridization is complete in 2.5 h. Coating of the wells with BSA–$(dT)_{30}$ provides a universal solid phase for capturing both the target and IS.

Real-time PCR offers simplicity and automation but requires specialized and expensive equipment and reagents. The photoprotein-based quantitative PCR methods are endpoint assays (post-PCR detection) that use DNA competitors and are performed in a high-throughput format. The cost of the reagents and equipment are considerably lower than real-time PCR.

Contrary to competitive quantitative PCR methods that are based on electrophoretic separation of the amplification products, in bioluminometric quantitative PCR methods the DNA competitor is identical in size to the target sequence but is distinguishable by hybridization because it differs in a short (usually 20–25 bp) internal segment. DNA competitors may be prepared either by using appropriate vectors and standard cloning procedures or by faster and simpler approaches that employ PCR as a synthetic tool. The latter employs the target DNA sequence as a starting template and generates two short and overlapping DNA fragments through PCR. Subsequently, the two fragments are subjected to a PCR-like joining reaction to create the sequence of the DNA competitor. The procedure is illustrated in Fig. 5.3.

PCR-A and PCR-B are performed using primer sets a1, a2, and b1, b2. The downstream primer a2 and the upstream primer b1 carry extensions at the 5′ end that are complementary to each other and represent the new sequence that will be introduced into the DNA competitor. PCR-A gives a product that is identical to the left part of the target sequence and carries a new extension downstream. The product of PCR-B is homologous to the right part of the target DNA and carries, upstream, a new extension. Upon mixing, denaturation, and annealing of products A and B, two new types of hybrids are formed. Subsequently, DNA polymerase uses the one strand as a primer and the other as a template and synthesizes DNA in the 5′ to 3′ direction only. Thus, only one of the hybrids is extended and provides the sequence of the internal standard.

In the case of RNA determination, the use of RNA competitors is advantageous compared to DNA competitors because they allow compensation for any variation

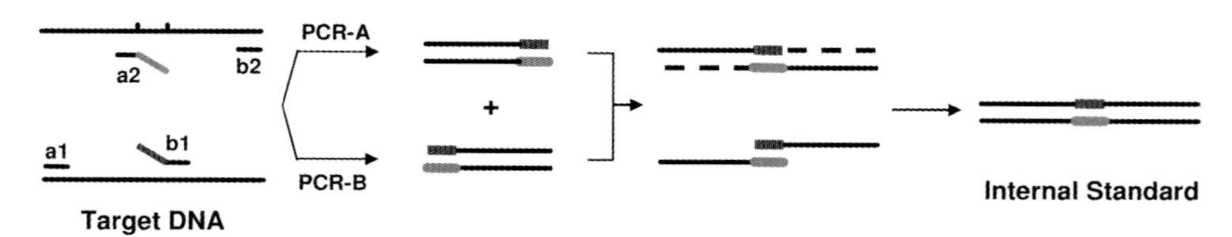

Fig. 5.3 Outline of the reactions used for the synthesis of a DNA internal standard (competitor) for quantitative PCR. Reprinted with permission from ref. 12.

in the efficiency of both the reverse transcription and the polymerase chain reaction. The RNA competitor is prepared by first synthesizing a DNA internal standard as described above and then introducing the T7 promoter. This is accomplished by subjecting the DNA competitor to PCR amplification using a primer with a T7 promoter sequence at its 5′ end. The DNA fragment is transcribed, *in vitro*, by T7 RNA polymerase to produce the RNA competitor [13].

Automation and throughput of quantitative PCR can be further enhanced by exploiting the variation in the kinetics of light emission from bio(chemi)luminescent reactions, thus allowing the development of dual hybridization assays for determination of target DNA/RNA and DNA/RNA competitor in the same reaction vessel (e.g., microtiter well) [11]. A rapid flash of light is generated from the aequorin reaction with a decay half-life of about 1 s, whereas a much slower emission (glow-type) that lasts from minutes to hours is produced by enzyme-catalyzed chemiluminescent reactions. Figure 5.4 presents the configuration of a dual-analyte assay for target and competitor that uses both aequorin and alkaline phosphatase as reporters. The two biotinylated PCR products from target DNA and IS are captured on a single microtiter well coated with streptavidin. The non-biotinylated strand is dissociated with NaOH and removed by washing. The immobilized single-stranded target DNA and IS hybridize simultaneously with their corresponding probes. The target- and IS-specific probes are labeled with the haptens digoxigenin and fluorescein, respectively. A solution containing aequorin-labeled anti-digoxigenin antibody and alkaline phosphatase-labeled anti-fluorescein antibody is added to the well and allowed to bind to the corresponding haptens. The excess of reagents is removed and the aequorin reaction is triggered with Ca^{2+}. The signal from the aequorin reaction is integrated for 3 s, followed by the addition of a chemiluminogenic substrate (CSPD) for ALP. The ALP reaction

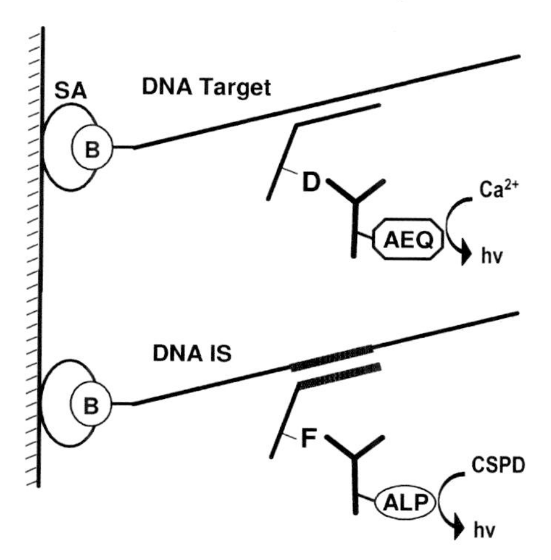

Fig. 5.4 Principle of quantitative competitive PCR based on a dual-analyte bio(chemi)luminometric assay for target and competitor (internal standard). SA: streptavidin; B: biotin; IS: internal standard; D: digoxigenin; F: fluorescein; Aeq: aequorin; ALP: alkaline phosphatase.

is allowed to proceed for 20 min followed by integration of the signal for 10 s. The ratio of the luminescence values for aequorin and ALP reactions is a linear function of the initial amount of the DNA (or RNA) in the sample prior to PCR. The linear range extends from 430 to 315 000 target DNA molecules and depends on the number of PCR cycles and the amount of DNA IS.

Besides facilitating high-throughput and automation, bioluminometric quantitative PCR methods eliminate a series of drawbacks of electrophoretic methods (including slab gel electrophoresis and capillary electrophoresis). For example, contrary to electrophoresis, hybridization methods provide sequence confirmation. In addition, it has been reported that PCR efficiency is inversely related to the size of the DNA [14]. This might create problems with electrophoretic methods that depend on size differences. The bioluminometric QPCR, however, employs an IS of a size identical to the target sequence. Finally, because of the high homology of the target and IS sequences, their co-amplification in the same reaction mixture leads to the formation of heteroduplexes, comprising a strand from the target DNA and a strand from the IS. The heteroduplexes usually migrate in a different way than the target and IS, causing errors in the determination of the amplification products, especially if they cannot be resolved from homoduplexes. Heteroduplex formation also may interfere with the digestion of the DNA in those QPCR methods that use competitors differing from the target sequence by a restriction site followed by electrophoresis of the digestion products. Heteroduplex formation is not a concern for the photoprotein-based QPCR methods described above, because they all rely on the denaturation of the amplified DNA and quantification of only one strand (the immobilized one).

5.3 Genotyping of Single-nucleotide Polymorphisms

Single-nucleotide polymorphisms (SNPs) can change gene function through amino acid substitution, modification of gene expression, or alteration of splicing. In recent years SNPs have emerged as a new generation of markers for disease susceptibility, prognosis, and response to therapy [15, 16]. The development of DNA microarray technology has enabled genome-wide studies of SNPs and correlation with various diseases. This technology, however, is designed to analyze thousands of polymorphisms from only a few samples (one sample per chip) by hybridization of the interrogated DNA to the probes of the microarray. Consequently, the DNA microarray is a valuable technique for establishing which SNPs are strongly associated with certain phenotypes, but it is not suitable for the screening of large numbers of samples. In the clinical laboratory, however, a small number of SNPs for each disease will be analyzed on a routine basis, and the results from disease-related genes or genes encoding drug-metabolizing enzymes will be utilized for disease prevention or for the design of effective therapeutic strategies. In this context, photoproteins may serve as reporters for high-throughput SNP genotyping performed in microtiter wells [17, 18].

The general procedure for SNP genotyping includes (1) the isolation of genomic DNA; (2) amplification, by PCR or other exponential amplification techniques, using primers flanking the locus of interest; (3) a genotyping reaction and; (4) detection of the products. The most commonly used genotyping reactions are the oligonucleotide ligation and the primer extension. The photoprotein aequorin has been used as a reporter in both genotyping reactions.

Figure 5.5 presents the configuration of a dual-analyte assay that combines flash- and glow-type reactions for the detection of the products of the oligonucleotide ligation reaction. The discrimination by DNA ligase against mismatches at the ligation site in two adjacently hybridized oligonucleotides (probes) is the basis for genotyping of SNPs by oligonucleotide ligation. The ligation methods make use of two recognition events between oligonucleotides and their targets, thus allowing these methods the required specificity for allele-specific SNP detection. Following PCR amplification, the DNA is denatured and hybridized simultaneously with three probes. Probes N and M are labeled at the 5′ end with biotin and digoxigenin, respectively. The last nucleotide at the 3′ end of N and M is specific for the normal and mutant allele, respectively. Probes N and M hybridize to the target DNA at a position adjacent to probe C. Upon perfect complementarity, a thermostable ligase joins the adjacent probes, giving two possible products depending on the allele that is present in the sample. If the normal allele is present, then biotinylated N-C is formed. Digoxigenin-labeled M-C is formed when the mutant allele is present. Probe C contains at the 3′ end an extension (about 17 nucleotides) that is irrelevant

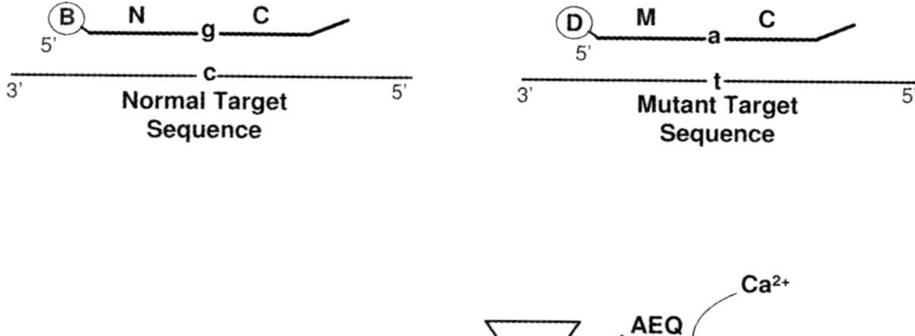

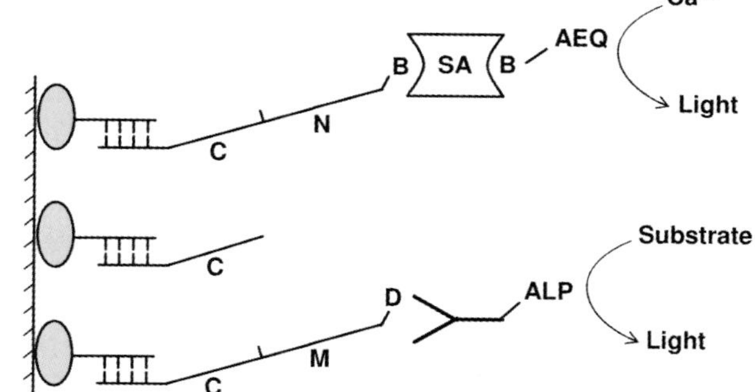

Fig. 5.5 Genotyping of SNPs by the oligonucleotide ligation reaction using the photoprotein aequorin and alkaline phosphatase as reporters. B: biotin; N: normal; M: mutant; C: common; SA: streptavidin; D: digoxigenin. Reprinted with permission from ref. 17.

to the target and enables the capture of the ligation products. Ligation products are heat-denatured and captured on the surface of a microtiter well through hybridization with an immobilized oligonucleotide that is complementary to the characteristic extension of the C probe. Aequorin-labeled streptavidin is added for detection of the N-C (normal allele) and ALP–anti-digoxigenin conjugate is used for detection of the mutant allele (M-C). The ratio of the luminescence signals obtained from aequorin and ALP gives the genotype for each sample. The microtiter well assay format is highly automatable and enables high-throughput genotyping of a large number of samples.

The principle of a bioluminometric assay that is based on a primer extension (PEXT) genotyping reaction and uses aequorin as a reporter is illustrated in Fig. 5.6. The distinction between the genotypes is based on the high accuracy of nucleotide incorporation by DNA polymerase. Two PEXT reactions are performed for each locus. Two allele-specific primers that hybridize with the target DNA adjacent to the mutation and have, at the 3′ end, a nucleotide complementary to the allelic variant are used in PEXT reactions. Only the primer with a perfectly matched 3′ end is extended by a thermostable DNA polymerase. Biotin is incorporated in the extended primer through the use of biotin-dUTP along with the dNTPs. Both genotyping primers N and M have at their 5′ end a $(dA)_{30}$ segment to enable affinity capture of extension products onto microtiter wells coated with a bovine serum albumin (BSA)-$(dT)_{30}$ conjugate. Prior to the bioluminometric assay, PEXT

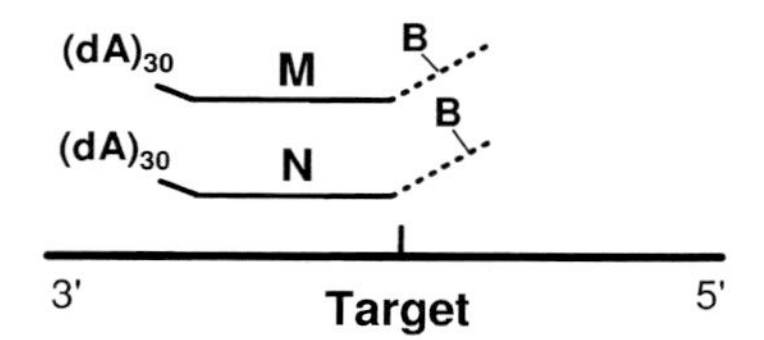

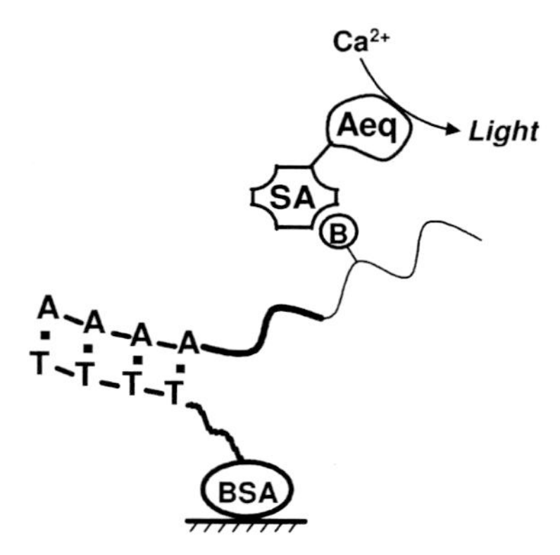

Fig. 5.6 Genotyping of SNPs by the primer extension (PEXT) assay using the photoprotein aequorin as a reporter. N: normal; M: mutant; B: biotin; SA: streptavidin; Aeq: aequorin; BSA: bovine serum albumin. Reprinted with permission from ref. 18.

products are heat-denatured to ensure separation of the extended strand of the genotyping primer from any biotinylated strands generated from extension of PCR primers. This denaturation step eliminates the need for purification of amplified fragments from unincorporated PCR primers. If primer extension has occurred, then the product carries incorporated biotin moieties that are detected by adding a streptavidin–aequorin conjugate. The PEXT reaction is completed in 10 min and the detection of the products takes less than 40 min.

The methods described above have been applied to the genotyping of the beta globin gene and the mannose-binding lectin gene.

5.4
Conjugation of Aequorin to Oligodeoxynucleotide Probes

A crucial step in the development of methods for detection and/or quantification of DNA/RNA is the linking between the recognition molecule (complementary DNA probe) and the reporter molecule (photoprotein). This linkage is carried out "directly" by chemical conjugation or "indirectly". In the indirect approach, a ligand (e.g., biotin or a hapten) is attached to the DNA probe and the hybrids are detected via a specific binding protein (streptavidin or an antibody), which is conjugated or complexed to the photoprotein. The advantage of the direct labeling approach lies in the fact that it eliminates an incubation step and a washing step, thus reducing considerably the assay time. The chemical attachment of aequorin to DNA is accomplished through primary amino groups or sulfhydryl groups of the photoprotein employing homo- or heterobifunctional cross-linking reagents [1, 19, 20]. Two conjugation strategies of aequorin to oligonucleotide probes are outlined in Fig. 5.7. The oligonucleotide probe is modified with an -NH_2 group at the 5′ end. The first reaction scheme involves activation of the probe by reacting with a large excess of the homobifunctional cross-linking reagent bis(sulfosuccinimidyl)suberate (BS^3) in order to avoid formation of dimers. The derivatized oligo carries a free succinimide group for subsequent reaction with -NH_2 groups of aequorin.

The second strategy entails the introduction of protected sulfhydryl groups to aequorin with *N*-succinimidyl-*S*-acetylthioacetate (SATA). The amino group of the DNA probe is converted to a maleimide group by reacting with the heterobifunctional cross-linker sulfosuccinimidyl 4-[*N*-maleimidomethyl]-cyclohexane-1-carboxylate (sulfo-SMCC). The SATA-modified photoprotein is then mixed with the maleimide-oligo and the conjugation reaction is initiated by the addition of hydroxylamine to deprotect the sulfhydryl group.

A successful conjugation procedure, besides maintaining the functionalities of the photoprotein and the DNA probe, requires the removal of the unreacted DNA probe which otherwise competes with the aequorin–DNA conjugate for hybridization to the target sequence and deteriorates the performance of the hybridization assays. The removal of the probe is usually accomplished by laborious chromatographic procedures followed by concentration steps.

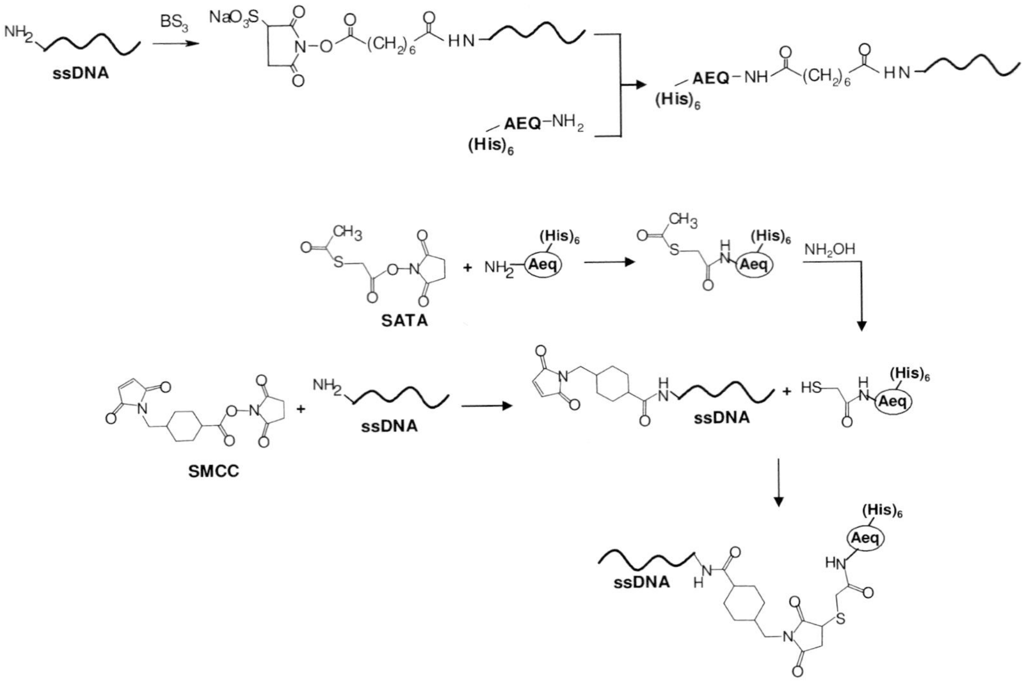

Fig. 5.7 Outline of two strategies used for conjugation of the photoprotein aequorin to oligonucleotide probes. Reprinted with permission from ref. 20.

In order to facilitate both the purification of recombinant aequorin from crude bacterial cultures and the preparation of aequorin conjugates, a suitable plasmid was constructed [21] in which a hexahistidine-coding sequence was fused upstream of the apoaequorin cDNA (Fig. 5.8). The $(His)_6$-apoaequorin fusion protein was overexpressed in *E. coli* under the control of the *tac* promoter. The inclusion bodies were solubilized by treating with 6 M urea, and the $(His)_6$-apoaequorin was purified in a single step by immobilized metal-ion affinity chromatography using a Ni^{2+}-nitrilotriacetate agarose column. Proper refolding of the photoprotein was achieved

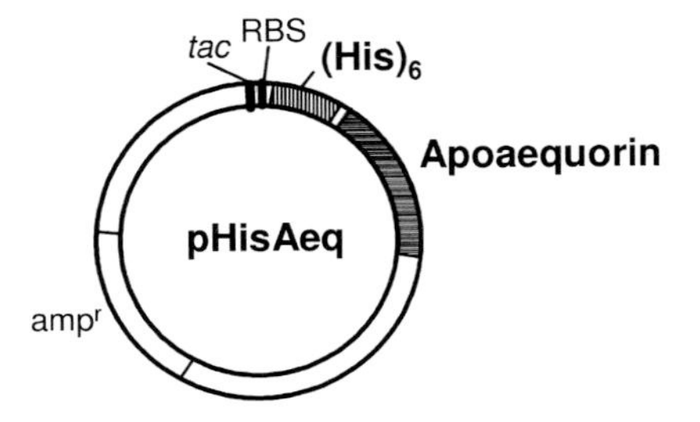

Fig. 5.8 Structure of the plasmid used for bacterial overexpression of recombinant hexahistidine–aequorin fusion protein.

by slow removal of urea using a gradient of 6–0 M. The fusion protein was eluted by imidazole and used for preparation of conjugates with DNA probes. Following the conjugation reactions described above, the $(His)_6$-apoaequorin was captured on a Ni^{2+}-nitrilotriacetate agarose column, the unreacted probe was washed away, and the conjugate was eluted as above. The presence of free aequorin in the conjugate solution does not interfere with the hybridization assay.

5.5
Development of New Recombinant Bioluminescent Reporters

The investigation and exploitation of new bioluminescent proteins as reporters in nucleic acid analysis (as well as in other applications) is an active area of research. The process of development and evaluation of new bioluminescent reporters involves the following steps:

1. isolation and cloning of the cDNA encoding the bioluminescent protein,
2. construction of suitable vectors for large-scale expression of the protein in a heterologous system (e.g., bacteria),
3. development of a purification method for the protein, and
4. linking of the bioluminescent protein to a DNA probe in order to prepare a recognition and detection reagent for nucleic acid analysis.

The luciferase of the marine copepod *Gaussia princeps* (*Gaussia* luciferase, GL) is a typical example of this process [22, 23]. GL is a single polypeptide chain of 185 amino acids (MW 19 900) and catalyzes the oxidative decarboxylation of coelenterazine to produce coelenteramide and light (470 nm). The cloning of the cDNA of GL was accomplished recently. In order to facilitate the purification of the protein and its subsequent use as a reporter in hybridization assays, a suitable vector was constructed (Fig. 5.9) that drives the expression of *in vivo* biotinylated GL in *E. coli*. The plasmid contains the sequence coding for the biotin acceptor peptide (bap) from *Propionibacterium shermanii* transcarboxylase positioned downstream of the *tac* promoter and a ribosome-binding site (RBS). The structure

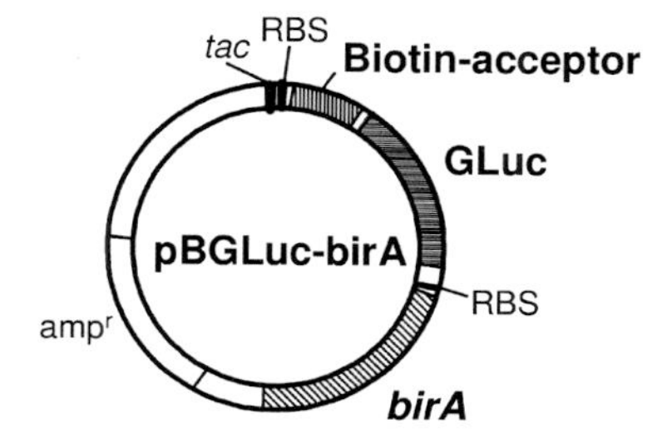

Fig. 5.9 Structure of the plasmid used for overexpression of *in vivo* biotinylated *Gaussia* luciferase in bacteria.

of the biotin domain of *P. shermanii* transcarboxylase is very similar to that of *E. coli* acetyl-CoA carboxylase, which is the physiological substrate of biotin protein ligase (BPL), the enzyme that is responsible for the *in vivo* biotinylation of bap at a unique site. The GL cDNA is positioned downstream of bap. Consequently, the expressed fusion is an *in vivo* biotinylated bap–GL sequence. Because the endogenous activity of BPL was found to be low, the gene birA, which codes for BPL, was also introduced in the same plasmid, thus achieving overexpression of both the BPL and the bap–GL fusion.

The *in vivo* biotinylated bap–GL was purified from the crude cellular extract by affinity chromatography using a monomeric avidin resin. Monomeric avidin has a much lower affinity for biotin ($k_d = 10^{-7}$) than tetrameric avidin ($k_d = 10^{-15}$), thereby allowing the binding of biotinylated GL and subsequent elution with free biotin. Moreover, biotinylation facilitates the linking of GL to streptavidin, avoiding chemical conjugation that may inactivate the protein. The complex of streptavidin (SA) and biotinylated GL (BGL) is prepared by precise optimization of the BGL: SA molar ratio. If BGL is present in excess, then the four biotin-binding sites of SA become saturated. If there is an excess of SA, then free SA competes with the SA-BGL complex for binding to the biotinylated hybrids on the well. The fact that a single biotin moiety is attached to the protein outside of the GL sequence is advantageous over chemical biotinylation methods that may alter functional groups of the luciferase. A typical (model) hybridization assay (Fig. 5.10) involves denaturation of a biotinylated DNA target and hybridization with an immobilized specific probe. The hybrids are detected by the addition of the SA-BGL complex. Coelenterazine is then added as a substrate. BGL is detectable down to 1 amol following light emission integration for 20 s, with linearity extending over 5 orders of magnitude. The hybridization assay has a linearity range of 1.6–800 pM of target DNA.

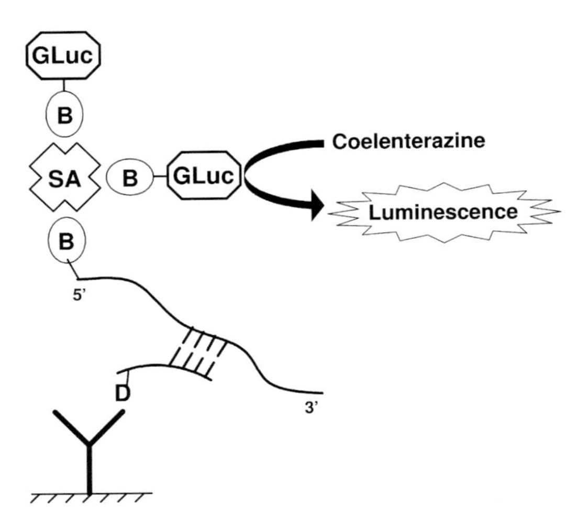

Fig. 5.10 Model hybridization assay using *Gaussia* luciferase as a reporter. B: biotin; SA: streptavidin; GLuc: *Gaussia* luciferase; D: digoxigenin.

5.6 Signal Amplification by in Vitro Expression of DNA Reporters Encoding Bioluminescent Proteins

Cell-free expression of DNA fragments entails transcription of DNA to RNA followed by translation of RNA to protein. Several RNA molecules are synthesized from each DNA template during transcription, and more than one protein molecule is generated from each transcript. As a consequence, gene expression forms the basis of a signal amplification system. The detectability is further improved if the DNA template encodes a bioluminescent protein. Contrary to previously described hybridization assays in which only one reporter molecule is linked to the hybrid, a photoprotein-coding DNA fragment upon expression generates multiple bioluminescent molecules in solution [24–27].

DNA templates were engineered that contain all necessary elements to enable *in vitro* expression in wheat germ- or rabbit reticulocyte-coupled (one-step) transcription–translation systems under the control of the bacteriophage T7 RNA polymerase. The DNA template contains a T7 promoter, the cDNA of the bioluminescent protein, and a $(dA/dT)_{30}$ extension that enhances translation efficiency by facilitating translation initiation. When apoaequorin cDNA is used as a template, coelenterazine is added to the expression mixture to enable the formation of fully functional aequorin. The *in vitro* expression reaction typically proceeds for 90 min and then Ca^{2+} is added to trigger light emission. The wheat germ extract is preferred for expression of aequorin because the rabbit reticulocyte extract absorbs a significant portion of the emitted light. *In vitro* expression experiments have shown that although the transcription–translation process consists of a series of complex and not completely understood reactions that require the concerted action of numerous factors (RNA polymerase, initiation, elongation and termination factors, ribosomal subunits, aminoacyl-tRNA synthetases, etc.), the final outcome is a simple and reproducible linear relationship between the bioluminescence signal and the amount of the DNA template. As low as 5×10^3 molecules of aequorin DNA are detectable and the linearity extends up to 10^8 molecules. For application to hybridization assays, the DNA template (reporter) must be linked to a specific probe. This is carried out by tailing the template with dATP using terminal deoxynucleotidyl transferase. The probe is tailed with dTTP. In a typical expression hybridization assay configuration (Fig. 5.11), biotinylated target DNA is captured on streptavidin-coated wells and, after removal of one strand by NaOH, is hybridized with the probe. The poly(dA)-tailed expressible DNA fragment is then hybridized to the poly(dT) tail of the probe. The transcription–translation "cocktail" is then added (one step) followed by a 90-min incubation during which the photoprotein is synthesized from the immobilized DNA template. These studies [24–27] showed for the first time data from *in vitro* expression of immobilized DNA templates. Alternatively, an unlabeled target DNA can be sandwiched between an immobilized probe and a poly(dT) probe followed by the addition of the expressible DNA fragment and *in vitro* expression. The linear range of these hybridization assays extends from

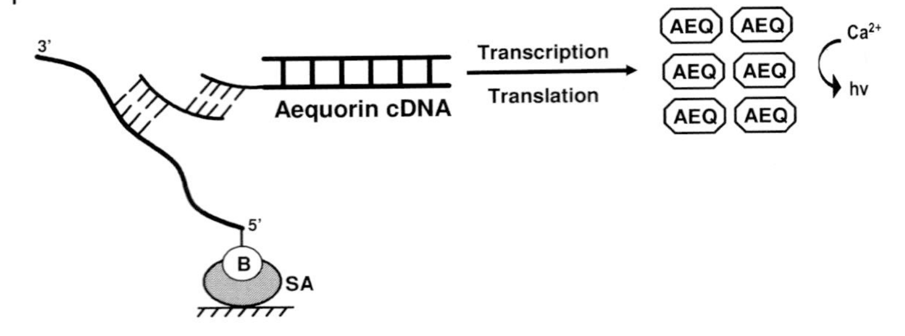

Fig. 5.11 Representative hybridization assay using, as a reporter, a DNA template encoding the photoprotein aequorin.

0.25 amol to 1500 amol target DNA. The above assays demonstrate that DNA, besides its wide use as a recognition molecule (probe), can also serve as a signal-generating molecule (reporter) providing an effective means for the generation of multiple aequorin molecules in solution (more than 150 molecules of the photoprotein per DNA template).

Luciferases, although highly detectable, have found only limited use as labels in DNA hybridization assays because of their significant loss of activity upon conjugation to other molecules. A distinct advantage, however, of using the luciferase-coding DNA as a reporter instead of the enzyme itself is that inactivation problems are avoided because the synthesized luciferase remains free in solution. The cDNA of firefly luciferase (FL, MW = 62 000) and *Renilla* luciferase (RL, MW = 36 000) have been used as reporters in expression hybridization assays. Furthermore, a microtiter well–based hybridization assay that allows simultaneous determination of two target DNA sequences has been developed [27]. The target DNAs were heat-denatured and hybridized with specific capture and detection probes. The capture probes were immobilized by physical absorption in the form of conjugates with bovine serum albumin. One detection probe was biotinylated and the other was tailed with dTTP. The hybrids were reacted with a mixture of SA-FL DNA complex and poly(dA)-tailed RL DNA. Subsequently, the transcription–translation mixture was added to initiate expression of the two immobilized reporter DNAs. It was shown that although the two DNA templates used the same promoter and shared the same transcription and translation machinery, they were expressed independently, i.e., the expression of one template does not affect the expression of the other in a large range of concentrations. Furthermore, the two luciferases (FL and RL) were co-determined in the expression mixture by a dual assay, exploiting the fact that the presence of Mg^{2+} is necessary for the activity of FL but is not required for RL. Thus, following expression, the substrate of FL was added (containing luciferin, ATP, and Mg^{2+}) and the luminescence was measured for 30 s. The FL reaction then was stopped by the addition of EDTA, the substrate of RL was injected, and the luminescence was measured for 10 s.

5.7 Conclusions

Bioluminescent reporters can be detected within a few seconds at the atto mol level using simple instrumentation. This is the basis for their application as non-radioactive reporters in nucleic acid analysis by hybridization with detectabilities in the low pmol range of target DNA. This is more than a 100-fold improvement over current hybridization assays that use fluorescent labels. In addition, photoproteins can be combined with glow-type chemiluminogenic reactions for the development of dual-analyte hybridization assays. The hybridization assays can be performed in microtiter wells, thus enabling automation and high-throughput analysis. Major application fields include the detection and determination of PCR products and the genotyping of single-nucleotide polymorphisms. One of the most important applications of photoprotein reporters is in the development of high-throughput quantitative competitive PCR methods that use synthetic DNA or RNA internal standards (competitors).

References

1. Stults, N. L., Stocks, N. F., Rivera, H., Gray, J., McCann, R. O., O'Kane, D., Cummings, R. D., Cormier, M. J., Smith, D. F. Use of recombinant biotinylated aequorin in microtiter and membrane-based assays: purification of recombinant aequorin from Escherichia coli. *Biochemistry* **1992**, *31*, 1433–1442.
2. Galvan, B., Christopoulos, T. K. Bioluminescence hybridization assays using recombinant aequorin. Application to the detection of prostate-specific antigen mRNA. *Anal. Chem.* **1996**, *68*, 3545–3550.
3. Glynou, K., Ioannou, P. C., Christopoulos, T. K. Detection of transgenes in soybean by polymerase chain reaction and a simple bioluminometric assay based on a universal aequorin-labeled oligonucleotide probe. *Anal. Bioanal. Chem.* **2004**, *378*, 1748–1753.
4. Xiao, L., Yang, C., Nelson, C. O., Holloway, B. P., Udhayakumar, V., Lal, A. A. Quantitation of RT-PCR amplified cytokine mRNA by aequorin-based bioluminescence immunoassay. *J. Immun. Methods* **1996**, *199*, 139–147.
5. Mullis, K. B., Faloona, F. A. Specific synthesis of DNA in vitro via a polymerase-catalyzed chain reaction. *Methods Enzymol.* **1987**, *155*, 335–350.
6. Christopoulos, T. K. Polymerase chain reaction and other amplification systems. Encyclopedia of Analytical Chemistry (Mayers, R. A., Ed.), pp. 5159–5173, John Wiley & Sons, Chichester UK, **2000**.
7. Laios, E., Ioannou, P. C., Christopoulos, T. K. Enzyme amplified aequorin-based bioluminometric hybridization assays. *Anal. Chem.* **2001**, *73*, 689–692.
8. Gilliland, G., Perrin, S., Blanchard, K., Bunn, H. F. Analysis of cytokine mRNA and DNA: Detection and quantitation by competitive polymerase chain reaction. *Proc. Natl. Acad. Sci. USA* **1990**, *87*, 2725–2729.
9. Wittwer, C. T., Herrmann, M. G., Moss, A. A., Rasmussen, R. P. Continuous fluorescence monitoring of rapid cycle DNA amplification. *BioTechniques* **1997**, *22*, 130–138.
10. Nurmi, J., Wikman, T., Lovgren, T. High performance real-time quantitative RT-PCR using lanthanide probes and a dual-temperature hybridization assay. *Anal. Chem.* **2002**, *74*, 3525–3532.

11 Verhaegen, M., Christopoulos, T. K. Quantitative polymerase reaction based on a dual-analyte chemiluminescence hybridization assay for the target DNA and the internal standard. *Anal. Chem.* **1998**, *70*, 4120–4125.

12 Mavropoulou, A. K., Koraki, T., Ioannou, P. C., Christopoulos, T. K. High-throughput double quantitative competitive polymerase chain reaction for determination of genetically modified organisms. *Anal. Chem.* **2005**, *77*, 4785–4791.

13 Verhaegen, M., Ioannou, P. C., Christopoulos, T. K. Quantification of prostate-specific antigen mRNA by coamplification with a recombinant RNA internal standard and microtiter well-based hybridization. *Clin. Chem.* **1998**, 44, 1170–1176.

14 McCulloch, R. K., Choong, C. S., Hurley, D. M. An evaluation of competitor type and size for use in the determination of mRNA by competitive PCR. *PCR Methods Appl.* **1995**, *4*, 219–226.

15 Carlson, C. S., Newman, T. L., Nickerson, D. A. SNPing in the human genome. *Curr. Opin. Chem. Biol.* **2001**, *5*, 78–85.

16 Syvanen, A.-C. Accessing genetic variation: Genotyping single nucleotide polymorphisms. *Nature Genetics* **2001**, *2*, 930–942.

17 Tannous, B., Verhaegen, M., Christopoulos, T. K., Kourakli, A. Combined flash- and glow-type chemiluminescent reactions for high-throughput genotyping of biallelic polymorphisms. *Anal. Biochem.* **2003**, *320*, 266–272.

18 Zerefos, P. G., Ioannou, P. C., Traeger-Synodinos, J., Dimissianos, G., Kanavakis, E., Christopoulos, T. K. Photoprotein aequorin as a novel reporter for SNP genotyping by primer extension-Application to the variants of mannose-binding lectin gene. *Hum. Mutat.* **2006**, *27*, 279–285.

19 Stults, N. L., Rivera, H. N., Burke-Payne, J., Ball, R. T., Smith, D. F. Preparation of stable conjugates of recombinant aequorin with proteins and nucleic acids. In *Bioluminescence and Chemiluminescence: Molecular Reporting With Photons* (Hastings, J. W., Kricka, L. J., Stanley, P. E., Eds.), p. 243, John Wiley & Sons, Chichester UK, **1997**.

20 Glynou, K., Ioannou, P. C., Christopoulos, T. K. Affinity capture-facilitated preparation of aequorin-oligonucleotide conjugates for rapid hybridization assays. *Bioconj. Chem.* **2003**, *14*, 1024–1029.

21 Glynou, K., Ioannou, P. C., Christopoulos, T. K. One-step purification of recombinant photoprotein aequorin by immobilized metal-ion affinity chromatography. *Prot. Expr. Purif.* **2003**, *27*, 384–390.

22 Szent-Gyorgyi, C., Ballou, B. T., Dagmal, E., Bryan, B. Cloning and characterization of new bioluminescent proteins. Part of the SPIE Conference on Molecular Imaging: Reporters, Dyes, Markers, and Instrumentation. San Jose, CA. Proc. SPIE *3600*, 4–11, **1999**.

23 Verhaegen, M., Christopoulos, T. K. Recombinant Gaussia luciferase. Over-expression, purification and analytical application of a bioluminescent reporter for DNA hybridization. *Anal. Chem.* **2002**, *74*, 4378–4385.

24 White, S. R., Christopoulos, T. K. Signal amplification system for DNA hybridization assays based on in vitro expression of a DNA label encoding apoaequorin. *Nucleic Acids Res.* **1999**, *27*, e25(i–viii)

25 Chiu, N. H. L., Christopoulos, T. K. Hybridization assays using an expressible DNA fragment encoding firefly luciferase as a label. *Anal. Chem.* **1996**, *68*, 2304–2308.

26 Laios, E., Ioannou, P. C., Christopoulos, T. K. Novel hybridization assay configurations based on *in vitro* expression of DNA reporter molecules. *Clin. Biochem.* **1998**, *31*, 151–158.

27 Laios, E., Obeid, P., Ioannou, P. C., Christopoulos, T. K. Expression hybridization assays combining cDNAs from firefly and renilla luciferases for simultaneous determination of two target sequences. *Anal. Chem.* **2000**, *72*, 4022–4028.

6
Bioluminescence Resonance Energy Transfer in Bioanalysis

Suresh Shrestha and Sapna K. Deo

6.1
Introduction

Nonradiative transfer of energy from a donor molecule to an acceptor molecule resulting from dipole–dipole coupling is termed resonance energy transfer (RET). When the donor molecule is a fluorescent probe, it is known as fluorescence resonance energy transfer (Fig. 6.1) [1]. Bioluminescence resonance energy transfer (BRET) is a phenomenon in which resonant energy is transferred from a bioluminescent donor protein to a fluorescent acceptor, which, in turn, emits light at its characteristic wavelength of emission (Fig. 6.1) [2].

BRET is a naturally occurring phenomenon observed in marine organisms such as the jellyfish *Aequorea victoria* and the sea pansy *Renilla reniformis* [3–5]. A pair of bioluminescent proteins and fluorescent proteins responsible for this phenomenon has been isolated from these organisms. In the jellyfish, this pair consists of the bioluminescent protein aequorin and the green fluorescent protein (GFP). The BRET pair in the sea pansy includes the bioluminescent protein *Renilla* luciferase (RLuc) and GFP.

This transfer of energy in marine organisms leads to an increase in the quantum efficiency of light emission [5]. Initial attempts to reproduce this BRET phenomenon *in vitro* using isolated aequorin and GFP in solution were unsuccessful.

However, BRET could be achieved upon co-adsorption of high concentration of these two proteins on a solid surface or if the two proteins were genetically conjugated.

Photoproteins in Bioanalysis. Edited by Sylvia Daunert and Sapna K. Deo

ISBN: 3-527-31016-9

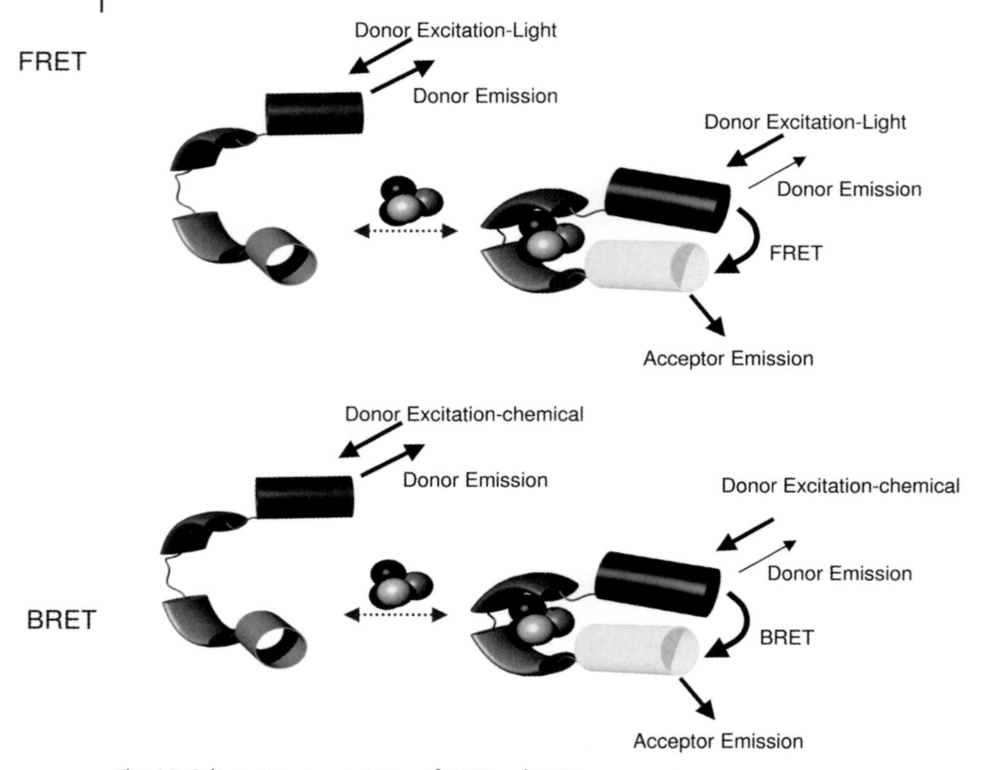

Fig. 6.1 Schematic representation of FRET and BRET.

6.2
BRET Principle, Efficiency, and Instrumentation

The efficiency of BRET is dependent upon factors such as spectral properties, relative distance, and orientation of donor and acceptor molecules [6, 7]. BRET normally occurs when the distance between the donor and the acceptor is within 100 Å, and the efficiency of BRET is inversely related to the distance to the sixth power. The rate of energy transfer is given by Förster's equation,

$$k_T = (1/\tau_d)\ (R_0/R)^6,$$

where k_T is the rate of energy transfer, τ_d is the fluorescence lifetime of the donor in the absence of the acceptor, R_0 is the Förster critical radius at which 50% of the excitation energy is transferred to the acceptor, and R is the distance between the centers of the donor and acceptor chromophores. Thus, the relative location between the donor molecule and the acceptor is critical for an efficient energy transfer.

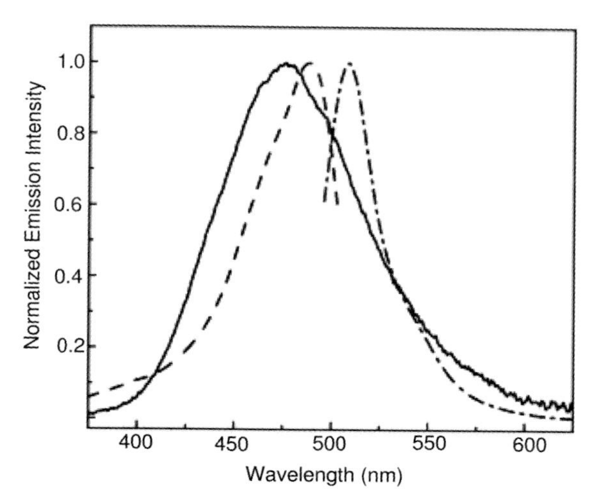

Fig. 6.2 Emission spectrum of Rluc (solid line) and excitation and emission spectrum of YFP (dotted line). The spectral overlap between the emission of RLuc and the excitation of YFP is evident from the spectra. Adapted from Ref. [25]. Reprinted with permission from Elsevier Science.

In addition to the distance, when designing fusion proteins for BRET assays, consideration should be given to the relative orientation of the donor and acceptor molecules. There should be sufficient flexibility between the donor and the acceptor in a fusion protein. It is necessary that the spectrum of donor emission and acceptor excitation overlap significantly in order to achieve a BRET of high efficiency (Fig. 6.2). Spectral resolution is lost if the donor and acceptor emission spectrum overlap, resulting in a low signal-to-noise ratio.

Luminescence and fluorescence measurements are performed in many laboratories routinely. Therefore, there is interest in developing instruments that allow sensitive and easy detection of light from emission-based methods. A large number of instruments capable of high-throughput measurements have been developed in recent years. These instruments are capable of working with traditional sample volumes, such as 96-well plate readers, as well as with low sample volumes, which are based on 384-well and 1536-well plates [6]. Some of these instruments have the ability to perform measurement in multiple modes such as absorbance, fluorescence, luminescence, fluorescence polarization, and anisotropy [6]. Typically, BRET measurements are performed by using either a microtiter plate reader or a scanning spectroscopy-type of instrument. In microtiter plate readers, the light is collected using fiber optics and a filter wheel that allows selection of emission from the fluorophore, which is kept in the path of the light. A highly sensitive PMT is used as a detector. Careful consideration should be given while selecting the bandpass filter, such that the signal corresponding to the donor emission is not collected. To overcome this drawback, some companies have made available BRET-optimized filters that allow longer bandpass for excitation and smaller bandpass for fluorophore emission measurement [6].

6.3 Comparison of BRET and FRET

Fluorescence resonance energy transfer (FRET), in which both the donor and acceptor molecules are fluorescent proteins or chemical probes, is a very popular biochemical technique employed in a variety of applications [8]. Several FRET pairs have been well studied and allow for a range of selection in terms of fluorescence excitation–emission wavelengths. In FRET studies, there is no need for addition of substrate to generate a signal. High signal levels can be achieved by selecting suitable fluorescent probes. In addition, because of the availability of extensive literature in this field, FRET has become a method of choice for many applications such as drug screening [9], *in vivo* biochemical analysis [10], receptor activation studies [11], metabolic screening [12], protein folding, [13] etc. In recent times, BRET has slowly gained in popularity over FRET. The former has several advantages as well as some limitations when compared to FRET. The main advantage of BRET over FRET is that it does not require an excitation source and therefore is a better option for analysis of cells that are damaged by excitation light or are photoresponsive. The problem of photobleaching of fluorophores, as in the case of FRET, is avoided when employing BRET. In addition, the contribution of cell autofluorescence to the background becomes insignificant when the cells are assayed using BRET. While spectral separation between donor and acceptor excitation is needed in FRET to avoid problems with the possibility of exciting both the fluorophores, this problem is eliminated in BRET because the light emission from the donor occurs as a result of a chemical reaction. It should be noted that BRET biosensing systems can be designed by employing molecular biology tools, such as gene fusion, to create protein chimeras where both the donor and the acceptor molecule are part of the same protein molecules [6]. In these cases, a fusion protein is constructed using molecular biology tools such that the bioluminescent donor is fused to one of the termini of the protein under study. At the other terminus the fluorescent protein is genetically fused. This also allows measurement of expression levels of donor and acceptor fusion partners in a BRET pair independently, in contrast to FRET, where the acceptor can get excited to some extent with donor excitation. This is especially true in cases where green fluorescent protein (GFP) mutants are employed as FRET pairs. The knowledge of relative levels of expression of the fusion partner allows for a proper comparison between all the experiments performed, which in turn allows for the examination of other factors that may affect the background and thus the protein expression levels.

A major disadvantage of BRET over FRET is that the BRET signal can be weak. This can be overcome by using highly sensitive instruments in conjunction with the integration over a long period of time of the BRET signal generated as a result of the assay. There is no question that BRET is a powerful technique; however, to date only a few BRET donors have been identified and characterized. Unfortunately, this limits the choice of acceptors and the wide applicability of the method.

6.4 Examples of BRET Donor–Acceptor Pairs

Renilla luciferase (RLuc) was originally isolated from the anthozoan coelenterate *Renilla reniformis* [4, 14]. It catalyzes the oxidative decarboxylation of the substrate coelenterazine to produce coelenteramide and light (wavelength maxima 480 nm) (Fig. 6.3). Renilla luciferase has been used as a reporter gene for studying *in vitro* and *in vivo* gene regulation. *Native* coelenterazine is the natural substrate for *Rluc*; however, it can also utilize analogues of coelenterazine as substrates yielding different properties in terms of emission wavelength and quantum efficiency. RLuc has a broad bioluminescence emission spectra with a peak around 480 nm. It was the first reported bioluminescence energy donor in a BRET application. RLuc was chosen because of the similarity of its emission spectrum with the cyan mutant of GFP (CFP) routinely employed in FRET pairs with the yellow mutant of GFP (YFP). The emission spectrum overlap between RLuc and YFP is poor [6], and hence other substrates of RLuc are now employed that allow the efficient transfer of energy with other mutants of GFP. A few examples of these substrates are given in Table 6.1.

Another BRET pair that has found a good number of applications in bioanalysis is that of aequorin and GFP. Aequorin, a bioluminescent photoprotein consisting of an apoprotein, the chromophore coelenterazine, and bound oxygen, undergoes a conformational change in the presence of calcium. This leads to oxidation of coelenterazine to coelenteramide, resulting in an excited-state coelenteramide.

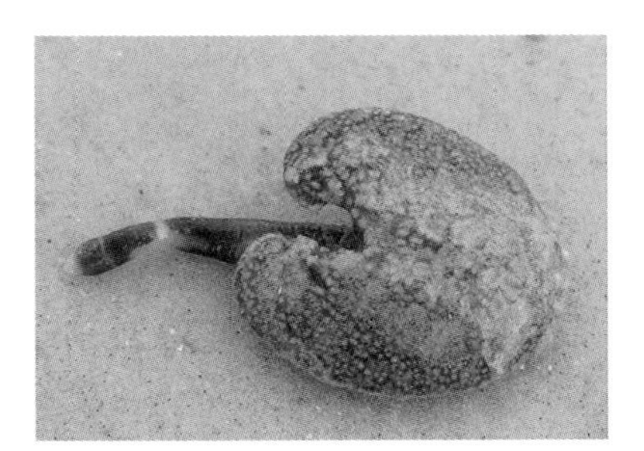

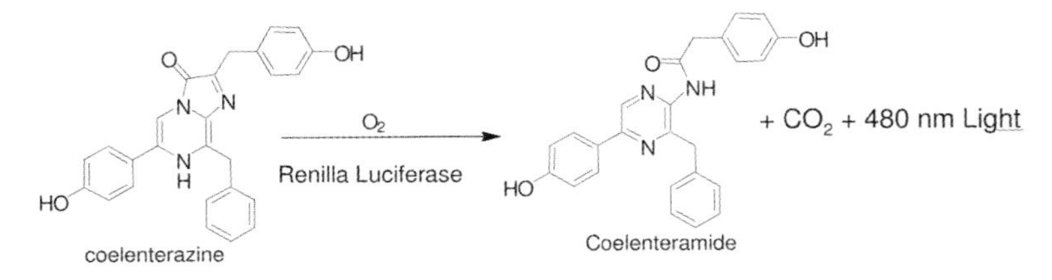

Fig. 6.3 The sea pansy *Renilla reniformis* adapted from (http://www.jaxshells.org/1113bb.htm) and a schematic showing the bioluminescence reaction of *Renilla* luciferase.

Table 6.1 Examples of BRET pairs and their spectral characteristics.

Donor	*Substrate*	*Donor emission (λ, nm)*	*Acceptor*	*Acceptor excitation/ Emission (λ, nm)*
RLuc	Coelenterazine	420–530	EYFP	513/527
RLuc	Coelenterazine h	420–530	Topaz	514/527
RLuc	Coelenterazine DeepBlueC	385–420	GFP	405/510
Aequorin	Coelenterazine	430–500	GFP	489/510
Firefly luciferase	D-luciferin	560–580	Red fluorescent protein	553/583

This excited-state coelenteramide relaxes to ground state with a concomitant emission of light at 469 nm and release of CO_2. The bioluminescence emission spectrum of aequorin is broad (approximately 430–500 nm), allowing an overlap with the excitation spectrum of EGFP (wavelength maximum at 489 nm). The gene for aequorin has been known and cloned for many years, and thus it is a natural one for use in BRET applications [15, 16].

Firefly luciferase has also been employed in one study as a BRET donor [10]. Its emission spectrum is in the range of 560–580 nm [17]. The red fluorescent protein isolated from the *Discosoma* species, which has an excitation maximum of 553 nm, is a suitable acceptor when firefly luciferase is employed as a donor. Firefly luciferase has a higher quantum yield than Rluc; hence, when employed in conjunction with a suitable acceptor, it can potentially yield higher sensitivity in the detection of target analytes. Chromophore maturation in the red fluorescent protein is slow, which limits its application. The cloning of monomeric red fluorescent protein with faster maturation may help overcome this problem.

In an alternative strategy, a bioluminescence energy transfer from aequorin to synthetic fluorophores has been demonstrated (Fig. 6.4) [18]. In this work, the photoprotein aequorin was modified by attaching fluorophores through a unique cysteine introduced site-specifically on the protein. The fluorophores were selected such that the excitation spectrum of the fluorophore overlapped with the emission spectrum of aequorin. Upon addition of calcium, the emitted bioluminescence of aequorin was transferred to the fluorophore, leading to the excitation of the fluorophore and subsequent emission at the characteristic wavelength of the fluorophore. These modified proteins were termed “artificial jellyfish” because of their similarity to the naturally occurring phenomenon in the marine jellyfish *Aequorea aequorea* in which under certain conditions (i.e., deep dark waters or cold waters) energy transfer from aequorin to GFP occurs [19]. The strategy developed here allows for the construction of bioluminescent labels with different emission maxima by selecting different fluorophores for applications in the simultaneous detection of multiple analytes. By examining the X-ray crystal structure of the protein, four different sites were evaluated for introduction of the unique cysteine residue. In this study, two fluorophores, IANBD ester (λ_{ex}= 478 nm, λ_{em}= 536 nm) and Lucifer yellow (λ_{ex} = 428 nm, λ_{em}= 531 nm) with differing emission maxima were attached individually to the aequorin mutants through the sulfhydryl group of the cysteine molecule. Two of the fluorophore-labeled mutants showed a peak corresponding to fluorophore emission, indicating resonance energy transfer from aequorin to the fluorophore.

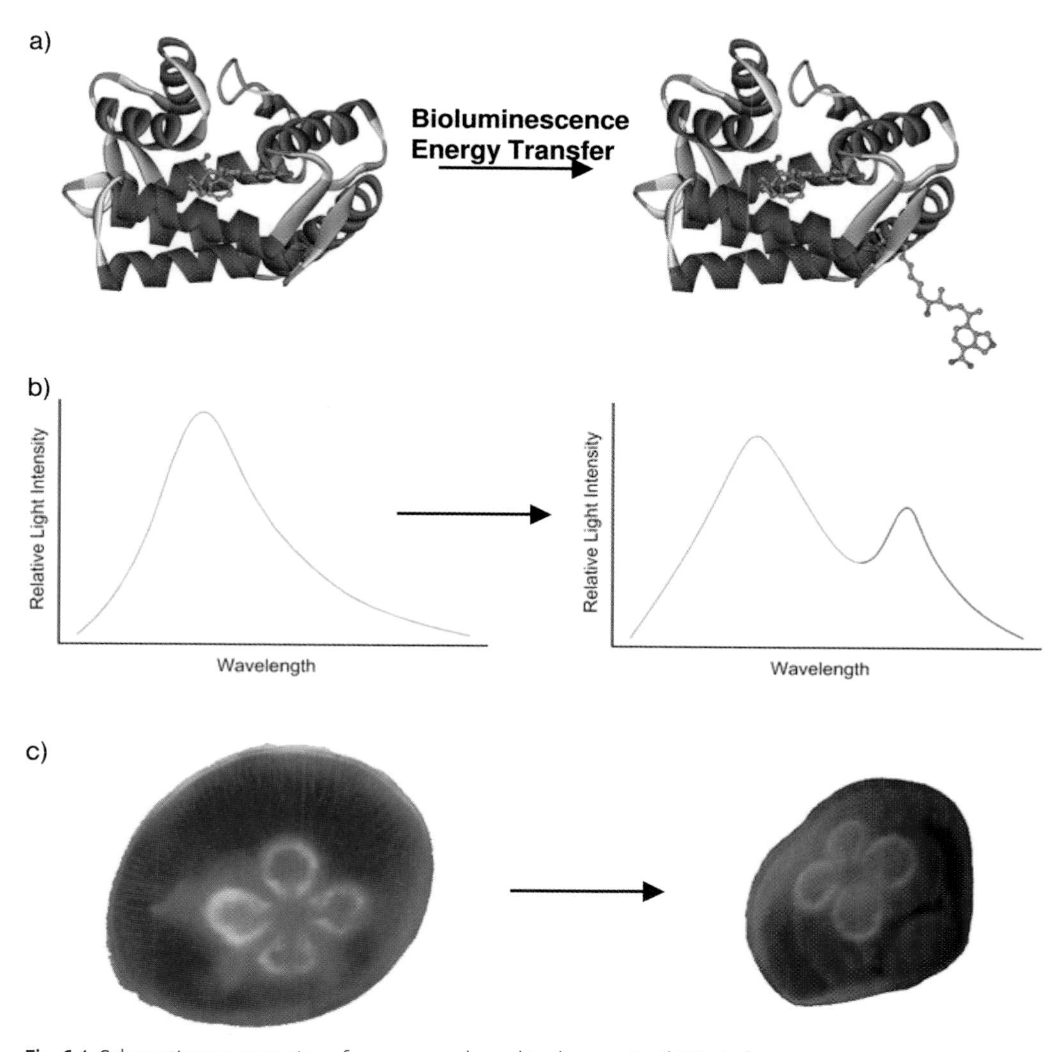

Fig. 6.4 Schematic representation of bioluminescence energy transfer from the protein aequorin to the fluorophore (reprinted from Ref. [18]).
(a) Representation of the three-dimensional structure of aequorin with the fluorophore bound to the protein. (b) Typical spectra showing the bioluminescence emission curve and the emission from the fluorophore.
(c) Photographs of different color jellyfish. Reprinted with kind permission from Springer Science and Business Media.

6.5
Applications of BRET in Bioanalysis

6.5.1
Homogeneous Assays

BRET technology has found applications in the development of homogeneous assays for the detection of biomolecules. These homogeneous assays present advantages over traditional heterogeneous assays because they are simple and fast, involve single-step analysis (there are no separation steps), and are amenable to automation and incorporation into miniaturized microfluidics systems. In addition, BRET reporters are proteinaceous in nature, which allows for their genetic manipulation (i.e., gene fusion, mutagenesis, etc.) and direct *in vivo* analysis. In that regard, Campbell et al. developed detection methods for proteases using aequorin as a donor and GFP as an acceptor in a BRET-type assay [20]. These fusion proteins were called rainbow proteins because their emission profile changed as a result of energy transfer in the presence of the analyte of interest. These rainbow proteins had an emission spectrum identical to that of GFP and a quantum yield corresponding to free aequorin. A linker length of at least six amino acids was needed to allow energy transfer from aequorin to GFP. Linkers shorter than six amino acids showed no energy transfer, whereas the efficiency of transfer increased with longer linkers up to a linker length of 50 amino acids. Two proteases namely, α-thrombin and caspase-3, were employed as model proteases in this study. Caspase-3 is an intracellular protease involved in apoptosis. α-Thrombin is an extracellular protease involved in the blood coagulation cascade. Fusion proteins were engineered with aequorin as the energy donor and GFP as the acceptor, with a protease linker recognition site between the donor and acceptor parts of the fusion protein molecule. A recognition site, DEVD, was employed for caspase-3 detection, and LVPRGS was used as the recognition site for α-thrombin. In the absence of proteases, addition of calcium triggered bioluminescence emission from aequorin, which was transferred to GFP, yielding spectra corresponding to that of GFP. Upon addition of the protease, it cleaved the fusion protein at the recognition site, yielding bioluminescence spectra corresponding to that of the aequorin. Analysis of proteolytically cleaved fragments showed that the ratio of green to blue light indicated the extent of proteolysis. The effect of protease on energy transfer was found to be dependent on dose and time. α-Thrombin (3.2–13.2 units mL^{-1}) showed a change from 2 to 4 units in the ratio of light emission (500/450 nm). Caspase-3 activity could be detected from 500 to 5000 units mL^{-1}. The rainbow protein for caspase-3 showed that the recognition site employed in the study (DEVD) could not distinguish between caspase-3 and caspase-9. However, these caspases could be distinguished using specific inhibitors of each of the proteases. These rainbow proteins could be employed as *in vitro* and *in vivo* indicators of protease activity. Furthermore, these rainbow proteins could be targeted to different organelles by engineering targeting signal peptides on these proteins. For example, the rainbow protein for caspase-3 was employed in the endoplasmic reticulum (ER) to study apoptosis induced by ER stress. The target event to be studied triggered by caspase-3 was imaged using a camera that could image four different colors simultaneously (Photek, East Sussex, United Kingdom, 512×512 pixels, at 60 Hz).

In another study, Mouland et al. developed a BRET-based assay to measure the activity of human immunodeficiency virus type 1 (HIV-1) protease *in vivo* [21]. In this work, the HIV-1 Gag-p2/Gag-p7 (p2/p7) protease recognition site was inserted between a human codon-optimized GFP and RLuc. This gene fusion (RLuc-p2/p7-GFP) was co-expressed with an HIV-1 codon-optimized protease. Expression of the hRLuc-p2/p7-hGFP alone generated a BRET signal resulting from energy transfer from RLuc to GFP, indicating that the protease recognition site was not cleaved (Fig. 6.5). In the presence of HIV-1 protease, a significant reduction in the BRET signal was observed as a consequence of the cleavage caused by the protease on the fusion protein (see Fig. 6.5). Upon addition of inhibitors of HIV-1 protease such as saquinavir or amprenavir, the cleavage was blocked, thus

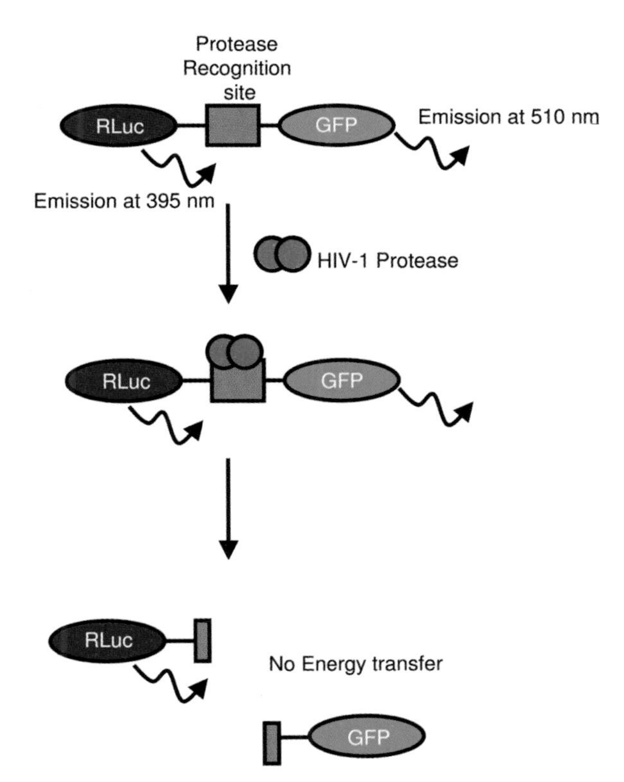

Fig. 6.5 Schematic of the HIV-1 protease assay based on BRET. In the absence of protease, energy transfer occurs from RLuc to GFP. In the presence of the protease, it cleaves at the recognition site, separating RLuc and GFP, which prevents transfer of energy. Adapted from Ref. [21] and reprinted with permission from Elsevier B. V.

resulting in the generation of BRET emission. This assay was further validated by co-expressing the HIV-1 auxiliary protein Vif, which inhibits cleavage at the HIV-1 protease recognition site p2/p7 when overexpressed. In this study, the generation of BRET signal was observed in a dose-dependent manner. This BRET assay for the detection of protease can be adapted to high-throughput screening of inhibitors of protease activity because it is a one-step assay performed in a homogeneous manner. The technology developed is noninvasive since the fusion protein between Rluc–protease recognition site–GFP can be genetically encoded for production in the cell, and assays can be performed *in vivo*, making it suitable for the identification of bioactive inhibitory molecules.

Calcium is a regulator of several intracellular processes; therefore, its measurement is of great biological significance. Many different methods of calcium detection have been developed. These methods are mostly based on calcium-sensitive fluorophore probes, microelectrodes, FRET assays, etc. Brulet et al. utilized aequorin's sensitivity toward calcium in order to develop a BRET-based assay for the determination of calcium in single neuroblastoma cells (Fig. 6.6) [22]. In nature and under certain conditions, the jellyfish *Aequorea victoria* emits green fluorescence in addition to bioluminescence alone. This phenomenon occurs when there is a stimulus that triggers the release of a bolus of calcium within the photocyte cells in the umbrella of the jellyfish that contain the two photoproteins native to the jellyfish, namely aequorin and the green fluorescence protein. Green fluorescence is observed because of a radiationless energy transfer event from aequorin to GFP. On the basis of this observation, a fusion protein between aequorin and GFP was constructed by fusing the C-terminus of GFP to the N-terminus of aequorin. Different linker lengths of 5–50 amino acids were engineered between these two proteins. The fusion proteins were then transiently transfected into neuronal cells. Using an intensified CCD camera, photon emission was monitored after the addition of a calcium solution. A signal integration time of 1 s was enough to record the fluorescence signal emitted by GFP. When aequorin was expressed alone or when co-expressed with the free GFP, no signal could be observed. The aequorin–GFP fusion protein showed improved calcium-triggered bioluminescent activity compared to aequorin alone. This is attributed to increased stability of the fusion protein. This fused reporter protein should allow calcium monitoring in cellular (Fig. 6.6) and subcellular compartments [22].

In another study, an immunoassay for the determination of an antigen from hen egg lysozyme was developed by Nagamune et al. based on BRET technology [23]. For that, two fusion proteins, an antibody heavy-chain fragment (V_H)-RLuc, and an antibody light-chain fragment, (V_L)-enhanced yellow fluorescent protein (EYFP), were constructed. Upon addition of the antigen, a re-association of the antibody's variable domains occurred in an antigen concentration-dependent manner, resulting in the generation of a BRET signal (Fig. 6.7). The BRET signal was monitored in the form of luminescence ratio I_{525}/I_{475}. An increase in the luminescence ratio was observed with increasing antigen concentration. The assay was performed in a single step with a range of detection spanning from 0.1 to 10 μg mL^{-1} of hen egg lysozyme.

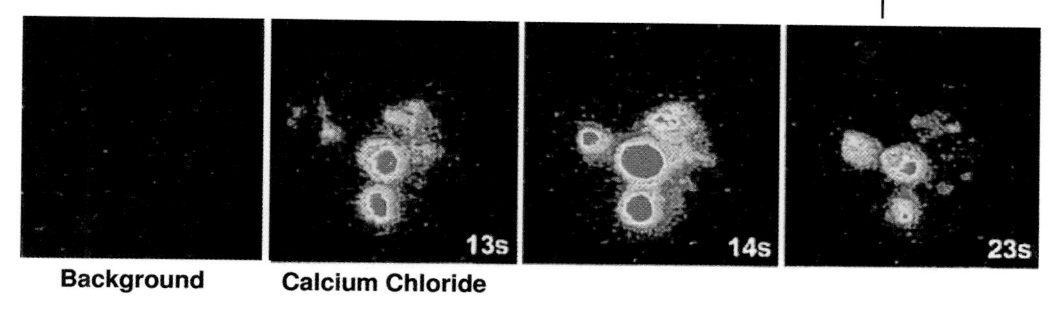

Fig. 6.6 Photographical images of neuro 2A cells showing calcium-induced bioluminescence (reprinted from Ref. [22]). Background is collected before addition of calcium. Reprinted with permission from PNAS.

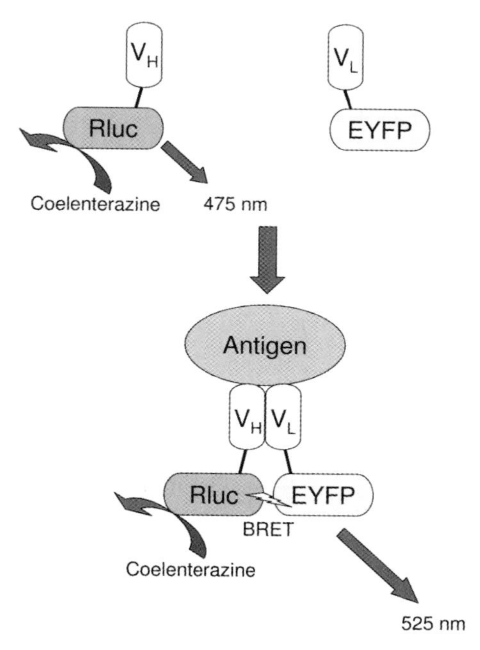

Fig. 6.7 Schematic representation of a homogeneous immunoassay based on energy transfer (adapted from Ref. [23]). In the absence of antigen, the two fusion proteins remain separated. Addition of antigen leads to heterodimerization of antibody chains, leading to energy transfer from Rluc to EYFP.
Reprinted with permission from Academic Press.

A BRET-based competitive assay for the detection of biotin was also developed by Adamczyk et al., using aequorin as a donor molecule and a dye quencher as the acceptor [24]. For that purpose, conjugates of avidin, which binds biotin with high specificity and affinity, were prepared containing the covalently attached dyes QSY-7 (λmax 560 nm) and dabcyl (λ_{max} 470 nm). These dyes were chosen because of the overlap between their absorption spectra and the emission of aequorin (430–500 nm). Conjugates were prepared using different ratios of avidin to dye. When avidin conjugates were added to biotinylated aequorin, a quenching in the bioluminescence of aequorin was observed. A maximum quenching value of 51% and 28% was obtained for QSY-7 and dabcyl-avidin conjugates, respectively. Quenching efficiency was shown to be dependent on the label-to-avidin ratio, as well as on the concentration of the avidin conjugate. A dose–response curve was generated by setting up a competition between free biotin and biotin conjugated to aequorin. Using the dose–response curve obtained, a biotin concentration as low as 0.1 nM could be detected.

In another study by Vinokurov et al., a BRET assay was developed for the detection of biotin using aequorin and EGFP as a BRET pair [25]. As part of this work, fusion proteins between aequorin and streptavidin (SAV) and between EGFP and the biotin carboxyl carrier protein (BCCP) were prepared. The spectral shape and intensity of the aequorin–SAV and EGFP–BCCP were found to be slightly different when compared to the spectra obtained with native aequorin and EGFP. Nevertheless, a sufficient spectral overlap existed for BRET to occur. As expected, when SAV–aequorin and EGFP–BCCP fusion proteins were mixed, a decrease in bioluminescence intensity at 470 nm and an increase in EGFP emission at 510 nm were observed. The maximum luminescence ratio (I_{510}/I_{450}) of 3.5 was obtained at saturating concentrations of EGFP–BCCP. Free biotin was shown to inhibit BRET in a dose-dependent manner due to its competition with EGFP–BCCP for the binding to SAV–aequorin. A dose–response curve was generated by monitoring the luminescence ratio at different biotin concentrations. Using this dose–response curve, a detection limit of 120 pM for biotin was obtained. The assay was performed using ratiometric dual-color measurement. This method is advantageous over others published in the literature because it demonstrates that for a given molar ratio of Aeq–SAV and EGFP–BCCP, the measured luminescence ratio is not dependent upon the amount of sample injected to generate bioluminescence. In addition, by using the luminescent ratio the method also overcomes the generation of potential errors resulting from partial deactivation of aequorin luminescence or from other natural or sample-related fluctuations.

6.5.2
Protein–Protein Interactions and High-throughput Screening

BRET technology has become a very useful tool in the study of protein–protein interactions. In this method, the donor and acceptor proteins are genetically fused to proteins of interest. Co-expression of these fusion proteins in cells allows real-time monitoring of protein–protein interactions in a quantitative manner. When proteins of interest interact with each other, energy transfer from a donor bioluminescent protein to an acceptor fluorescent protein can occur, yielding a BRET signal. The BRET signal depends upon the strength and stability of protein–protein interactions. In addition, the signal depends on the relative orientation, spectral properties, distance, and ratio of donor/acceptor molecules. Several accounts describing the use of BRET technology in the study of protein–protein interactions have been reported. In addition, several detailed reviews on this topic have been published [26–31]. Therefore, application of BRET technology in studying protein–protein interactions will not be discussed in this chapter.

Another aspect of studying protein–protein interactions is its usefulness in screening for agonists or antagonists of these interactions, making this technology especially useful in drug discovery. This is of potential interest to pharmaceutical, biotech, and chemical companies that are involved in identifying drug molecules that affect protein–protein interactions. For example, this method has been applied in the screening of agonist-induced interactions between G protein-coupled receptors (GPCRs) and β-arrestins. In general, GPCR and β-arrestin are tagged with energy donor and acceptor molecules. Activation of GPCR by an agonist leads to phosphorylation of GPCR and subsequent interaction with β-arrestin, which is monitored by energy transfer. This approach has been employed by several pharmaceutical companies to identify new drugs that act upon this biochemical signaling pathway. Reviews on the use of BRET technology in drug screening at the GPCR receptor have been published [32, 33].

BRET technology is also useful in the determination of the kinetics of agonist-induced changes in interactions between receptors. For example, when agonists of thyrotropin-releasing hormone receptors were studied, an increase in BRET was observed, reaching a maximum in 20 min [34]. In another study, BRET between cholecystokinin receptors in the presence of agonists decreased after only 2 min [35]. Such studies can provide information about the kinetic profile of interactions, but in this case, data are collected at predetermined time intervals. Alternatively, real-time monitoring of interactions can be performed, provided that the substrate of the BRET donor receptor molecule is stable. Coelenterazine h, which is a substrate for Rluc, is not stable and allows real-time measurement only up to 10 min. Another substrate for RLuc introduced by Promega, EnduRen, is a protected form of coelenterazine h that is metabolized by endogenous esterases to free substrate, which allows monitoring of BRET for several hours.

BRET technology has been employed in the development of a screening assay for estrogen-like compounds [36]. The assay is based on the homodimerization of the estrogen receptor (ERα) upon binding of an estrogen-like compound. In this study, the ER α-receptor monomer was genetically fused to either *Renilla* luciferase (RLuc) or enhanced yellow fluorescent protein (EYFP). Dimerization of the ER α-receptor in the presence of estrogen-like compounds caused RLuc and EYFP to come in close proximity, leading to energy transfer. Using purified fusion proteins, an *in vitro* BRET assay was developed employing 17-β-estradiol as the lead compound. This was followed by demonstrating the application of this assay for *in vivo* screening of estrogen-like compounds. To achieve that, the two fusion proteins (ERα-RLuc and ERα-EYFP) were co-expressed in live HepG2 cells. The assay was performed in a typical 96-well microplate format with a 30-min incubation time. This assay demonstrated a detection limit of 1 nM of 17-β-estradiol (Fig. 6.8). This method may find applications in the screening of environmental pollutants that act like xenoestrogens. Some examples of xenoestrogens include pollutants

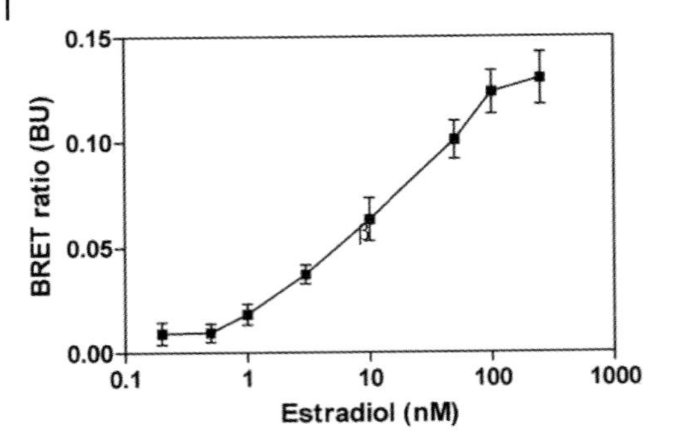

Fig. 6.8 Dose–response curve generated for estradiol by plotting the BRET ratio against the concentration of analyte. Reprinted from Ref. [36] with permission from ACS.

such as dioxins and polychlorinated biphenyls that bind to estrogen receptors and disrupt endocrine function [37].

A major problem with BRET-based protein interaction studies performed *in vivo* is that if protein expression is high, then it can lead to random collision of BRET donor and BRET acceptor fusion proteins, thereby causing a BRET emission that is not related to the detection of the target molecule. This can give a false indication of protein–protein interaction events that, in reality, are not occurring. This was observed by Mercier et al. in a study where dimerization of β_2-adrenergic (AR) receptor was monitored using BRET. The BRET data generated corresponded to different expression levels of the fusion proteins β_2-AR-Rluc and β_2-AR-GFP. It was observed that for a total concentration of protein between 1.4 and 26.3 pmol mg^{-1} of protein, the BRET signal did not increase [38]. However, at protein expression levels higher than 26.3 pmol mg^{-1}, an enhancement in BRET signal was observed. This was attributed to random collision between β_2-AR-Rluc and β_2-AR-GFP. Therefore, when *in vivo* BRET assays are designed, careful consideration should be given to the protein expression levels and the potential for nonspecific energy transfer.

6.6 Conclusions

In conclusion, BRET technology offers advantages in terms of sensitivity and suitability for cellular analysis. Careful experimental design and development of new instrumentation should aid in the improvement of this technology. Instruments that allow *in situ* real-time imaging and high-throughput measurement capabilities are needed. Advances in the identification and characterization of new organisms that are bioluminescent will increase the availability of additional donor and acceptor proteins. This should result in systems with enhanced spectral separation and higher quantum yields. In addition, bioluminescent substrates that have better stability are still desired. These advances in BRET technology should expand its applications in mainstream bioanalysis. It is envisioned that BRET technology, along with complementary technologies such as high-throughput mass spectrometry, could potentially transform the field of receptor interaction research. A unique feature of BRET lies in the fact that it allows for the *in vivo* study of organelle-specific protein interactions. In addition, drug discovery programs are incorporating BRET assays into their drug screening processes, which should open new areas of research.

Acknowledgement

We thank the American Cancer Society Indiana University Institutional Research Grant #IRG 84-002-22.

References

1 Uhl, J. R., Cockerill III, F. R. The fluorescence resonance energy transfer system. *Molecular Microbiology* **2004**, 295–306.
2 Pontier, P. J., Charest, S., Pascale, G., Muriel, A., Bouvier, M. Real-time monitoring of ubiquitination in living cells by BRET. *Nature Methods* **2004**, *1(3)*, 203–208.
3 Ward, W. W., Bokman, S. H. Reversible denaturation of Aequorea green-fluorescent protein: physical separation and characterization of the renatured protein. *Biochemistry* **1982**, *21(19)*, 4535–4540.
4 Hart, R. C., Matthews, J. C., Hori, K., Cormier, M. J. Renilla reniformis bioluminescence: luciferase-catalyzed production of nonradiating excited states from luciferin analogues and elucidation of the excited state species involved in energy transfer to Renilla green fluorescent protein. *Biochemistry* **1979**, *18(11)*, 2204–2210.
5 Morise, H., Shimomura, O., Johnson, Frank, H., Winant, J. Intermolecular energy transfer in the bioluminescent system of Aequorea. *Biochemistry* **1974**, *13(12)*, 2656–2662.
6 Xu, Y., Kanauchi, A., Piston, David, W., Johnson, C. H. Resonance energy transfer as an emerging technique for monitoring protein–protein interactions *in vivo*: BRET vs. FRET. *Luminescence Biotechnology* **2002**, 529–538.
7 Dionne, P., Caron, M., Labonte, A., Carter-Allen, K., Houle, B., Joly, E. Taylor, S. C., Menard, L. BRET2: efficient energy transfer from *Renilla* luciferase to GFP2 to measure protein–protein interactions and intracellular signaling events in live cells. *Luminescence Biotechnology* **2002**, 539–555.
8 Arun, K. H. S., Kaul, C. L., Ramarao, P. Green fluorescent proteins in receptor research: An emerging tool for drug discovery. *Journal of Pharmacological and Toxicological Methods* **2005**, *51(1)*, 1–23.
9 Rodems, S. M., Hamman, B. D., Lin, C., Zhao, J., Shah, S., Heidary, D., Makings, L., Stack, J. H., Pollok, B. A. A FRET-based assay platform for ultra-high density drug screening of protein kinases and phosphatases. *Assay and Drug Development Technologies* **2002**, *1(1–1)*, 9–19.
10 Cremazy, F. G. E., Manders, E. M. M., Bastiaens, P. I. H., Kramer, G., Hager, G. L., van Munster, E. B., Verschure, P. J., Gadella, T. W. J., van Driel, R. Imaging *in situ* protein–DNA interactions in the cell nucleus using FRET-FLIM. *Experimental Cell Research* **2005**, *309(2)*, 390–396.

11 Hoffmann, C., Gaietta, G., Buenemann, M., Adams, S. R., Oberdorff-Maass, S., Behr, B., Vilardaga, J., Tsien, R. Y., Ellisman, M. H., Lohse, M. J. A FlAsH-based FRET approach to determine G protein-coupled receptor activation in living cells. *Nature Methods* **2005**, *2(3)*, 171–176.

12 Looger, L. L., Lalonde, S., Frommer, W. B. Genetically encoded FRET sensors for visualizing metabolites with subcellular resolution in living cells. *Plant Physiology* **2005**, *138(2)*, 555–557.

13 Wozniak, A. K., Nottrott, S., Kuehn-Hoelsken, E., Schroeder, G. F., Grubmueller, H., Luehrmann, R., Seidel, C. A. M., Oesterhelt, F. Detecting protein-induced folding of the U4 snRNA kink-turn by single-molecule multiparameter FRET measurements. *RNA* **2005**, *11(10)*, 1545–1554.

14 Hori, K., Wampler, J. E., Matthews, J. C., Cormier, M. J. Identification of the product excited states during the chemiluminescent and bioluminescent oxidation of Renilla (sea pansy) luciferin and certain of its analogs. *Biochemistry* **1973**, *12*, 4463.

15 Prasher, D., McCann, R. O., Cormier, M. J. Cloning and expression of the cDNA coding for aequorin, a bioluminescent calcium-binding protein. *Biochemical and Biophysical Research Communications* **1985**, *126(3)*, 1259–1268.

16 Akagi, Y., Sakaki, Y., Inoue, S., Noguchi, M., Iwanaga, S., Miyata, T., Tsuji, F. A. Plasmid vector containing aequorin gene from Aequorea victoria. *Jpn. Kokai Tokkyo Koho* **1986**, 11 pp.

17 Arai, R., Nakagawa, H., Kitayama, A., Ueda, H., Nagamune, T. Detection of protein–protein interaction by bioluminescence resonance energy transfer from firefly luciferase to red fluorescent protein. *Journal of Bioscience and Bioengineering* **2002**, *94(4)*, 362–364.

18 Deo, S. K., Mirasoli, M., Daunert, S. Bioluminescence resonance energy transfer from aequorin to a fluorophore: an artificial jellyfish for applications in multianalyte detection. *Analytical and Bioanalytical Chemistry* **2005**, *381(7)*, 1387–1394.

19 Morise, H., Shimomura, O., Johnson, F. H., Winant, J. Intermolecular energy transfer in the bioluminescent system of Aequorea. *Biochemistry* **1974**, *13(12)*, 2656–2662.

20 Waud, J. P., Bermudez Fajardo, A., Sudhaharan, T., Trimby, A. R., Jeffery, J., Jones, A., Campbell, A. K. Measurement of proteases using chemiluminescence-resonance-energy-transfer chimaeras between green fluorescent protein and aequorin. *Biochem. J.* **2001**, *357(Pt 3)*, 687–697.

21 Hu, K., Clement, J. F., Abrahamyan, L., Strebel, K., Bouvier, M., Kleiman, L., Mouland, A. J. A human immunodeficiency virus type 1 protease biosensor assay using bioluminescence resonance energy transfer. *J. Virol. Methods* **2005**, *128(1–2)*, 93–103.

22 Baubet, V., Le Mouellic, H., Campbell, A. K., Lucas-Meunier, E., Fossier, P., Brulet, P. Chimeric green fluorescent protein-aequorin as bioluminescent Ca2+ reporters at the single-cell level. *Proc. Natl. Acad. Sci. USA* **2000**, *97(13)*, 7260–7265.

23 Arai, R., Nakagawa, H., Tsumoto, K., Mahoney, W., Kumagai, I., Ueda, H., Nagamune, T. Demonstration of a homogeneous noncompetitive immunoassay based on bioluminescence resonance energy transfer. *Anal. Biochem.* **2001**, *289(1)*, 77–81.

24 Adamczyk, M., Moore, J. A., Shreder, K. Quenching of biotinylated aequorin bioluminescence by dye-labeled avidin conjugates: application to homogeneous bioluminescence resonance energy transfer assays. *Org. Lett.* **2001**, *3(12)*, 1797–1800.

25 Gorokhovatsky, A. Y., Rudenko, N. V., Marchenkov, V. V., Skosyrev, V. S., Arzhanov, M. A., Burkhardt, N., Zakharov, M. V., Semisotnov, G. V., Vinokurov, L. M., Alakhov, Y. B. Homogeneous assay for biotin based on Aequorea victoria bioluminescence resonance energy transfer system. *Anal. Biochem.* **2003**, *313(1)*, 68–75.

26 Kroeger, K. M., Eidne, K. A. Study of G-protein-coupled receptor-protein

interactions by bioluminescence resonance energy transfer. *Methods Mol. Biol.* **2004**, *259*, 323–333.

27 Subramanian, C., Xu, Y., Johnson, C. H., von Albrecht, G. *In vivo* detection of protein–protein interaction in plant cells using BRET. *Methods in Molecular Biology* **2004**, *284*, 271–286.

28 Eidne, Karin, A., Kroeger, Karen, M., Hanyaloglu, Aylin, C. Applications of novel resonance energy transfer techniques to study dynamic hormone receptor interactions in living cells. *Trends in Endocrinology and Metabolism* **2002**, *13(10)*, 415–421.

29 Kroeger, K. M., Hanyaloglu, A. C., Eidne, K. A. Applications of BRET to study dynamic G-protein coupled receptor interactions in living cells. *Letters in Peptide Science* **2002**, *8(3–5)*, 155–162.

30 Pfleger, K. D. G., Eidne, K. A. New technologies: bioluminescence resonance energy transfer (BRET) for the detection of real time interactions involving G-protein coupled receptors. *Pituitary* **2003**, *6(3)*, 141–151.

31 Gomes, I., Jordan, B. A., Gupta, A., Rios, C., Trapaidze, N., Devi, L. A G protein coupled receptor dimerization: implications in modulating receptor function. *Journal of Molecular Medicine* **2001**, *79(5–6)*, 226–242.

32 Boute, N., Jockers, R., Issad, T. The use of resonance energy transfer in high-throughput screening: BRET versus FRET. *Trends in Pharmacological Sciences* **2002**, *23(8)*, 351–354.

33 Milligan, G. Applications of bioluminescence- and fluorescence resonance energy transfer to drug discovery at G protein-coupled receptors. *Eur. J. Pharm. Sci.* **2004**, *21(4)*, 397–405.

34 Kroeger, K. M., Hanyaloglu, A. C., Seeber, R. M., Miles, L. E. C., Eidne, K. A. Constitutive and agonist-dependent homo-oligomerization of the thyrotropin-releasing hormone receptor. Detection in living cells using bioluminescence resonance energy transfer. *Journal of Biological Chemistry* **2001**, *276(16)*, 12736–12743.

35 Cheng, Z., Miller, L. J. Agonist-dependent dissociation of oligomeric complexes of G protein-coupled cholecystokinin receptors demonstrated in living cells using bioluminescence resonance energy transfer. *Journal of Biological Chemistry* **2001**, *276(51)*, 48040–48047.

36 Michelini, E., Mirasoli, M., Karp, M., Virta, M., Roda, A. Development of a bioluminescence resonance energy-transfer assay for estrogen-like compound *in vivo* monitoring. *Anal. Chem.* **2004**, *76(23)*, 7069–7076.

37 Layton, A. C., Sanseverino, J., Gregory, B. W., Easter, J. P., Sayler, G. S., Schultz, T. W. *In vitro* estrogen receptor binding of PCBs: measured activity and detection of hydroxylated metabolites in a recombinant yeast assay. *Toxicology and Applied Pharmacology* **2002**, *180(3)*, 157–163.

38 Mercier, J., Salahpour, A., Angers, S., Andreas, B. M. Quantitative assessment of β1- and β2-adrenergic receptor homo- and heterodimerization by bioluminescence resonance energy transfer. *Journal of Biological Chemistry* **2003**, *278(20)*, 18704.

7
Photoproteins as *in Vivo* Indicators of Biological Function

Rajesh Shinde, Hui Zhao, and Christopher H. Contag

7.1
Overview

A number of whole-body imaging techniques have been developed for the *in vivo* study of mammalian biology. Optical imaging methods are particularly well suited for the study of small animal models, and the use of light-emitting enzymes as reporters of biological activity has become one of the cornerstone technologies for studying biology and pharmacology in rodent models. This approach has been significantly advanced by continued development of modified genes encoding photoproteins, coupled with advances in detection technologies that have improved the sensitivity and quality of the experimental data obtained noninvasively from laboratory animals.

Proteins with optical signatures that are bioluminescent or fluorescent have been used for nearly a decade in biological assays and in cells. Imaging experience gained from cultured cells has led to the application of reporter genes in living animals. Optical techniques such as fluorescence and bioluminescence imaging have been utilized in a wide range of animal models for studying oncology, stem cell biology, and infectious diseases. In general, noninvasive imaging reduces the number of animals needed for a given experiment while providing a rapid means of accurate quantification of biological processes. The use of photoproteins as *in vivo* reporters further enhances these approaches by enabling sensitive detection with relatively inexpensive instrumentation and an ease of use that permits fine temporal resolution. This is in contrast to the traditional approach where spatiotemporal information is obtained by sacrificing a statistically significant number of animals for each time point and the temporal information for individual animals is lost.

In order to develop effective imaging modalities with molecular reporters, consideration should be given to (1) selection of the appropriate probe for assaying the targeted cellular or molecular process throughout the entire study, (2) the sensitivity of detection of small changes in cell function and/or distribution, and (3) the effects of the reporter on the biological process itself. These approaches can

Photoproteins in Bioanalysis. Edited by Sylvia Daunert and Sapna K. Deo

ISBN: 3-527-31016-9

be enriched through modifications to the exogenous molecules that impart unique distinct optical signatures on products of reporter genes when expressed in specific cells and tissue types. Several new reporter systems based on photoproteins try to overcome these issues. Advances in optical detection technologies over the past decade have allowed for assessment of low levels of light emission from cells and tissues in animals. This coupled with advances in reconstruction algorithms and development of reporter molecules has helped target specific molecular events.

Studies that incorporate noninvasive animal imaging techniques to investigate questions in biology and advance development of new pharmaceuticals have increased dramatically in the last decade. In this chapter, we discuss the different methods of using photoproteins for *in vivo* imaging of biology in small animals, the effects of optical properties of tissues in imaging, and examples of where these approaches have shed light on biology. We also look at possible future developments in the biochemistry of bioluminescent reporter proteins and their substrates used for *in vivo* imaging. This includes the development of activable substrates for protease sensors and sensors of other enzymatic activities. These imaging methods have attained significance in biomedical research and drug development, and additional developments will increase their utility and versatility [1–6].

7.2
Probes Used for in Vivo Bioluminescence Imaging

In vivo bioluminescence imaging (BLI) is a powerful tool that can be used to study biological events temporally and noninvasively in live animals. Biological entities such as bacteria, tumors, or other cells or genes are labeled with bioluminescent reporters and light is detected externally using a highly sensitive cooled charged-couple device (CCD) camera [7]. As a bioluminescent probe, used in *in vitro* and *in vivo* assays, luciferase can generate visible light by oxidation of an enzyme-specific substrate in the presence of oxygen and, usually, ATP as a source of energy. Luciferase has the advantages of rapid turnover and an inherently low background, giving real-time detection and near absence of endogenous light from mammalian cells and tissues and leading to a low signal-to-noise ratio (SNR) [8].

Luciferase enzymes have been found in many species, including bacteria, marine crustaceans, fish, and insects [9]. However, only a subset of luciferase-substrate combinations have been characterized and utilized as *in vivo* reporter genes, including those from firefly, click beetle, and *Renilla*. Luciferases from firefly and click beetle use D-luciferin as a substrate, while *Renilla* luciferase uses coelenterazine as a substrate. Luciferase genes have been modified in many aspects to optimize its applications in BLI. Codon usage was optimized for use of these reporter genes in mammalian cells, and any peroxisome-targeting sequences were removed [10]. More recently, potential binding sites for mammalian transcription factors have been eliminated from some of these genes [10]. These modifications have resulted in significantly higher levels of transcription in mammalian cells

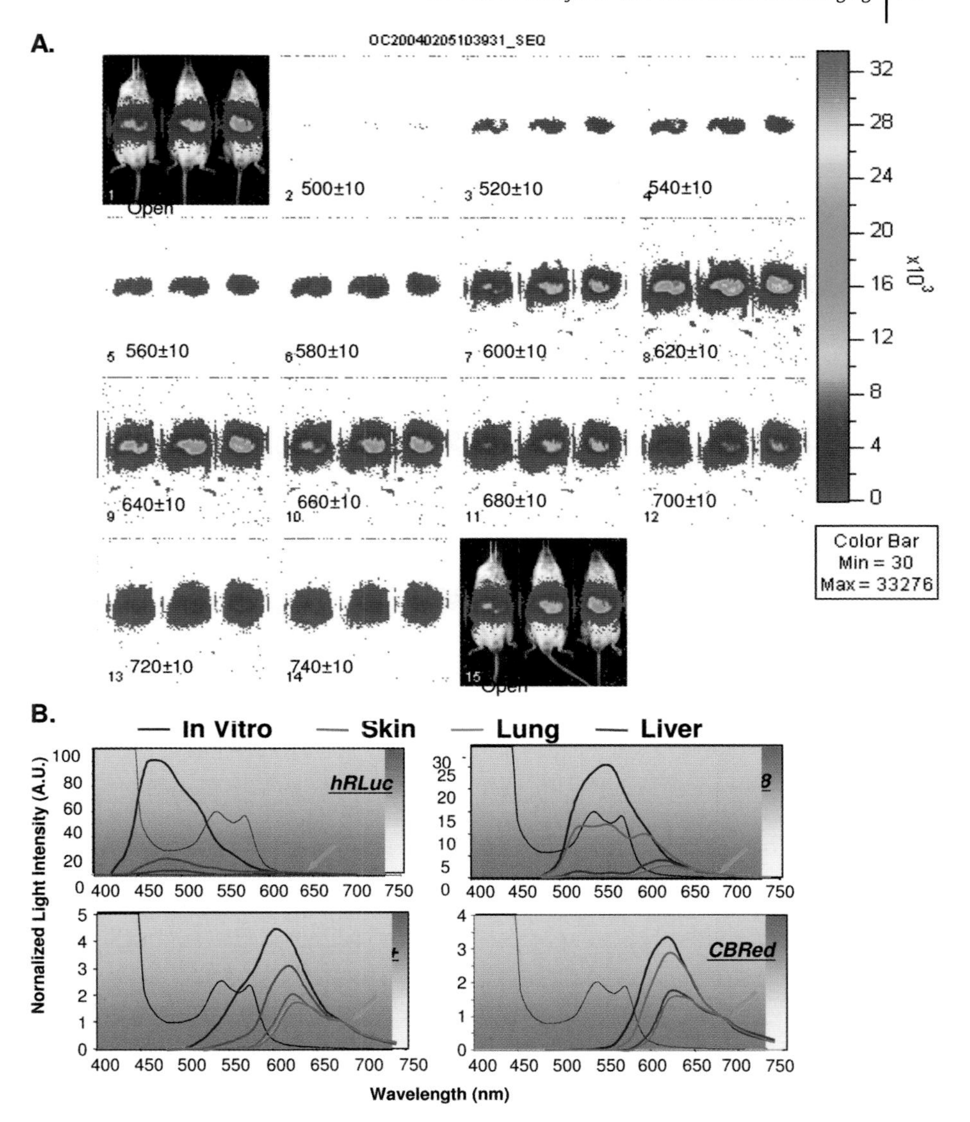

Fig. 7.1 Transmission of bioluminescent signals through mammalian tissues. Spectral scanning was performed on three mouse models – tumor cells under the skin, tumor cells in the lungs and liver transfections using four different bioluminescent markers: *Renilla* luciferase (hRluc), firefly luciferase (Fluc), and the green (CBGr68) and red (Cb red) luciferases from the click beetle (Promega Corp., Madison WI). (A) Example of the liver model showing the 20-nm windows that were used to obtain the plots in panel B. (B) Spectral data were plotted over the hemoglobin absorption spectrum (gray). These data reveal that light greater than 600 nm (red) is transmitted through mammalian tissues more efficiently than shorter wavelengths of light (blue) (adapted from ref. 14).

in culture as well as in animals. On the other hand, site-directed mutagenesis and chimerization of different luciferase genes have been performed, and various colors were obtained from amino acid substitutions [11, 12].

Bioluminescence images of live animals are a diffuse projection at the body surface from bioluminescent sources that can be located deep inside the body. Only light that escapes the absorbing and scattering environment of mammalian tissues can be detected externally. Rice et al. have demonstrated that, consistent with the optical properties of mammalian tissue, luciferases with emission in the red to near-infrared range (> 600 nm) may have higher transmission efficiency through tissue than those of shorter wavelengths [13]. Light in the blue and green region of the spectrum (500–600 nm) is strongly absorbed by macromolecules such as hemoglobin, melanin, and other pigments [13]. This was further confirmed recently by Zhao et al., who evaluated four luciferases that emit with peaks across the visible region in three different animal models using spectral imaging [14] (Fig. 7.1). The influence of tissue on detection and spectral profiles was studied by comparing light emission from the models with the sources located at different depths in the body [14]. Each of the four luciferases studied (firefly, click beetle green, click beetle red, and *Renilla* luciferase) has unique emission spectra. They observed that the spectra of firefly luciferase were red-shifted at 37 °C relative to emission at room temperature. The spectra of the other luciferases were not affected by changes in temperature, but the activity of the click beetle luciferases was increased at 37 °C relative to 25 °C.

The influence of tissue is largely due to strong hemoglobin absorption of light in the visible region of the spectrum (Fig. 7.1). The spectra in the absence of tissue (cell culture) revealed emission in the blue and green regions, as did spectra at superficial sites in the body (subcutaneous). In contrast, transmission from bioluminescent sources located at deep tissue sites, lung and liver, was differentially attenuated, with the red region dominating the spectra [14]. Luciferases with a higher proportion of light emitted above 600 nm, such as those from firefly and the red variant of the click beetle enzyme, demonstrated greater tissue penetration.

7.3
Probes Used for in Vivo Fluorescence Imaging

In vivo fluorescence imaging involves labeling cells with coding sequences for fluorescent proteins or addition of exogenous dyes that fluoresce. As with luciferase, a coding sequence for a fluorescent reporter (e.g., green fluorescent protein, GFP) can be fused to a gene's regulatory region and the expression patterns studied in cells or animals [15]. A fluorescent dye-binding domain can be incorporated into cells or tissues such that an exogenously added dye fluoresces only after binding to this domain [16]. Alternatively, fluorescence reporter gene–substrate combinations have been created from enzymes that are typically used with chromogenic substrates. For example, beta-lactamase or beta-galactosidase has been used to convert an exogenously added, non-fluorescent substrate into a fluorescent derivative and shift the wavelength of a fluorescent substrate emission [17, 18]. The light output per molecule of reporter proteins in the first two strategies is usually lower than that of enzymatically produced signals because a single enzyme can catalyze many substrate molecules, resulting in amplification. Enzymes that process fluorescent substrates may generate a greater photon output per molecule of reporter protein, providing significant signals for *in vivo* fluorescence imaging.

Fluorescence requires an external light source for illuminating the fluorophores, and a majority of reporters with fluorescent signals require excitation in the blue-green region of the visible spectrum where tissue absorption of light is high. Many fluorophores also emit light in this wavelength range. There can be substantial background in fluorescence imaging resulting from a variety of endogenous molecules such as hemoglobin and cytochromes, leading to low SNR [15, 18]. To improve SNR, Weissleder et al. have used biocompatible autoquenched near-infrared (NIR) fluorescent compounds to target specifically to cells or tissues. These compounds can be activated by tumor-specific proteases for cathepsins B and H and emit a signal above 600 nm [19–21]. The dyes are tethered to peptides and change their emission properties upon cleavage of this tethered peptide chain. The peptide sequence can be selected for specificity for desired proteases. This is an effective strategy that is gaining currency for molecular imaging because it allows for targeting specific molecules and can be easily adapted for studying pro-drugs.

7.4 Detection Technologies

Advances in detector technologies have greatly expanded *in vivo* imaging capabilities, with many of these advances occurring in charge-coupled device (CCD) cameras and their configurations for *in vivo* imaging. CCD cameras come in various architectures, including intensified and cooled CCDs [22]. Intensified detectors use photocathodes to convert captured photons to electrons that are then amplified and converted back to photons using a phosphor screen, and these photons are finally detected on the CCD. Photocathodes have defined spectral sensitivity to regions of the light spectrum and tend to be blue sensitive. Amplified systems that utilize bialkali photocathodes work optimally in the blue-green region and lack sensitivity for detection of wavelengths over 600 nm. Photocathodes made of gallium arsenide extend the spectral range of the intensifiers into the red and may be suitable for some *in vivo* applications. Cooled CCDs are designed to reduce noise by cooling. Cooling to –120 °C greatly enhances the SNR of the detector while retaining its spectral sensitivity. Unfortunately, cooling below –120 °C can also reduce the ability to read the chip, and current designs do not offer any additional advantages. Many new features have been implanted into imaging systems that use CCD camera detectors, and these have served to advance molecular imaging applications available to biologists interested in studying *in vivo* processes. Advances such as installation of a series of band-pass filters into these systems have enabled *in vivo* spectral imaging and allowed depth information to be determined based on the differential transmission of blue vs. red light. Spectral unmixing algorithms and tunable filters are being applied to *in vivo* imaging and provide extraordinary images of bright fluorescent markers [23]. The challenge for spectral imaging approaches is the limited signals that are available from inside living animals and the efficiency of filters that are used.

One approach to deal with the loss of photons resulting from scattering and absorption has been developed in the field of intravital microscopy, where dorsal skin-fold window chambers are employed and microscopes are used to directly examine normal tissues and tumors [24–26]. In this technique, tissues of interest, which are labeled with fluorescent dyes or photoproteins, are surgically implanted into a skin pocket and observed with fluorescence microscopes. Recent developments in this technique involve high-speed intravital multiphoton laser scanning microscopy to enhance the depth of tissue that can be interrogated and to accurately project three-dimensional structures [26]. Though this method produces high-resolution images of tumor growth, vascularity, and immune cell migration, whole-body images cannot be obtained because photon detection is restricted to the transparent window [26–29]. A commonly used approach in intravital microscopy is to label cells or tissues with fluorophores that are then injected into the animal [30, 31]. A potential drawback of this approach is that upon cell division, the detectable emission signal reduces, thus limiting temporal measurements to early time points. A way around this is to encode cells with genes for photoproteins, such as GFP [32–35].

7.5 Current Applications

Optical methods that can detect exogenously administered photoproteins noninvasively permit serial studies over long time periods and interrogation of all sites in the body. These nondestructive approaches are well suited for metastasis models and other models where the location of the reporter could be at any number of sites in the body. Among these methods, bioluminescent reporters obviate the need for an external light source, and the absence of autoluminescence offers a tremendous SNR that enables very sensitive detection of biological processes. In this section, biological applications of optical imaging using photoproteins will be discussed.

7.5.1 Oncology

The study of cancer biology and animal models of malignancy has benefited significantly from advances in *in vivo* imaging of small animals. Bioluminescent and

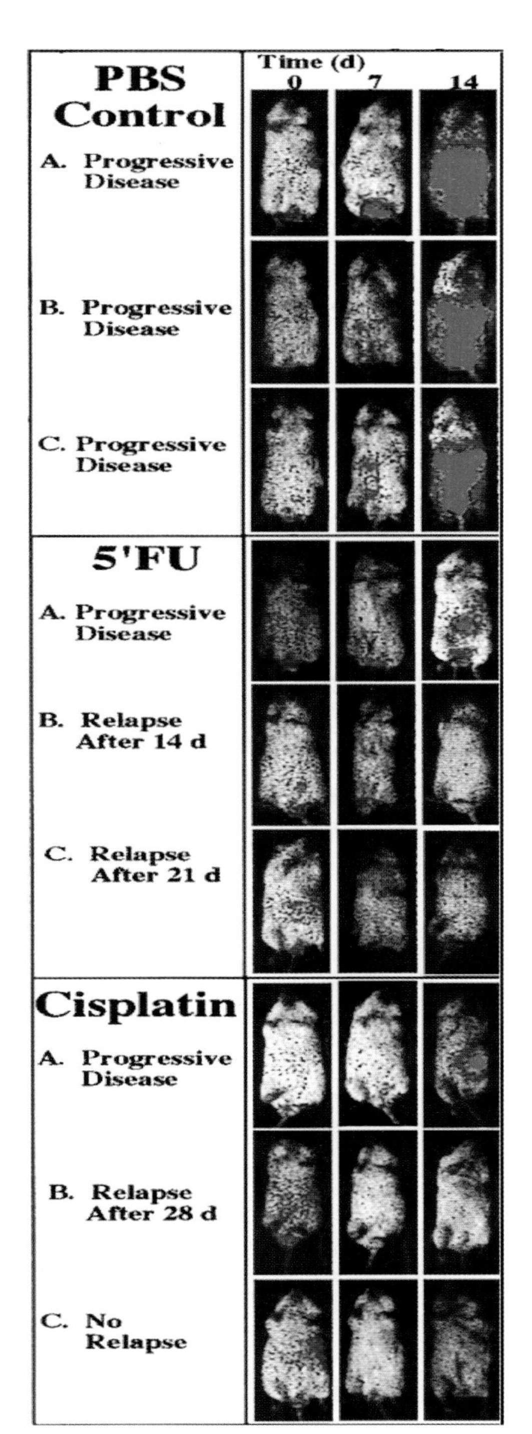

Fig. 7.2 Imaging therapeutic responses. Constitutive expression of photoproteins in tumor cells enables noninvasive measures of tumor burden and assessment of response to the therapies. Here, two chemotherapies (5′fluorouracil [5 FU] and cisplatin) were tested using labeled human cervical carcinoma cells as xenografts in *scid* mice. The cells were labeled by expression of firefly luciferase and the images were taken using a Hamamatsu C2400 intensified CCD camera. Adapted from Ref. [5].

fluorescent reporters have been used in a variety of mouse models of human cancer and have included xenograft, allograft, spontaneous-transgenic, and orthotopic models. Imaging tools that can noninvasively quantify the spatiotemporal changes throughout the disease course have proven invaluable for refining the study of cancer biology and for accelerating the assessment of therapeutic outcome for experimental therapies. Imaging enables real-time analysis of tumor burden for lesions from cells implanted orthotopically, intraperitoneally, intravenously, or intracranially, similar to how calipers are used for subcutaneous models. Moreover, the reporters can be introduced into tumors such that their expression and activity are linked to biological processes; thus, the *in vivo* assays can go beyond mere measures of tumor burden. Such models would otherwise be difficult to monitor without sacrificing the animal, and analyses of these animal models are becoming more refined through imaging [36–40].

Noninvasive imaging techniques reduce the number of animals that need to be sacrificed and provide a more accurate quantification of the growth kinetics. Numerous papers in the past few years have demonstrated the potential of bioluminescence and fluorescence optical imaging techniques in tumor growth and metastasis [4, 5, 34, 41, 42]. *In vivo* bioluminescence imaging for the purpose of visualizing the kinetics of tumor clearance from living animals in response to novel cell therapies was first demonstrated in 1999 (Fig. 7.2) [5]. In this paper the authors evaluated both conventional chemotherapies and immune cell therapy that used cytokine-induced killer (CIK) cells, a tumoricidal population of cells that are enriched from primary cells. The spatiotemporal distribution of bioluminescent human-derived tumor cells in immunodeficient mice was observed and their response to therapy measured. In the absence of therapy, there were increases in the signal intensities indicating increased tumor burden. Mandl et al. used a multimodality imaging approach to reveal spatiotemporal patterns of tumor cell loss and annexin V uptake, as an indication of extent of apoptosis. Annexin V is being investigated as a marker of therapeutic efficacy [43]. Weissleder et al. developed techniques that used biocompatible near-infrared fluorescence probes to image tumors in living animals [44]. These methods indicate that optical methods will provide excellent tools for the study of *in vivo* biology and that photoproteins will be a significant component of these approaches.

7.5.2
Infectious Disease

Decades of studies on pathogens have enhanced our understanding of their virulence properties and of the host cell response to microbial insults. Many of these studies have been conducted in cell culture or animal models where the readout is illness and death. Indeed, these studies form the first stage of any meaningful investigation, but they may not provide clues to the complex interplay between the pathogens and host cells and tissues in intact biological systems. Information is usually obtained by using laboratory animals wherein the host is infected with the pathogen and the information on the disease is assessed by external analyses on biopsy or postmortem samples. Unfortunately, such techniques do not provide the opportunity for real-time analyses and often lead to data being obtained from single points in time. To generate data with temporal information, a statistically significant number of animals need to be sacrificed for each time point, and even then the time course information in individual animals is lost. *In vivo* imaging techniques have revolutionized the approach towards research in infectious diseases.

7.5.3
Bacterial Infections

The *Lux* operons from bioluminescent bacteria are ideal for labeling bacterial pathogens, as they encode all the genes involved in light production [45] and appear to be compatible with microbial physiology. The five lux genes on the *Photorhabdus luminescens* operon (*luxA-E*) are sufficient for a bioluminescence signal. These genes encode the heterodimeric luciferase (Lux A and B) and the enzymes for substrate biosynthesis (Lux C, D, and E), which eliminate the need for exogenous supply of substrate in cell cultures and animals [46–48]. Contag and coworkers provided the first demonstration of BLI for the investigation of bacterial pathogens in living hosts [7]. A *Salmonella typhimurium* infection model was used to monitor disease progression in living mice using BLI. Three different strains of the bacteria were used to infect the mice, via the oral route, each expressing the *lux* genes from a plasmid. Bioluminescent signals were detected in the mice throughout the entire course of the infection, and they correlated well with the bacterial burden, which was measured using standard culture methods for colony-forming units.

Imaging has utility in the study of opportunistic infections that can take place in the complex immunosuppressive environment of cancer therapies. Myelotoxic treatments for oncological diseases are complicated by neutropenia, which renders patients susceptible to infections. BitMansour et al. used a labeled strain of *P. aeruginosa* that constitutively expressed a bacterial luciferase to demonstrate that lineage-restricted progenitors such as common myeloid progenitors (CMP) and granulocyte myeloid progenitors (GMP) protected mice from lethal doses of the pathogens [49]. Hematopoietic stem cells and CMP/GMP subpopulations of the HSCs were isolated and transplanted into irradiated host mice. They were then infected with lethal doses of the gram-negative bacteria *P. aeruginosa*. The intensity of the bioluminescent signal correlated with clinically apparent illness and survival of the mice. BLI offers the opportunity to study various parameters of immune reconstitution and the protection that they offer to transplant patients. In addition, infections often occur as a result of complex interactions between tissues and artificial surfaces that are implanted clinically. It is necessary to develop models that approximate these conditions, and developing imaging markers will refine and accelerate the analysis of these models.

Artificial surfaces provide an ideal substrate for bacterial colonization. A number of medical practices involve implantation of diverse prosthetic devices.

The colonization of these devices leads to formation of biofilms of microbes that are difficult to treat because of their increased resistance to host defenses and exogenous therapies. Kadurugamuwa et al. developed a rapid and continuous method for monitoring the efficacy of antibacterial agents used against such pathogens [50]. They developed a mouse model of chronic biofilm infection by using bioluminescence imaging of the bacterium *Staphylococcus aureus*, which was modified using a gram-positive adapted *lux* operon. Traditional methods for predicting the outcome of *in vivo* antibiotic treatments against microbial biofilms have involved culture methods that suffer from sampling limitations and time delays in data acquisition. As such, they may not be very predictive of the therapeutic outcome [51, 52]. These techniques traditionally lack a nondestructive and reproducible monitoring system that could assess antibiotic efficacy. Imaging enables temporal analysis of the same animal throughout the duration of the study. Using BLI, the authors were able to evaluate antibacterial effects and re-growth of the bacterial biofilm in living animals.

Recently, Hardy et al. used BLI to demonstrate that the bacterium *Listeria monocytogenes* can replicate in the murine gall bladder and that the replication is extracellular and intraluminal [53]. The spatiotemporal distribution of *L. monocytogenes* in infected BALB/c mice was revealed using BLI (Fig. 7.3), and further study provided evidence that the bacterium grew extracellularly in the lumen of the gall bladder. This location and method of growth were not previously reported for this organism, and as such this is an excellent example of what noninvasive assays can reveal about pathogenesis. Whole-body analysis was the only method by which this would be revealed, since the timing of colonizing the gall bladder was not inherently obvious and more labor-intensive assays could have been unable to provide the temporal resolution to reveal this location of replication. The significance of this observation is that replication of *Listeria* in the gall bladder could constitute a considerable source of infection in the human population.

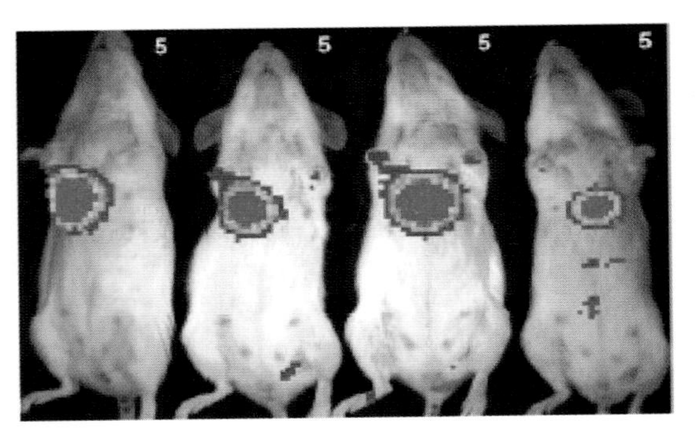

Fig. 7.3 Monitoring persistent infection of *Listeria in vivo*. The lux operon from *Photorhabdus luminescens* was modified for expression in gram-positive bacteria and used to label a reduced-virulence strain of *Listeria*. After intravenous inoculation of this strain into BALB/c mice, persistence was observed in the gall bladder. This is a previously undescribed tissue site of infection that could have been identified only by using imaging. Adapted from Ref. [53].

7.5.4 Viral Infections

Cook and Griffin applied BLI to monitor Sindbis virus (SV) infection in mice [54]. Earlier studies had suggested that SV enters the brain by replicating in cerebral capillary endothelial cells, but subsequent studies suggested otherwise. Other viruses that also cause encephalitis can enter the central nervous system (CNS) by retrograde axonal transport from the olfactory epithelium or peripheral sites of replication. In the case of SV, the simultaneous presence of a high amount of virus in the blood at the time of CNS entry complicates the analysis. BLI post peripheral inoculation showed significant replication in the lower spinal cords, nose, and brains of different animals, as well as in the abdominal regions of some animals. The authors were able to support the hypothesis that the SV entered the CNS by first replicating in the nasal neuroepithelium and then spread through the blood to the olfactory epithelium prior to retrograde transport to the brain.

7.5.5 Viral-mediated Gene Transfer

Reporter genes have been used to monitor gene transfer and translation of this approach to animal models, and whole-body analysis offers the opportunity to localize and quantify transfer methods. Lipshutz et al. reported noninvasive BLI measurements of viral-mediated gene therapy using an adeno-associated viral vector (AAV) [55]. This study revealed that intraperitoneal injection of viral vectors resulted primarily in transduction of cells lining the peritoneal wall and not the liver.

Adams et al. used BLI to demonstrate metastasis of prostate cancer by employing a prostate-specific adenovirus vector, AdPSE-BC-luc [56]. Endocrine therapy is the current treatment for prostate cancer. It involves androgen ablation, but the disease typically relapses within 18–36 months. Viral-mediated gene therapy is a potential alternative to existing treatments. Strong constitutive promoters such as cytomegalovirus (CMV) have high expression but lack specificity. Hence, enhancing the specificity of such constructs can offer additional advantages. In this study, the authors developed a viral construct driven by an enhanced prostate-specific antigen (PSA) promoter. The construct was 20-fold more active than the native PSA promoter and exhibited enhanced specificity and restricted expression. The virally mediated gene therapy concept was tested in *scid* mice that were implanted with several human prostate cancer models (LAPC series). They observed that their vector transduced the xenografts. Using BLI, it was also possible to track the tumors as they progressed from androgen dependence to androgen independence. They also showed that the AdPSE-BC-luc vector could be used as a tool to interrogate endogenous androgen receptor pathway.

7.5.6
Cell Biology

Developing an understanding of cellular and molecular biology is increasingly driven by our ability to visualize and quantify molecular events in living animals. A high spatiotemporal resolution is required for such studies, and *in vivo* bioluminescence and fluorescence imaging provide the necessary tools to biologists. For instance, cell migration is important in the development of an effective immune response to foreign antigens. For decades, immune cell migration has been studied without a good experimental tool for *in vivo* studies. Blood and tissue samples were collected from test subjects and analyzed by flow cytometry and histology techniques at certain time intervals after an immunologic event had been started. Neither approach could provide spatiotemporal information of these events. Recently, BLI has been used for elucidating the spatiotemporal trafficking patterns of lymphocytes in living animals [42, 57–60]. In these studies the kinetics of cell migration was apparent, and the function of these cells for tumor cell killing and migration to sites of autoimmune destruction was revealed.

Murugan et al. developed a bioluminescence imaging strategy to image protein–protein interactions in living mice using the split reporter technique [61]. Protein–protein interactions are important for describing many cellular processes. The authors adapted the split reporter protein approach to demonstrate the interaction between MyoD and Id, two members of the helix-loop-helix (HLH) family of nuclear proteins. MyoD is a myogenic regulatory protein that is expressed in the skeletal muscle. Id is a negative regulator of myogenic differentiation that can associate with, among others, MyoD. In the split reporter technique, a reporter gene such as the firefly luciferase is separated into N-terminus and C-terminus halves. Each of these halves is attached to two proteins whose interactions are under study. In one methodology, interaction between the two proteins reconstitutes the complete luciferase enzyme activity that can be imaged in cells and mice. In another approach, the interaction between the two proteins mediates the splicing of the N- and C-termini of the luciferase, which is then reconstituted to give an active luciferase enzyme. In another study, the authors used this technique to evaluate heterodimerization of the human proteins FRB and FKBP12 mediated by rapamycin [62]. They used synthetic *Renilla* luciferase protein fragment-assisted complementation to assess the heterodimerization of the two proteins. The system could be titrated by changing the concentration of the interacting molecules in cells and living mice, which is very essential in preclinical drug screening and validation processes. Hence, the development of this technique can lead to the possibility of screening agonists and antagonists of protein–protein interactions in cancer models and appraising the efficacy of small pharmaceutical molecules with all the pharmacokinetic issues.

7.5.7 Stem Cell Biology

Stem cell biology is a rapidly developing field that has recently been aided by whole imaging of photoprotein expression [63]. Hematopoietic reconstitution from hematopoietic stem cells (HSC) has previously been observed by flow cytometric measurements of the bone marrow. For these studies animals need to be sacrificed to harvest the bone marrow cells. Although the reconstitution could be assessed with peripheral blood, the localization of stem cells remained a difficult problem. With photoproteins, such as GFP and luciferase, it has become possible to measure engraftment levels and the relative contribution of the donor HSCs to various hematopoietic lineages. Recently, Crooks et al. used BLI to quantitate different levels of chimerism of human donor cells within the marrow space that provided a dynamic profile of engraftment and proliferation in live recipient animals [64]. They showed that late-term engraftment predominated when purified $CD34^{+}CD38^{-}$ cells were transplanted. Lentiviral vectors were used to transfer the luciferase reporter into human HSCs that were then imaged using BLI, thus providing a dynamic profile of progenitor and HSC engraftment. Kim et al. imaged recruitment of neural progenitor stem cells (NPC) to ischemic infarcts in murine models using BLI. They transfected a clonal multipotent murine neural precursor cell line to stably express luciferase and GFP [65]. Their work could potentially lead to novel therapy for ischemic damage and replace dead neurons and supporting tissue.

Studies have shown improved functional outcome after stroke using stem cell therapy and fetal graft transplants [66, 67]. Stem cells migrate to sites of cerebral pathology, including stroke and numerous other central nervous system pathologies. Current methods of studying stem cell mobility involve largely histological experiments. Approaches using labeling of these cells with magnetic particles, isotopes, or gadolinium chelates have not been successful [68–72]. It is difficult to quantify and observe the proliferation of the stem cells because the markers are degraded, diluted, or excreted as cells divide. Kim et al. found that NPCs migrated to an ischemic lesion as compared to control animals without infarcts [73]. Using imaging, they also demonstrated that both methods of administration of the cells, intraparenchymal or intraventricular, yielded the same results.

BLI also identified heterogeneous differences that otherwise would not be identified. In a study by Cao et al., a transgenic donor mouse was described where all of the cells in the body could potentially express luciferase [74]. This donor animal provided a uniformity of integration site and more uniform expression than would be expected from retroviral transduction. Using stem cells isolated from this donor line, Cao et al. used BLI to reveal, in real time, the early events and dynamics of hematopoietic reconstitution in living animals. They imaged the engraftment from single luciferase-labeled HSCs in irradiated recipients (Fig. 7.4). Discrete foci of these transplanted HSCs were found in the spleen and bone marrow at a frequency that correlated with the bone marrow compartment size.

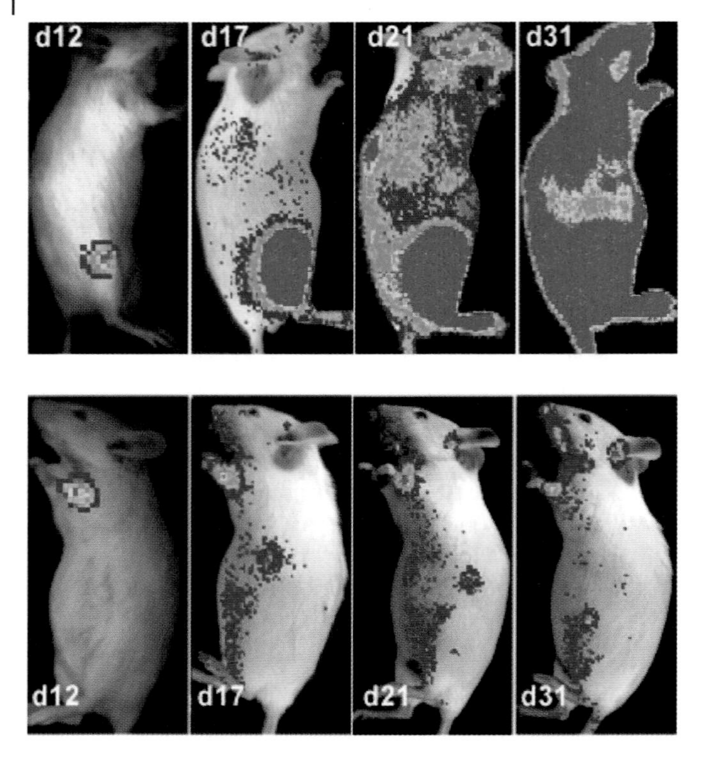

Fig. 7.4 Visualizing hematopoiesis from single hematopoietic stem cells (HSCs). Single *luc*$^+$ or *luc*$^+$ and GFP$^+$ KTLS HSCs were injected intravenously into lethally irradiated FVB recipients along with unlabelled 3×10^5 stem cell-depleted BM cells as a radioprotective population. Animals were imaged daily and initial foci were observed at 12 d post transplant. In one of the two animals, the single-labeled cell participated fully in hematopoiesis (upper), and in the other limited participation was observed. Adapted from Ref. [74].

7.6 Protease Sensors

Proteases are key enzymes that are implicated in a number of cardiovascular, oncological, neurodegenerative, and inflammatory diseases. For instance, in cancer, proteases are thought to be involved in the invasion, metastasis, and angiogenesis of tumor cells [75]. Cancer treatments are being sought that target these proteases and inhibit their activity. For example, matrix metalloproteases (MMP) and cathepsins have been found to enable malignant cells to cross the basement membrane and digest the extracellular matrix (ECM), thus leading to metastasis. Unfortunately, development of drugs that target these proteases has been handicapped by a lack of proper *in vivo* imaging techniques.

Bioluminescence and fluorescence imaging will become useful tools for these studies. Weissleder et al. first demonstrated the use of near-infrared fluorescent dyes as probes for tumors in living animals [44]. They used biocompatible, optically quenched, near-infrared fluorescence (NIRF) imaging probes with which tumor-specific proteases such as cathepsin B, -D, -H, -L, trypsin, and other hydrolases were assayed. When cysteine protease inhibitors (for proteases cathepsin B, -H, and -L) were added, they inhibited all signal from the NIRF probes. Protease-activable NIRF probes have been developed to study endogenous proteases implicated in cancer and cardiovascular diseases [44, 76]. These NIRF probes offer higher signal because of the lower absorption by hemoglobin and tissue, and because of low autofluorescence from nontarget tissue. The role of cathepsin B as a protease involved in inflammatory response and atherosclerotic plaque rupture has been demonstrated using such NIRF probes [76]. More recently, Shah and Weissleder et al. demonstrated the use of NIRF probes for detecting HIV-1 proteases in human glioblastoma cells [77]. The cells were infected with a human simplex virus amplicon vector (HSV) expressing HIV-1. The HIV protease belongs to a family of aspartic acid proteases and specifically cleaves VSQNYPIV and other related peptide sequences. The HIV-1PR protease was virally delivered to glioblastoma tumors in mice and assayed using the NIRF probes. The *in vivo* imaging modality also allowed for assessing the therapeutic effects of the HIV-1PR protease inhibitor indinavir.

Laxman et al. developed a recombinant luciferase reporter that had attenuated activity levels in mammalian cells but showed increased levels of activity under apoptotic conditions [78]. This was achieved by silencing the activity of the reporter molecule via fusion with an estrogen receptor (ER) regulatory domain. A caspase-3-specific cleavage site, DEVD, was inserted between the luciferase and ER domain. Caspase-3 is the protease that is activated by numerous cells during apoptosis. Thus, mammalian cells expressing this construct emitted light as a result of higher luciferase activity only during apoptosis. The protease assay was used to assay D54 human glioma cells that were stably transfected to produce the construct. The cells were implanted into athymic nude mice. Apoptosis was induced in the D54 cells by the tumor necrosis factor α-related apoptosis-inducing ligand (TRAIL). Using a different strategy, Luker et al. fused ubiquitin and luciferase (Ub-Luc) for imaging the activity of the 26S proteasome [79]. Using this probe, they could image proteasome activity, because the proteasome degraded the activity of Ub-Luc. *In vivo* imaging of the proteasome activity was shown by implanting cells that expressed Ub-Luc and wild-type luciferase. The efficacy of the proteasome inhibitor bortezomib could also be determined by this approach. Thus, this technique could become a powerful method for monitoring the therapeutic efficiency of small-molecule drugs rapidly and noninvasively.

7.7 Conclusions

Developments in *in vivo* imaging in the past decade have made it possible to visualize various aspects of cell biology in living animals. It is now also possible to study a wide spectrum of diseases in physiologically relevant settings, including oncology, cardiovascular diseases, and infectious diseases. Developments in the design of novel reporters for enzymatic activity have helped to better target markers in tumors and other diseases. Use of *in vivo* imaging techniques in animal models has the potential to greatly enhance the process of validating pharmacokinetic and pharmacodynamic models and assessing drug candidates.

References

1 Nakajima, A. et al. *J. Clin. Invest.* **2001**, *107*, 1293–1301.
2 Costa, G. L. et al. *J. Immunol.* **2001**, *167*, 2379–2387.
3 Shi, N. et al. *Proc. Natl. Acad. Sci.* **2000**, *87*, 14709–14714.
4 Edinger, M. et al. *Neoplasia* **1999**, *1*, 303–310.
5 Sweeney, T. J. et al. *Proc. Natl. Acad. Sci.* **1999**, *96*, 12044–12049.
6 Reynolds, P. N. et al. *Mol. Ther.* **2000**, *2*, 562–578.
7 Contag, C. H. et al. *Mol. Microbiol.* **1995**, *18*, 593–603.
8 Contag, C. H., Bachmann, M. H. *Annu. Rev. Biomed. Eng.* **2002**, *4*, 235–260.
9 Hastings, J. W. et al. *Methods Enzymol.* (Bioluminescence and chemiluminescence) **2003**, *360*, 75–104.
10 *Promega* **2003**, Chroma-GLo™ luciferase assay system (Technical Manual 062) and Chroma-Luc™ series reporter vectors (Technical Manual 059).
11 Branchini, B. R. et al. *Biochemistry* **2001**, *40*, 2410–2408.
12 Eames, B. et al. Construction and characterization of a red-emitting luciferase in 'Biomedical Imaging: reporters, dyes, and instrumentation', *Int. Sic. Opt. Eng.* **1999**, San Jose, CA.
13 Rice, B. W. et al. *J. Biomed. Opt.* **2001**, *6*, 432–440.
14 Zhao, H. et al. *J. Biomed. Opt.* **2005**, *10* (4), 41210.
15 Tsein, R. Y. *Annu. Rev. Biochem.* **1998**, *67*, 509–544.
16 Griffin, B. A. *Science* **1998**, *281*, 269–272.
17 Nolan, G. P. et al. *Proc. Natl. Acad. Sci.* **1988**, *85*, 2603–2607.
18 Raz, E. et al. *Dev. Biol.* **1998**, *203*, 290–294.
19 Bremer, C. et al. *Radiology* **2001**, *221*, 523–529.
20 Tung, C. H. *Bioconjugate Chem.* **1999**, *10*, 892–896.
21 Mahmood, U. et al. *Radiology* **1999**, *213*, 866–870.
22 Oshiro, M. *Methods Cell Biol.* **1998**, *56*, 45–62.
23 Gao, X. et al. *Nat. Biotech.* **2004**, *22*, 969–976.
24 Qian, H. et al. *Nature Biotech.* **1999**, *17*, 1033–1035.
25 *Tumor Models in Cancer Research* **2001** (Teicher, B. A., Ed.), pp. 34, 647–672, Totowa, NJ, Humana.
26 Padera, T. P. et al. *Mol. Imaging* **2001**, *1*, 9–15.
27 Becker, M. D. et al. *J. Immuno. Methods* **2000**, *240*, 23–37.
28 Fujimoro, H. et al. *Digestion* **2001**, *1*, 97–102.
29 Koseki, S. et al. *Int. Immunol.* **2001**, *13*, 1165–1174.
30 Ntziachristos, V. et al. *Proc. Natl. Acad. Sci.* **2000**, *97*, 2767–2772.
31 Hawrysz, D. J. et al. *Neoplasia* **2000**, *2*, 388–417.
32 Contag, C. H. et al. *Photochem. Photobiol.* **1997**, *66*, 523–531.
33 Contag, P. R. et al. *Nat. Med.* **1998**, *4*, 245–247.

34 Yang, M. et al. *Proc. Natl. Acad. Sci.* **2000**, *97*, 1206–1211.
35 Yang, M. et al. *Proc. Natl. Acad. Sci.* **2000**, *97*, 12278–12282.
36 Pasqualini, R. et al. *Nature Biotech.* 1997, *15*, 542–546.
37 Buchsbaum, D. J. *Cancer* **1997**, *80*, 2371–2377.
38 Rusckowski, M. J. et al. *Pept. Res.* **1997**, *50*, 393–401.
39 Jain, R. *Science* **1996**, *271*, 1079–1080.
40 Muldoon, L. L. et al. *Am. J. Pathol.* **1995**, *147*, 1840–1851.
41 Rehemtulla, A. et al. *Neoplasia* **2000**, *2*, 491–495.
42 Vooijs, M. et al. *Cancer Res.* **2002**, *62*, 1862–1867.
43 Mandl, S. et al. *Molecular Imaging* **2004**, *3*, 1–8.
44 Weisleder, R. et al. *Nat. Biotech.* **1999**, *17*, 375–378.
45 Corbisier, P. et al. *FEMS Microbiol. Lett.* **1993**, *110*, 231–238.
46 Engebrecht, J. et al. *Cell* **1983**, *32*, 773–781.
47 Engebrecht, J. et al. *Proc. Natl. Acad. Sci.* **1984**, *81*, 4154–4158.
48 Meighen, E. A. *Annu. Rev. Microbiol.* **1988**, *42*, 151–176.
49 BitMansour, A. et al. *Blood* **2002**, *100*, 4660–4667.
50 Kadurugamuwa, J. L. et al. *Anti. Agents Chemo.* **2003**, *47*, 3130–3137.
51 Brown, M. R. W. et al. *Annu. Rev. Microbiol.* **1985**, *39*, 527–556.
52 *J. Infect. Diseases* **1990**, *162*, 96–102.
53 Hardy, J. et al. *Science* **2004**, *303*, 851–853.
54 Cook, S. H., Griffin, H. D. E. *J. of Virology* **2003**, *77*, 5333–5338.
55 Lipshutz, G. S. et al. *Mol. Ther.* **2001**, *3*, 284–292.
56 Adams, J. Y. *Nat. Med.* **2002**, *8*, 891–896.
57 Costa, G. L. et al. *J. Immunol.* **2001**, *167*, 2379–2387.
58 Hardy, J. et al. *Exp. Hematol.* **2001**, *29*, 1353–1360.
59 Tarner, I. H. et al. Clin. Immunol. **2002**, *105*, 304–314.
60 Edinger, M. et al. Blood **2003**, *101*, 640–648.
61 Paulmurugan, R. et al. *Proc. Natl. Acad. Sci.* **2002**, *99*, 15608–16613.
62 Paulmurugan, R. et al. *Cancer Research* **2004**, *64*, 2113–2119.
63 Morrison, S. J., Weissman, I. L. *Immunity* **1994**, *1*, 661–673.
64 Case, S. S. et al. *Proc. Natl. Acad. Sci.* **1999**, *96* (6), 2988–2993.
65 Kim, D.-E. et al. *Stroke* **2004**, *35*, 952–957.
66 Chen, J. et al. *Stroke* **2001**, *32*, 2682–2688.
67 Mattsson, B. et al. *Stroke* **1997**, *28*, 1225–1232.
68 Bulte, J. W. M. et al. *Proc. Natl. Acad. Sci.* **1999**, *96*, 15256–15261.
69 Lewin, M. et al. *Nat. Biotech.* **2000**, *18*, 410–414.
70 Modo, M. et al. *Neuroimage* **2002**, *17*, 803–811.
71 Kornblum, H. I. et al. *Nat. Biotech.* **2000**, *18*, 655–660.
72 Honigman, A. et al. *Mol. Ther.* **2001**, *4*, 239–249.
73 Wang, X. et al. *Blood* **2003**, *102*, 3478–3482.
74 Cao, Y. A. et al. *Proc. Natl. Acad. Sci.* **2004**, *101*, 221–226.
75 Kim, J. et al. *Cell* **1998**, *94*, 353–362.
76 Chen, J. et al. *Circulation* **2002**, *105*, 2766–2771.
77 Shah, K. et al. *Cancer Research* **2004**, *64*, 273–278.
78 Laxman, B. et al. *PNAS* **2002**, *99*, 16551–16555.
79 Luker, G. D. et al. *Nat. Med.* **2003**, *9*, 969–973.

8
Photoproteins as Reporters in Whole-cell Sensing

Jessika Feliciano, Patrizia Pasini, Sapna K. Deo, and Sylvia Daunert

8.1
Introduction

Reporter genes code for proteins that produce a detectable signal, which can be differentiated in a mixture of endogenous intra- or extracellular proteins [1, 2]. They have found applications in studies of transcription control, gene expression, cell-signaling mechanisms, drug target discovery, cell-based analyte biosensing, and environmental monitoring of specific microorganisms [3], among others. Photoproteins can serve as reporters in two different ways: via genetically engineered systems that are selective toward specific analytes or by using a constitutively bioluminescent/fluorescent cell line that measures overall toxicity. This chapter will focus on the applications of photoproteins as reporters in genetically engineered whole-cell sensing systems.

8.1.1
Biosensors Using Intact Cells

A biosensor combines a biological component (the sensing element), which is responsible for the selectivity of the device, with a detection system (the transducer) for measuring the reaction of the biological component with the substance (analyte) being monitored. The biological component can be an enzyme, an antibody, a receptor, or whole cells, while the biological reaction that is monitored can be the activity of an enzyme, the binding of an analyte to an antibody or receptor, the induction of gene expression within cells, or even cell death [4, 5]. A number of detection systems have been employed to monitor these biological events, including electrochemical, optical, and piezoelectric systems [6, 7].

There are a number of advantages in using intact microorganisms, rather than isolated biological components, in biosensors. Microorganisms are usually more tolerant of assay conditions that would be detrimental to an isolated enzyme or protein. Because microorganisms have mechanisms that help them regulate their internal environment, they are also more tolerant of suboptimal pH and

Photoproteins in Bioanalysis. Edited by Sylvia Daunert and Sapna K. Deo

ISBN: 3-527-31016-9

temperatures than purified enzymes and are less likely to be inhibited by solutes in the sample being assayed. Microorganisms are cheaper to use in biosensors because the active biological component does not have to be isolated and unlimited quantities can be prepared relatively inexpensively [8–10].

Because an analyte, such as a pollutant, must be taken up by the microorganism before a response is produced, using intact microorganisms can provide information about the bioavailability of the analyte. Classical analytical methods of detection cannot distinguish between bioavailable pollutants from those that are unavailable. In that regard, sensing systems using intact cells complement the more conventional sensors where isolated biological components are being used. These systems may provide physiologically relevant data and give an indication of the bioavailability of a pollutant, which is especially important when decisions regarding remediation of contaminated environments have to be made [9]. While only the bioavailable amount can be obtained, cell-based biosensors are preferred over other biological biosensors because one can acquire other types of information such as toxicity, mutagenicity, and pharmacological activity.

Because of all these characteristics, cell-based biosensing systems have found applications in biotechnology, pharmacology, molecular biology, and clinical and environmental chemistry. Among the advances over the years is the constant wish to improve the systems. Improved high-throughput assays for gene function, automation to enhance screening and capacity, and miniaturization to reduce costs have been developed. Systems for the correction of changes in signal that are due to the physiological state of the cell have also been developed. For a good review on how the bioreporter design can be improved by mutagenizing the sensing and the regulatory proteins involved, readers are encouraged to review Ref. [9].

Several reviews have been written on cell-based biosensors, including by our group in 2000 [10]. This chapter will present a general view of the photoproteins with an emphasis on their application specifically to bacterial whole-cell sensing systems. The most advanced technology from our 2000 review will be highlighted, as well as advances in reporter gene technology in cell-based assays since then.

8.1.2
Reporter Genes in Genetically Engineered Whole-cell Sensors

Genetically engineered whole-cell sensing systems can be developed by coupling a sensing element, which recognizes the analyte and confers selectivity to the system, with a reporter gene, which produces the detectable signal and determines the sensitivity of the system. In this chapter whole-cell sensors that employ receptors as sensing elements will not be discussed. Such biosensing systems are usually based on eukaryotic cells, either yeast or mammalian cells, and are routinely used in drug screening assays and for the environmental detection of dioxin- and estrogen-like compounds [11–13]. Instead, we will focus on inducible operon-based whole-cell sensors, in which the sensing element is composed of regulatory proteins that are capable of molecular recognition and promoter regions (O/P) of DNA.

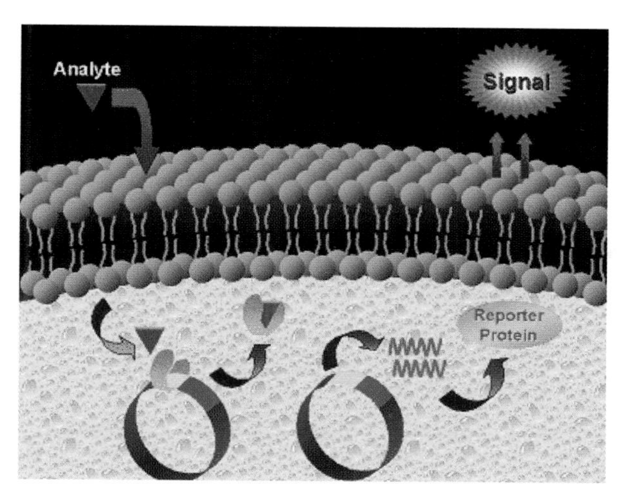

Fig. 8.1 Schematic representation of a whole-cell sensing system based on reporter genes.

In a whole-cell sensing system, the regulatory protein controls the transcription of the fused genes. When the target analyte is present, the regulatory protein is capable of recognizing the analyte and will unbind itself from the operator/promoter region, causing transcription to occur. Thus, when the induction takes place in the presence of the target analyte, the reporter gene is co-expressed along with the other genes of the operon. Consequently, the concentration of the inducer can be quantified by measuring the signal generated by the reporter protein (Fig. 8.1).

When a reporter is being selected, one has to take into consideration which type of application the reporter is needed for, including how sensitive the assay should be, in what concentration range one needs to detect, and the nature of the target analyte. In the case of sensitivity, when the reporter gene codes for an enzyme, internal amplification cascades can be used to increase the sensitivity of the device. The stability of the reporter plays a very important role as well. For instance, green fluorescent protein variants that are unstable have different half-lives and are more appropriate for kinetic studies than the red fluorescent protein, which takes days to maturate. A good reporter should be environmentally safe, easy to use, highly sensitive, relatively small in size (for inclusion in the vector), and nontoxic, so that expression is related only to the gene under study and not to overall toxicity (biologically inert and benign). Additionally, the reporter should produce a wide dynamic response, its signal should be distinguishable over background, and the expressed product should be enough so that it can be detected directly or indirectly. Table 8.1 summarizes the advantages and disadvantages of the reporter photoproteins. Details about specifics on each of the reporter proteins have been the subject of reviews elsewhere and will not be discussed here. The following section covers the specific characteristics of each photoprotein when used as a reporter in a cell-based assay.

Table 8.1 Advantages and disadvantages of photoproteins in whole-cell sensing applications.

Reporter protein	*Advantages*	*Disadvantages*
Bacterial luciferase	High sensitivity. Does not require addition of a substrate. No endogenous activity in mammalian cells. No light source requirement.	Heat labile and therefore of limited use in mammalian cells. Narrow linear range.
Firefly luciferase	High sensitivity. Broad linear range. No endogenous activity in mammalian cells. High quantum yield. Spectral variants.	Requires addition of a substrate. Requires an aerobic environment and ATP. Requires addition of solubilizers for the substrate to permeate the cell. Flash-type emission.
Sea pansy luciferase	No endogenous activity in bacterial and mammalian cells. No light source requirement. Its substrate, coelenterazine, is membrane permeable.	Requires addition of a substrate.
Aequorin	High sensitivity. No endogenous activity in mammalian cells. No light source requirement. Detection down to subattomole levels.	Requires addition of a substrate and the presence of Ca^{2+}. Flash-type emission.
Green fluorescent protein	Autofluorescent and thus does not require addition of a substrate or cofactors. Spectral variants. No endogenous homologues in most systems. Stable at biological pH.	Moderate sensitivity (no signal amplification). Requires post-translational modification. Background fluorescence from biological systems may interfere. Potential cytotoxicity in some cell types.

8.2 The Luciferases

Bioluminescence is the term used to describe the light produced by chemical excitation within living organisms. In many cases light emission is based on luciferase-mediated oxidation reactions. Luciferase is a generic name for an enzyme that catalyzes a light-emitting reaction [14]. Luciferases can be found in both terrestrial and aquatic organisms, including bacteria, algae, fungi, jellyfish, insects, shrimps, squids, and fireflies [15]. The most widely used luciferase genes in reporter gene assays are the bacterial luciferase genes (*lux* genes of terrestrial *Photorhabdus luminescens* and marine *Vibrio harveyi* bacteria) and the eukaryotic *luc* and *ruc* genes from firefly (*Photinus pyralis*) and sea pansy (*Renilla reniformis*).

8.2.1
Bacterial Luciferases

Bacterial luciferases have found applications in gene transfer and expression, promoter analysis, gene delivery, imaging of gene expression [16], drug discovery and signaling pathways, etc. However, because they are heat-labile dimeric proteins, their applications in mammalian cell-based systems have been limited [1].

Bacterial luciferase catalyzes the oxidation of a reduced flavin mononucleotide ($FMNH_2$) and a long-chain aldehyde to FMN and the corresponding long-chain carboxylic acid, emitting light at 490 nm (Table 8.2). The blue-green light produced in this reaction has a quantum yield between 0.05 and 0.15 [17]. The *lux* genes from bacteria luciferases have been isolated and extensively employed in the construction of bioreporters. They are five genes that are organized in the *luxCDABE* operon: *luxA* and *luxB* code for the α and β subunits of bacterial luciferase, while *luxC*, *luxD*, and *luxE* code for the synthesis enzymes for the long-chain aldehyde substrate. These five genes are conserved in all bacterial species identified to date. Depending on the combination of the genes employed, different types of bioluminescent sensing systems can be developed [18].

Table 8.2 Chemical reactions involving photoproteins.

Reporter protein	*Catalyzed reaction*
Bacterial luciferase	$FMNH_2 + R\text{-}CHO + O_2 \rightarrow FMN + H_2O + RCOOH + h\nu$ (490 nm)
Firefly luciferase	Firefly luciferin + O_2 + ATP → oxyluciferin + AMP + P_i + hν (550–575 nm)
Renilla luciferase	Renilla luciferase + Coelenterazine + O_2 → coelenteramide + CO_2 + hν (482 nm)
Aequorin	Coelenterazine + O_2 + Ca^{2+} → coelenteramide + CO_2 + hν (469 nm)
Green fluorescent protein	Post-translational formation of an internal chromophore

8.2.1.1 *luxAB* Bioreporters

These bioreporters contain only the *luxA* and *luxB* genes, which are sufficient for generating the bioluminescent signal; however, because the genes that code for the production of the substrate are not present, the substrate must be supplied to the cell. Typically, decanal is added at some point during the assay. This system has the advantage that one can control the moment when the reaction is triggered and the measurement be taken right after, dramatically reducing the basal bioluminescence resulting from "leaky" promoters. Various *luxAB* sensing systems have been developed in bacterial, yeast, insect, nematode, plant, and mammalian cell sensing systems. Nevertheless, the addition of a substrate makes it impractical for deployable applications, and platforms for biosensors have utilized the entire *luxCDABE* cassette instead because expression of all five genes has the advantage of not requiring addition of a substrate.

8.2.1.2 *luxCDABE* Bioreporters

These bioreporters contain the full *lux* operon cassette, thereby allowing for a completely independent bioluminescent system that requires no external addition of a substrate. In this biosensing system, the bioreporter is exposed to the target analyte and a quantitative change in bioluminescence is detected, often within an hour. The rapidity of the assay, its ease of use, and the substrate-free characteristic of this system make *luxCDABE* bioreporters ideal for real-time, online, field, and high-throughput applications. For that reason, *luxCDABE*-based bacterial luciferases are the most widely used photoproteins in whole-cell sensing applications. They have been incorporated into a diverse range of platforms and have found applications in environmental and clinical applications, among others.

8.2.1.3 Naturally Luminescent Bioreporters

One of the earliest applications of bacterial luciferase as a reporter was in the Microtox assay [19, 20]. This assay relies on the changes in light production of the natural luminescence of *Photobacterium phosphoreum* and is used to measure overall toxicity. Similarly, toxin-sensitive cells can be used in biosensors for the nonspecific detection of toxins. When the cells are exposed to critical amounts of toxins, either the metabolic activity of the cells decreases or the cells die. In either case, as the concentration of toxins in the environment increases there is a corresponding decrease in the signal produced by a reporter protein expressed constitutively within the cells. It should be noted that this type of biosensor will respond to any type of substance that is toxic to the cells. Based on this principle, biosensors have been developed for determining the biotoxicity of a sample [21, 22].

8.2.2
Eukaryotic Luciferases

8.2.2.1 Firefly Luciferase

Firefly luciferase, encoded by the *luc* gene, was first isolated from the North American firefly *Photinus pyralis* [23, 24]. The structure of this luciferase is different from that of bacterial luciferase, and the resulting bioluminescent reaction exhibits other characteristics as well. The oxygen-dependent bioluminescent reaction converts the substrate luciferin to oxyluciferin based on energy transfer from ATP to the substrate to yield AMP, carbon dioxide, and light in the 550–575 nm range. The bioluminescence produced by firefly luciferase is characterized by a peak after 300 ms, after which it exhibits decay. This decay is produced by the product oxyluciferin, which associates with the enzyme and inhibits light production. This can be overcome by addition of coenzyme A in the reaction mix, which will render dissociation of oxyluciferin from the enzyme and prolong emission over several minutes [23, 24]. The bioluminescence resulting from firefly luciferase exhibits the highest quantum yield known for a bioluminescent reaction (0.88, which is about 10 times larger than that of bacterial luciferase) [1]. Like bacterial luciferase, mammalian organisms do not exhibit endogenous firefly luciferase activity. However, the genes required to synthesize the substrate (luciferin) are not present; thus, addition of an exogenous substrate is always required for bioluminescence to occur. There has been controversy in this matter because the substrate cannot be permeated into the cell [25]. The addition of solubilizers such as DMSO has been required to facilitate entry into the cell. An interesting feature of some eukaryotic luciferases is that they can exhibit light in different wavelengths, ranging from 547 nm to 593 nm [26]. This characteristic has been employed only in colony detection on the same plate, but it can be further exploited in multi-analyte detection by fusing each mutant to a different regulatory gene. Firefly luciferase has found applications in whole-cell sensing systems for the detection of heavy metals and aromatic organics. Its high sensitivity (subattomole level) and broad dynamic range (eight orders of magnitude) make this reporter a favorite in mammalian cell-based sensing systems.

8.2.2.2 **Sea Pansy Luciferase**

The sea pansy *Renilla reniformis* is a coelenterate that, unlike other bioluminescent organisms, emits a bright green light. *In vivo, Renilla* luciferase oxidizes its substrate coelenterazine, and after energy transfer to the green fluorescent protein, light is emitted at 509 nm. *In vitro,* it is represented by the *Rluc* gene and produces a blue light of 482 nm [27]. Like other luciferases, *Renilla* luciferase has no endogenous activity in bacteria and thus can find applications in bacteria-based sensing systems. Studies have shown a sensitivity and detection range similar to the firefly luciferase. Despite its major advantage over firefly luciferase that coelenterazine is membrane permeable, it has limited applications in analytical chemistry. This is probably due to a flash half-life of several seconds. The main application of this eukaryotic luciferase has been in combination with firefly luciferase, in a dual luciferase-based reporter system [28]. Here, both eukaryotic luciferases show a linear response to the target analyte: five orders of magnitude for *Renilla* luciferase and seven orders of magnitude for firefly luciferase. In both cases, substrate addition makes it impractical for field applications.

8.3 Aequorin

Aequorin is a calcium-binding photoprotein isolated from the jellyfish *Aequorea victoria*. This protein is comprised of three components: apoaequorin (the precursor of the photoprotein aequorin), a coelenterazine chromophore, and molecular oxygen [29]. Within the protein, it contains three Ca^{2+}-binding sites that, once occupied, cause the protein to undergo a conformational change, which in turn triggers the oxidation of coelenterazine to a highly unstable intermediate, coelenteramide. Like *Renilla* luciferase, it catalyzes the oxidation of coelenterazine to yield a flash blue light of 469 nm that lasts five seconds. This flash-type light emission is accompanied by the release of CO_2, has a quantum yield between 0.15 and 0.20, and requires a Ca^{2+} concentration of 0.1–10 μM to induce light emission [30]. To date, aequorin is one of the most extensively studied calcium-binding photoproteins. It can be regenerated by dialyzing with EDTA to remove Ca^{2+} and adding fresh coelenterazine.

Among the advantages of using aequorin in reporter gene assays are its high sensitivity and low background. Aequorin can be detected at subattomole levels and is not toxic to various types of cells [31]. Furthermore, a complete absence of endogenous aequorin in mammalian cells can find applications in this type of cell-based sensing systems. Despite these advantages, aequorin is not a popular reporter in cell-based assays. This is probably due to its flash-type light emission (lasting less than five seconds), which requires a luminometer that can initiate and detect the reaction rapidly. In addition, the exogenous requirement of coelenterazine makes this impractical for out-of-the-laboratory applications. Currently, there are two ways of assessing aequorin bioluminescence: detection of aequorin itself or detection of the luminescence triggered by the addition of Ca^{2+} as the product of the signaling cascade. The latter is more commonly used.

Aequorin is widely used as a label in immunoassays and in nucleic acid probe assays, and, because it is a Ca^{2+}-binding protein, it is used to monitor intracellular levels of free calcium in both prokaryotes and eukaryotes. However, to our knowledge there is only one application in cell-based sensing assays. Rider et al. developed the CANARY (Cellular Analysis and Notification of Antigen Risks and Yields) assay for the detection of pathogenic bacteria and viruses [32, 33]. In this assay, B cells were genetically engineered to express membrane-form antibodies and apoaequorin. The cells were patterned on glass substrates using photolithography and packed into a flow chamber of a microfabricated device. Once the antigen binds, an intracellular signaling cascade is triggered, causing the release of Ca^{2+} and the activation of the luminescent reaction of aequorin. Detection of pathogenic bacteria and viruses was achieved in less than 1 min with high specificity and sensitivity. An advantage of this assay is that it can be engineered for any target analyte and can be specific to antigens of different pathogens.

8.4
Fluorescent Proteins

Fluorescent proteins absorb light at a certain maximum wavelength and emit it at a different maximum wavelength, normally a higher one. The green fluorescent protein (GFP) is by far the most popular fluorescent reporter protein in whole cell-based assays. GFP variants and the red fluorescent protein (DsRed) also have found applications in this field. Spectral properties of fluorescent proteins are reported in Table 8.3.

Table 8.3 Excitation and emission maxima of fluorescent photoproteins.

Reporter protein	***Excitation λ_{max} (nm)***	***Emission λ_{max} (nm)***
Green fluorescent protein (GFP)	395	509
Enhanced green fluorescent protein (EGFP)	488	509
Green fluorescent protein UV (GFPuv)	395	509
Blue fluorescent protein (BFP/EBFP)	380	440
Cyan fluorescent protein (CFP/ECFP)	433	475
Yellow fluorescent protein (YFP/EYFP)	513	527
Red fluorescent protein (DsRed)	558	583

8.4.1 Green Fluorescent Protein

The green fluorescent protein, encoded by the *gfp* gene, has been isolated and cloned from the jellyfish *Aequorea victoria. In vivo* the role of GFP is to shift the blue bioluminescence of aequorin to green fluorescence by means of a radiationless energy transfer from aequorin to GFP [34]. The wild-type GFP has an excitation maximum at 395 nm, with a minor peak at 475 nm, and an emission maximum at 509 nm with a small shoulder at 540 nm [35].

The main advantage of GFP as a reporter protein is that it does not require the addition of exogenous substrates or cofactors to produce light; therefore, its use is not limited by the accessibility of substrates [1]. Detection of GFP requires only irradiation by light, which makes GFP the reporter of choice for *in situ* detection of analytes. Induced bacteria expressing GFP can be detected at a single-cell level by using techniques such as epifluorescence microscopy, flow cytometry, or confocal laser scanning microscopy [36]. GFP is a very stable protein, which is advantageous because long-lasting GFP produced from weak promoters or under conditions of low metabolic activity can accumulate in the biosensor cells over time. GFP can be easily expressed in a wide range of organisms, in particular bacteria. This enabled extensive mutational studies of the native protein and the functions of individual amino acids, with the final goal of altering GFP stability and spectral properties (Fig. 8.2). Variants with improved fluorescence intensity, thermostability, and chromophore folding have been obtained. Shifts within the excitation and emission spectra have also been achieved, thus making available mutants such as the blue, cyan, and yellow variants that can be used for multi-analyte detection purposes [37]. Different strategies have been used to generate variants with shorter half-lives that are suitable for time-dependent induction studies [38].

Limitations to the use of GFP as a reporter include lower detectability compared to other reporter systems. This is due in part to lack of the amplification step occurring with enzyme reaction-based reporters. Additionally, despite the absence of endogenous GFP homologues in most target organisms, the sensitivity of GFP may be compromised by the presence of other fluorescent molecules found in

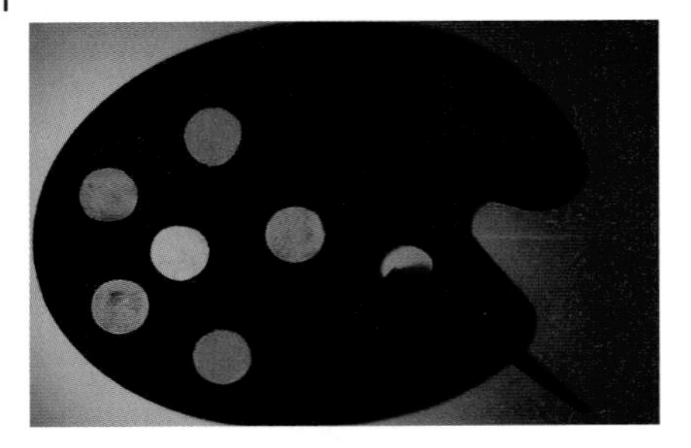

Fig. 8.2 Expression of fluorescent reporter proteins in *E. coli*.
Reprinted in part from Ref. [10] with permission from the American Chemical Society.

biological systems, which increase background signal [2]. The need for O_2 for folding and maturation also limits application to aerobic conditions.

Overall, GFP has unique characteristics that have led to its successful application as a reporter for various analytes [10, 38].

8.4.2
Red Fluorescent Protein

The *dsRed* gene encodes a red fluorescent protein (DsRed) with homology to GFP. The red fluorescent protein has been cloned from coral *Discosoma* sp. [39]. It has an excitation peak at 558 nm and an emission peak at 583 nm. These spectral properties are suitable for its use as a reporter protein. In fact, cellular autofluorescence is supposed to be lower at longer wavelengths, which should result in lower background signal with DsRed as compared to GFP [40]. However, the use of DsRed as a reporter is somewhat limited by the fact that it has a maturation period of several hours to days [41].

In one study the performance of *dsRed* was compared to that of other reporter genes (*gfp*, *luc*, and *luxCDABE*) in identical constructs in which the reporter genes were inserted under the control of mercury-responsive regulatory units (*mer*) [42]. As expected, both fluorescent proteins gave a slow response, with DsRed being much slower than GFP. DsRed was expected to give better sensitivity than GFP because of lower autofluorescence; however, that did not happen with the *mer–dsRed* construct used in the study.

8.5
Multiplexing

Many of these reporters have been used in combination. Multiple reporter genes are commonly utilized to provide multi-analyte detection or as viability control. Various combinations have been employed, including using different spectral variants of the same photoprotein and using a combination of a fluorescent photoprotein and bioluminescent one either within the same construct in the same cell [43, 44], within separate constructs within the same cell, or in different cell lines [45, 46]. For instance, the click beetle has four different luciferases that produce bioluminescence emission at different emission maxima (548 nm, 560 nm, 578 nm, and 590 nm) [26, 47]. If their emission spectra do not overlap, the genes encoding the 548-nm and 590-nm emission could be used in a dual-detection system. Not only would detection of more than one analyte be feasible, but also the assay would provide a much simpler system because only one substrate would be required. When using bioluminescent photoproteins, the signals are simultaneously generated and independently measured because of either different kinetics (half-life) or different emission spectrum. In contrast, the fluorescent photoproteins have to be independently excited and/or measured because of different excitation and emission spectra.

We developed a dual-analyte sensing system for the simultaneous detection of β-lactose and L-arabinose as model analytes. Two variants of the green fluorescent protein, BFP2 and GFPuv, were used as the signal-generating reporter proteins. The cells were excited at a given wavelength, and emission for the two reporters was collected at wavelengths characteristic of each reporter. The fluorescence emission thus obtained could be correlated with the amount of sugars present in the sample. The system proved to be highly selective for the two analytes [48]. Dual-reporter reagents using *Renilla* and firefly luciferase reporters are commercially available. For viability control, we developed a fluorescent dual-reporter system that utilizes two fluorescent proteins with different emission spectra (GFPuv and YFP) within the same genetic construct. One of the fluorescent proteins was used for the analytical signal, while the other was the internal reference of general cell viability. Consequently, any environmental conditions that resulted in a nonspecific effect on the analytical signal could be corrected. Such a system with internal response correction is useful for environmental samples whose complex matrices can affect cell viability [49].

8.6 Applications

Luciferases are the most widely used of the reporter proteins in whole-cell sensing systems. Unlike fluorescent photoproteins, there is no light scattering from other components in the mixture, as well as no background emission from the media because the media itself is not luminescent. This low background expression along with the enzymatic nature and high quantum efficiency of bioluminescent reactions account for the higher and faster detectability of bioluminescence when compared to fluorescence. Because luminometers do not contain a light source, whole-cell sensing systems based on bioluminescence, specifically bacterial luciferases, have been more widely used in real-sample analysis and field applications. The small number of instrument components required has made possible the construction of cheaper and more compact devices that are suitable for portability in contrast to fluorescence devices.

8.6.1
Stress Factors and Genotoxicants

Recombinant bacterial whole-cell sensing systems for the detection of general stress, oxidative stress, and genotoxic compounds have been developed. In these systems, reporter genes are fused to promoters of different stress-response regulons, which are activated by a broad range of compounds. A panel of these bacterial strains is usually employed in environmental monitoring studies [50]. Numerous bacterial strains have been developed that are able to respond to different stressful conditions with the production of a reporter photoprotein. Selected examples of such strains are reported in Table 8.4.

Table 8.4 Whole-cell sensing systems using photoproteins as reporters able to respond to different stress factors.

Stress inducer	*Promoter*	*Stress response*	*Reporter genes*	*References*
Protein damage	*GrpE*	Heat shock	*luxCDABE*	44, 51–56
Oxidative damage	*KatG*	Hydrogen peroxide	*luxCDABE*	43, 52–54, 56, 57
Oxidative damage	*MicF*	Superoxide	*luxCDABE*	52
Membrane synthesis	*FabA*	Fatty acid synthesis	*luxCDABE*	51–57
DNA damage	*RecA*	SOS	*luxCDABE* gfp_{uv}	43, 51–55, 57–60

8.6.2
Environmental Pollutants

A variety of bacterial whole-cell sensing systems based on the promoter–reporter gene concept have been developed for the specific or selective detection of single pollutants or classes of pollutants. Many microorganisms have evolved the ability to survive in suboptimal conditions, including contaminated environments. Such ability usually relies on genetically encoded resistance systems. In the presence of toxic compounds such as heavy metals and metalloids, particular bacteria can synthesize specific proteins that confer resistance to those substances. The mechanisms of resistance vary. Some microorganisms develop efflux pumps, loss-of-uptake systems, or chemical detoxification systems [61, 62]. Other bacterial strains living in contaminated sites can degrade organic xenobiotics and utilize them as carbon and energy sources [63]. Many of these resistance pathways are inducible, meaning that protein synthesis occurs only as required by the presence of given compounds, which makes their use possible for analytical purposes. The gene sequences of numerous regulatory systems have been cloned and plasmid reporter vectors have been constructed that feature the regulatory sequence in gene fusion with a reporter gene. The pioneering work of Sayler and coworkers, who constructed a whole cell-based sensing system for the detection of naphthalene and salicylate [64], prompted the development of a plethora of constructs responsive to organic and inorganic pollutants and including different reporter genes. A number of these constructs, along with the application of whole cell-based biosensors to different environmental samples, have been presented and discussed in recent reviews [65, 66]. Many whole-cell sensing systems for the detection of metals and metalloids have been developed, while smaller numbers of biosensing systems for organic pollutants are available. In Table 8.5 we list inorganic and organic analytes for which whole cell-based biosensors have been developed, along with the regulatory units and reporter genes used. Whole-cell sensing systems are suitable for the detection of metabolites that are formed along the pollutants degradation pathways. As an example, a whole-cell biosensor for the detection of

Table 8.5 Whole-cell sensing systems using photoproteins as reporters for the detection of environmental inorganic and organic analytes.

Analyte	Reporter genes	Regulatory unit	References
Arsenite, arsenate, antimonite	*luc* *luxAB* *gfp* gfp_{uv}	*ars*	67–71
Cadmium, lead	*luc* *luxAB*	*cad*	72
Cadmium, lead, zinc	*luc* *luxCDABE* *rs-GFP*	*znt*	73, 74
Chromate	*luxCDABE*	*chr*	75
Copper	*luxCDABE*	*copA*	73, 76, 77
Iron	*gfp*	*pvd*	78
Mercury, organomercurials	*luc* *luxAB* *luxCDABE* *gfp*	*mer*	79–84
Nickel	*luxCDABE*	*cnr*	85
Silver	*luxCDABE*	*copA*	77
Alkanes	*luxAB*	*alk*	86
Aromatic compounds	*luxCDABE*	*sep*	87
Benzene, toluene, ethylbenzene, xylene (BTEX)	*luc* *gfp* *luxCDABE*	*xyl* *tbu* *tod*	88–90
Naphtalene, salicylate	*luxCDABE*	*nah*	91
Phenolic compounds	*luxCDABE* *luxAB*	*mop* *dmp*	92
Polychlorinated biphenyls	*luxCDABE*	*bph*	93
Nitrate	*gfp*	*nar*	94, 95
Phosphate	*luxCDABE*	*pho*	96

hydroxylated polychlorinated biphenyls in biological and environmental samples is currently under development in our laboratory. Although whole-cell sensing systems have been mainly developed for environmental monitoring purposes, detection of biologically relevant molecules, such as sugars, has benefited from the analytical performance of reporter gene-based whole-cell biosensors [48, 97, 98]. Whole-cell sensing systems for the detection of quorum-sensing signaling molecules and antibiotics have been also developed. These are described in more detail in the following sections of this chapter.

8.6.3
Quorum-sensing Signaling Molecules

Quorum sensing is a phenomenon that enables certain bacteria to communicate with each other by producing and responding to secreted signaling molecules in proportion to cell density [99]. This process allows a population of bacteria to regulate expression of specialized genes in response to changes in its size. When cell population reaches a critical size, certain genes are expressed, hence the term quorum sensing. This phenomenon is responsible for many processes, including expression and secretion of virulence factors, formation of biofilms, competence, conjugation, antibiotic production, motility, sporulation, symbiosis, and bioluminescence [99]. Several quorum-sensing regulatory systems have been identified so far. The best-characterized are those of the LuxI/LuxR type found in most gram-negative bacteria. These quorum-sensing circuits use *N*-acyl-homoserine lactones (AHLs) as signaling molecules and contain regulatory proteins with homology to the regulatory proteins LuxI and LuxR, which control bioluminescence in the marine bacterium *Vibrio fischeri* [100]. The gene sequences of some of these regulatory systems have been cloned and inserted in plasmid reporter vectors, thus allowing the development of whole-cell sensing systems for the detection of quorum-sensing signaling molecules.

Whole cell-based sensors that contain the *luxCDABE* cassette as a reporter in gene fusion with quorum-sensing regulatory sequences have been employed to study structure-activity relationships of AHLs [101]. Such biosensing systems have limits of detection in the nanomolar range. They have been applied successfully to the study of AHLs in sputum of cystic fibrosis patients with bacterial lung infections [102]. Whole-cell sensing systems may also represent a tool to identify AHL-producing bacteria by detecting AHLs in their cell-free culture supernatants. Using this approach, it has been shown that *Porphyromonas gingivalis*, a gram-negative bacterium implicated in the etiology of human periodontal disease, actually regulates the production of virulence factors by quorum sensing but does not utilize AHL-mediated systems [103]. In another study, it has been reported that plant growth-promoting *Pseudomonas putida* produces cyclic dipeptides that potentially cross talk with the LuxI/LuxR quorum-sensing system [104]. In general, these biosensors can be used as a preliminary screening tool to identify active samples, which can be then subjected to mass spectrometric and NMR analysis in order to determine the chemical identity of individual molecules.

Recently, the molecular mechanisms of bacterial quorum sensing have been proposed as new drug target. In fact, the pharmacological inhibition of quorum sensing for the treatment of bacterial infections may be an alternative to traditional antibiotic treatments, whose therapeutic efficacy is currently decreasing as a result of the occurrence of multiple antibiotic-resistant pathogenic bacteria [105]. Although physical-chemical methods for the detection of AHLs have been developed [106], detailed investigations of quorum-sensing inhibition by methods such as HPLC, GC-MS, or TLC are not practical for routine screening. Whole-cell sensing systems are capable of easily, rapidly, and inexpensively investigating large numbers of compounds and thus are suitable for high-throughput screening of quorum-sensing inhibitors as potential antimicrobial drugs.

A GFP-based whole-cell sensing system was developed to perform single-cell analysis and online studies of AHL-mediated communication among bacteria [107]. In this application, a *gfp* mutant that encodes a protein with a relatively short half-life was used in order to monitor variations in AHL concentrations in real time. Such a biosensor represents a powerful tool for *in vivo* studies of cell communication and for *in vivo* efficacy studies of quorum-sensing inhibitors. Indeed, it has been used to detect the production of AHLs from *Pseudomonas aeruginosa* in a mouse infection model [108]. A similar biosensor allowed detection of AHL production in laboratory-based *P. aeruginosa* biofilms and showed the ability of a synthetic compound to penetrate microcolonies and block quorum-sensing signaling in most biofilm cells [109]. GFP-based whole-cell sensing systems combined with flow cytometry analysis have been successfully employed for *in situ* detection of AHLs in soil [110].

8.6.4 Antibiotics

Antibiotics belonging to the tetracycline family are extensively used in the therapy and prophylactic control of bacterial infections in human and veterinary medicine and as food additives for growth promotion in the farming industry. Intensive use of tetracyclines has led to widespread antibiotic resistance in bacterial species. The resistance mechanism genes are located in plasmids that can be efficiently transferred from one strain to another. In addition to the development of drug-resistant pathogens, the use of tetracycline has been associated with problems such as unacceptable levels of drug residues in food products for human consumption and release of drugs into the environment. Control of usage in animal farming is possible by monitoring antibiotic residues in different biological samples. Conventional methods for the detection of tetracycline residues include microbial inhibition tests, immunoassays, and chromatographic methods. Recently, whole-cell sensing systems have been proposed as a sensitive, simple, fast, and inexpensive method for measuring tetracyclines in biological fluids and food samples [111].

A bioluminescent bacterial strain containing the bacterial luciferase operon *luxCDABE* under the control of the tetracycline-responsive element, which is part of the regulation unit of the tetracycline resistance factor, has been developed.

It allows for the specific detection of different clinically relevant tetracyclines, with limits of detection at picomole levels and an induction time of 90 min [112]. Moreover, freeze-dried cells can detect tetracyclines as sensitively as freshly cultivated cells, thus envisioning the possibility of on-site use. The developed whole-cell sensing system has been tested on different biological samples. It has proved to be suitable to screen veterinary serum samples for tetracycline residues in real time [113] and to detect tetracyclines in raw bovine milk samples below the official limits set by the European Union and the U.S. Food and Drug Administration [114]. The biosensing strain was also applied to the detection of traces of tetracyclines in fish, after simple sample preparation. In this case, tetracycline residues were detected below official limits, and the results correlated with those obtained by conventional HPLC [115].

In another study, a GFP-based whole-cell sensor system for *in situ* detection of tetracyclines was developed. The biosensor was used for qualitative detection of oxytetracycline production by the bacterium *Streptomyces rimosus* in soil microcosms [116]. The tetracycline-induced GFP-producing biosensor cells were detected by using flow cytometry and fluorescence-activated cell-sorting (FACS) analysis. This study showed the biosensor potential for microbial ecology studies. The same sensor strain was employed in a separate study for *in vivo* detection and quantification of tetracyclines in rat intestine [117]. Bioavailable tetracycline concentration within the bacterial growth habitat of the intestine proved to be proportional to the intake concentration, but significantly lower. This finding, made possible by the use of the whole-cell sensing system, may help to clarify and optimize antimicrobial therapy in the intestinal environment.

8.7
Technological Advances

An emerging trend of research in luminescent whole-cell biosensors is moving toward integration of the bacterial reporter strain into the appropriate detection device. This goal is achieved mainly by immobilizing the cells onto optical fiber tips. Different immobilization methods that utilize membranes or alginate, solid agar, and sol-gel matrices have been developed and optimized. The obtained fiber-optic, whole-cell sensors have been proposed as self-contained, disposable biomonitoring devices for environmental analysis [59, 118–121]. When coupled to a portable photon-counting system, they allow performance of measurements in the field. Additionally, the use of multiple fibers enables multiple detection [122]. An interesting application of optical fiber led to the development of a high-density cell assay platform [123]. Here, fluorescent reporter cells were placed into ordered microwell arrays that were fabricated on an optical imaging fiber. Fluorescence signals from individual cells were detected and spatially located using an imaging detector. The main advantage of this cell array technology is the simultaneous analysis of multiple responses from a large number of different strains, thus showing potential for high-throughput screening applications.

A different approach to integration is the implementation of chip-based systems. An example of a whole-cell biochip is the bioluminescent bioreporter integrated circuit (BBIC), in which recombinant bioreporter cells are interfaced with an integrated circuit. In this application, the cells are either immobilized or encapsulated to integrate them into the circuit. Alternatively, a flow system is employed to bring the cell bioreporters in proximity with the photodiode-containing chip of a microluminometer optimized for very low-level luminescence detection [124, 125].

Further developments in whole-cell sensing aim toward system miniaturization, with consequent implementation of high-density assay platforms, including 96- and 384-well microtiter plates, microarrays, and microfluidics devices. The main advantage of these analytical formats is that they allow for high-throughput screening of samples. Not only can large numbers of samples be analyzed, but also large amounts of information from each sample can be assessed simultaneously and in a relatively short time. Additionally, miniaturized systems allow decreased reagent consumption, sample volume, and analysis time.

Several whole cell-based assay systems that use multi-well platforms have been developed. Reporter cells can be suspended in culture media within the wells [126], immobilized in agar matrices at the bottom of single wells [127], or encapsulated in sol-gel matrices that form thick silicate films [128]. The main advantage of immobilized cells is that, while maintaining the analytical characteristics of free cells, they can be used under continuous flow conditions for on-line monitoring and as multiple-use sensing elements. In that regard, the optimization of deposition and immobilization techniques can contribute significantly to the advancement of whole-cell sensing technology [129]. An example of whole-cell microarray is the Lux-Array, which includes hundreds of *E. coli* reporter strains, high-density printed to membranes on agar plates [130]. The system was developed for genome-wide transcriptional analysis by fusing a collection of *E. coli* functional promoters to a reporter gene. However, it shows the potential for assembling an array of reporter-sensing strains to be used for analytical purposes. Recently, we took advantage of microfluidics to develop a miniaturized whole cell-based sensing system [71]. A recombinant reporter strain for the detection of arsenite/antimonite was adapted to a microcentrifugal microfluidics platform in the shape of a compact disk. Centrifugal force controlled valving, reagent flow, and mixing and drove fluids through microfabricated channels and reservoirs. The force was generated by a motor that spun the disk at programmed velocities. Fluorescence from the reporter was detected using a fiber-optic probe. Despite miniaturization, the biosensing system retained its analytical performance and was able to perform quantitative analysis within short response times. Moreover, the developed system has the potential for portability and applicability in the field.

Acknowledgments

This work was supported in part by the National Institute of Environmental Health Sciences (Grant P42 ES 007380) and the National Science Foundation (Grant CHE-0416553).

References

1 Naylor, L. H. Reporter gene technology: the future looks bright. *Biochem. Pharmacol.* **1999**, *58*, 749–757.

2 Wood, K. V. Marker proteins for gene expression. *Curr. Opin. Biotechnol.* **1995**, *6*, 50–58.

3 Jansson, J. K. Marker and reporter genes: illuminating tools for environmental microbiologists. *Curr. Opin. Microbiol.* **2003**, *6*, 310–316.

4 Bousse, L. Whole cell biosensors. *Sens. Actuators B Chem.* **1996**, *34*, 270–275.

5 Pancrazio, J. J., Whelan, J. P., Borkholder, D. A., Ma, W., Stenger, D. A. Development and application of cell-based biosensors. *Ann. Biomed. Eng.* **1999**, *27*, 697–711.

6 *Biosensors Fundamentals and Applications* (Turner, A. P. F., Karube, I., Wilson, G. S., Eds.), Oxford University Press: Oxford, UK, **1987**.

7 *Biosensor Principles and Applications* (Blum, L. J., Coulet, P. R., Eds.), Marcel Dekker: New York, **1991**.

8 D'Souza, S. F. Microbial biosensors. *Biosens. Bioelectron.* **2001**, *16*, 337–353.

9 van der Meer, J. R., Tropel, D., Jaspers, M. Illuminating the detection chain of bacterial bioreporters. *Environ. Microbiol.* **2004**, *6*, 1005–1020.

10 Daunert, S., Barrett, G., Feliciano, J. S., Shetty, R. S., Shrestha, S., Smith-Spencer, W. Genetically engineered whole-cell sensing systems: coupling biological recognition with reporter genes. *Chem. Rev.* **2000**, *100*, 2705–2738.

11 Durick, K., Negulescu, P. Cellular biosensors for drug discovery. *Biosens. Bioelectron.* **2001**, *16*, 587–592.

12 Windal, I., Denison, M. S., Birnbaum, L. S., Van Wouwe, N., Baeyens, W., Goeyens, L. Chemically activated luciferase gene expression (CALUX) cell bioassay analysis for the estimation of dioxin-like activity: critical parameters of the CALUX procedure that impact assay results. *Environ. Sci. Technol.* **2005**, *39*, 7357–7364.

13 Rogers, J. M., Denison, M. S. Recombinant cell bioassays for endocrine disruptors: development of a stably transfected human ovarian cell line for the detection of estrogenic and anti-estrogenic chemicals. *In Vitr. Mol. Toxicol.* **2000**, *13*, 67–82.

14 Baldwin, T. O., Christopher, J. A., Raushel, F. M., Sinclair, J. F., Ziegler, M. M., Fisher, A. J., Rayment, I. Structure of bacterial luciferase. *Curr. Opin. Struct. Biol.* **1995**, *5*, 798–809.

15 Meighen, E. A. Molecular biology of bacterial bioluminescence. *Microbiol. Rev.* **1991**, *55*, 123–142.

16 Rutter, G. A., Kennedy, H. J., Wood, C. D., White, M. R. H., Tavare, J. M. Real-time imaging of gene expression in single living cells. *Chem. Biol.* **1998**, *5*, R285–R290.

17 Meighen, E. Bioluminescence, Bacterial. *Encyclo. Microbiol.* **1992**, *1*, 309–319.

18 Billard, P., DuBow, M. S. Bioluminescence-based assays for detection and characterization of bacteria and chemicals in clinical laboratories. *Clin. Biochem.* **1998**, *31*, 1–14.

19 Bulich, A. A practical and reliable method for monitoring the toxicity of aquatic samples. *Process Biochem.* **1982**, *17*, 45–47.

20 Hermens, J., Busser, F., Leeuwangh, P., Musch, A. Quantitative structure-activity relationships and mixture toxicity of organic chemicals in *Photobacterium phosphoreum*: the Microtox test. *Ecotoxicol. Environ. Saf.* **1985**, *9*, 17–25.

21 Gupta, R. K., Patterson, S. S., Ripp, S., Simpson, M. L., Sayler, G. S. Expression of the *Photorhabdus luminescens* lux genes (luxA, B, C, D, and E) in *Saccharomyces cerevisiae*. *FEMS Yeast Res.* **2003**, *4*, 305–313.

22 Kelly, C. J., Tumsaroj, N., Lajoie, C. A. Assessing wastewater metal toxicity with bacterial bioluminescence in a bench-scale wastewater treatment system. *Water Res.* **2004**, *38*, 423–431.

23 Dewet, J. R., Wood, K. V., Deluca, M., Helinski, D. R., Subramani, S. Firefly luciferase gene – structure and

expression in mammalian cells. *Mol. Cell Biol.* **1987**, *7*, 725–737.

24 Dewet, J. R., Wood, K. V., Helinski, D. R., DeLuca, M. Cloning of firefly luciferase cDNA and the expression of active luciferase in *Escherichia coli*. *Proc. Natl. Acad. Sci. USA* **1985**, *82*, 7870–7873.

25 Craig, F. F., Simmonds, A. C., Watmore, D., McCapra, F., White, M. R. Membrane-permeable luciferin esters for assay of firefly luciferase in live intact cells. *Biochem. J.* **1991**, *276 (Pt 3)*, 637–641.

26 Wood, K. V., Lam, Y. A., Seliger, H. H., McElroy, W. D. Complementary DNA coding click beetle luciferases can elicit bioluminescence of different colors. *Science* **1989**, *244*, 700–702.

27 Lorenz, W. W., McCann, R. O., Longiaru, M., Cormier, M. J. Isolation and expression of a cDNA encoding *Renilla reniformis* luciferase. *Proc. Natl. Acad. Sci. U S A* **1991**, *88*, 4438–4442.

28 Stables, J., Scott, S., Brown, S., Roelant, C., Burns, D., Lee, M. G., Rees, S. Development of a dual glow-signal firefly and *Renilla* luciferase assay reagent for the analysis of g-protein coupled receptor signalling. *J. Recept. Signal Transduct. Res.* **1999**, *19*, 395–410.

29 Inouye, S., Noguchi, M., Sakaki, Y., Takagi, Y., Miyata, T., Iwanaga, S., Tsuji, F. I. Cloning and sequence analysis of cDNA for the luminescent protein aequorin. *Proc. Natl. Acad. Sci. USA* **1985**, *82*, 3154–3158.

30 Ohmiya, Y., Hirano, T. Shining the light: the mechanism of the bioluminescence reaction of calcium-binding photoproteins. *Chem. Biol.* **1996**, *3*, 337–347.

31 Shimomura, O., Johnson, F. H., Saiga, Y. Extraction, purification and properties of aequorin, a bioluminescent protein from the luminous hydromedusan, *Aequorea*. *J. Cell Comp. Physiol.* **1962**, *59*, 223–239.

32 Rider, T., Petrovick, M., Young, A., Smith, L., Mathews, R., Palmacci, S., Hollis, M., Chen, J. NASA/NCI Biomedical Imaging Symposium, Bethesda, MD, **1999**.

33 Rider, T. H., Petrovick, M. S., Nargi, F. E., Harper, J. D., Schwoebel, E. D., Mathews, R. H., Blanchard, D. J., Bortolin, L. T., Young, A. M., Chen, J. Z., Hollis, M. A. A B cell-based sensor for rapid identification of pathogens. *Science* **2003**, *301*, 213–215.

34 Phillips, G. N. Jr. Structure and dynamics of green fluorescent protein. *Curr. Opin. Struct. Biol.* **1997**, *7*, 821–827.

35 Welsh, S., Kay, S. A. Reporter gene expression for monitoring gene transfer. *Curr. Opin. Biotechnol.* **1997**, *8*, 617–622.

36 Hansen, L. H., Sørensen, S. J. The use of whole-cell biosensors to detect and quantify compounds or conditions affecting biological systems. *Microb. Ecol.* **2001**, *42*, 483–494.

37 Lewis, J. C., Daunert, S. Photoproteins as luminescent labels in binding assays. *Fresenius J. Anal. Chem.* **2000**, *366*, 760–768.

38 March, J. C., Rao, G., Bentley, W. E. Biotechnological applications of green fluorescent protein. *Appl. Microbiol. Biotechnol.* **2003**, *62*, 303–315.

39 Matz, M. V., Fradkov, A. F., Labas, Y. A., Savitsky, A. P., Zaraisky, A. G., Markelov, M. L., Lukyanov, S. A. Fluorescent proteins from nonbioluminescent Anthozoa species. *Nat Biotechnol.* **1999**, *17*, 969–973.

40 Davey, H. M., Kell, D. B. Flow cytometry and cell sorting of heterogeneous microbial populations: the importance of single-cell analyses. *Microbiol. Rev.* **1996**, *60*, 641–696.

41 Gross, L. A., Baird, G. S., Hoffman, R. C., Baldridge, K. K., Tsien, R. Y. The structure of the chromophore within DsRed, a red fluorescent protein from coral. *Proc. Natl. Acad. Sci. USA* **2000**, *97*, 11990–11995.

42 Hakkila, K., Maksimow, M., Karp, M., Virta, M. Reporter genes *lucFF*, *luxCDABE*, *gfp*, and *dsred* have different characteristics in whole-cell bacterial sensors. *Anal. Biochem.* **2002**, *301*, 235–242.

43 Mitchell, R. J., Gu, M. B. Construction and Characterization of Novel Dual Stress-Responsive Bacterial Biosensors. *Biosens. Bioelectron.* **2004**, *19*, 977–985.

44 Molina, A., Carpeaux, R., Martial, J. A., Muller, M. A trans-

formed fish cell line expressing a green fluorescent protein-luciferase fusion gene responding to cellular stress. *Toxicol. In Vitro* **2002**, *16*, 201–207.
45 Sagi, E., Hever, N., Rosen, R., Bartolome, A. J., Premkumar, J. R., Ulber, R., Lev, O., Scheper, T., Belkin, S. Fluorescence and bioluminescence reporter functions in genetically modified bacterial sensor strains. *Sens. Actuators B Chem.* **2003**, *90*, 2–8.
46 Baumstark-Khan, C., Khan, R. A., Rettberg, P., Horneck, G. Bacterial Lux-Fluoro test for biological assessment of pollutants in water samples from urban and rural origin. *Anal. Chim. Acta* **2003**, *487*, 51–60.
47 Wood, K. V. The chemical mechanism and evolutionary development of beetle bioluminescence. *Photochem. Photobiol.* **1995**, *62*, 662–673.
48 Shrestha, S., Shetty, R. S., Ramanathan, S., Daunert, S. Simultaneous Detection of Analytes Based on Genetically Engineered Whole Cell Sensing Systems. *Anal. Chim. Acta.* **2001**, *444*, 251–260.
49 Mirasoli, M., Feliciano, J., Michelini, E., Daunert, S., Roda, A. Internal response correction for fluorescent whole-cell biosensors. *Anal. Chem.* **2002**, *74*, 5948–5953.
50 Pedahzur, R., Polyak, B., Marks, R. S., Belkin, S. Water toxicity detection by a panel of stress-responsive luminescent bacteria. *J. Appl. Toxicol.* **2004**, *24*, 343–348.
51 Gu, M. B., Gil, G. C., Kim, J. H. Enhancing the sensitivity of a two-stage continuous toxicity monitoring system through the manipulation of the dilution rate. *J. Biotechnol.* **2002**, *93*, 283–288.
52 Premkumar, J. R., Lev, O., Marks, R. S., Polyak, B., Rosen, R., Belkin, S. Antibody-based immobilization of bioluminescent bacterial sensor cells. *Talanta* **2001**, *55*, 1029–1038.
53 Kim, B. C., Gu, M. B. A bioluminescent sensor for high throughput toxicity classification. *Biosens. Bioelectron.* **2003**, *18*, 1015–1021.
54 Choi, S. H., Gu, M. B. A portable toxicity biosensor using freeze-dried recombinant bioluminescent bacteria. *Biosens. Bioelectron.* **2002**, *17*, 433–440.
55 Gu, M. B., Gil, G. C. A multi-channel continuous toxicity monitoring system using recombinant bioluminescent bacteria for classification of toxicity. *Biosens. Bioelectron.* **2001**, *16*, 661–666.
56 Choi, S. H., Gu, M. B. Toxicity biomonitoring of degradation byproducts using freeze-dried recombinant bioluminescent bacteria. *Anal. Chim. Acta* **2003**, *481*, 229–238.
57 Kim, B. C., Park, K. S., Kim, S. D., Gu, M. B. Evaluation of a high throughput toxicity biosensor and comparison with a *Daphnia magna* bioassay. *Biosens. Bioelectron.* **2003**, *18*, 821–826.
58 Gu, M. B., Min, J., LaRossa, R. A. Bacterial bioluminescent emission from recombinant *Escherichia coli* harboring a *reca::luxcdabe* fusion. *J. Biochem. Biophys. Methods* **2000**, *45*, 45–56.
59 Polyak, B., Bassis, E., Novodvorets, A., Belkin, S., Marks, R. S. Bioluminescent whole cell optical fiber sensor to genotoxicants: system optimization. *Sens. Actuators B Chem.* **2001**, *74*, 18–26.
60 Bartolome, A. J., Ulber, R., Scheper, T., Sagi, E., Belkin, S. Genotoxicity monitoring using a 2D-spectroscopic GFP whole cell biosensing system. *Sens. Actuators B Chem.* **2003**, *89*, 27–32.
61 Silver, S. Genes for all metals – a bacterial view of the Periodic Table. The 1996 Thom Award Lecture. *J. Ind. Microbiol. Biotechnol.* **1998**, *20*, 1–12.
62 Rosen, B. P. Transport and detoxification systems for transition metals, heavy metals and metalloids in eukaryotic and prokaryotic microbes. *Comp. Biochem. Physiol. A Mol. Integr. Physiol.* **2002**, *133*, 689–693.
63 Parales, R. E., Haddock, J. D. Biocatalytic degradation of pollutants. *Curr. Opin. Biotechnol.* **2004**, *15*, 374–379.
64 Burlage, R. S., Sayler, G. S., Larimer, F. Monitoring of naphthalene catabolism by bioluminescence with nah-lux transcriptional fusions. *J. Bacteriol.* **1990**, *172*, 4749–4757.
65 Belkin, S. Microbial whole-cell sensing systems of environmental pollutants. *Curr. Opin. Microbiol.* **2003**, *6*, 206–212.

66 Gu, M. B., Mitchell, R. J., Kim, B. C. Whole-cell-based biosensors for environmental biomonitoring and application. *Adv. Biochem. Engin./Biotechnol.* **2004**, *87*, 269–305.

67 Ramanathan, S., Shi, W., Rosen, B. P., Daunert, S. Sensing antimonite and arsenite at the subattomole level with genetically engineered bioluminescent bacteria. *Anal. Chem.* **1997**, *69*, 3380–3384.

68 Tauriainen, S., Karp, M., Chang, W., Virta, M. Recombinant luminescent bacteria for measuring bioavailable arsenite and antimonite. *Appl. Environ. Microbiol.* **1997**, *63*, 4456–4461.

69 Tauriainen, S., Virta, M., Chang, W., Karp, M. Measurement of firefly luciferase reporter gene activity from cells and lysates using *Escherichia coli* arsenite and mercury sensors. *Anal. Biochem.* **1999**, *272*, 191–198.

70 Roberto, F. F., Barnes, J. M., Bruhn, D. F. Evaluation of a gfp reporter gene construct for environmental arsenic detection. *Talanta* **2002**, *58*, 181–188.

71 Rothert, A., Deo, S. K., Millner, L., Puckett, L. G., Madou, M. J., Daunert, S. Whole-cell-reporter-gene-based biosensing systems on a compact disk microfluidics platform. *Anal. Biochem.* **2005**, *342*, 11–19.

72 Tauriainen, S., Karp, M., Chang, W., Virta, M. Luminescent bacterial sensor for cadmium and lead. *Biosens. Bioelectron.* **1998**, *13*, 931–938.

73 Riether, K. B., Dollard, M. A., Billard, P. Assessment of heavy metal bioavailability using *Escherichia coli zntAp::lux* and *copAp::lux*-based biosensors. *Appl. Microbiol. Biotechnol.* **2001**, *57*, 712–716.

74 Shetty, R. S., Deo, S. K., Shah, P., Sun, Y., Rosen, B. P., Daunert, S. Luminescence-based whole-cell-sensing systems for cadmium and lead using genetically engineered bacteria. *Anal. Bioanal. Chem.* **2003**, *376*, 11–17.

75 Peitzsch, N., Eberz, G., Nies, D. H. *Alcaligenes eutrophus* as a bacterial chromate sensor. *Appl. Environ. Microbiol.* **1998**, *64*, 453–458.

76 Vulkan, R., Zhao, F. J., Barbosa-Jefferson, V., Preston, S., Paton, G. I., Tipping, E., McGrath, S. P. Copper speciation and impacts on bacterial biosensors in the pore water of copper-contaminated soils. *Environ. Sci. Technol.* **2000**, *34*, 5115–5121.

77 Stoyanov, J. V., Magnani, D., Solioz, M. Measurement of cytoplasmic copper, silver, and gold with a lux biosensor shows copper and silver, but not gold, efflux by the CopA ATPase of *Escherichia coli*. *FEBS Lett.* **2003**, *546*, 391–394.

78 Joyner, D. C., Lindow, S. E. Heterogeneity of iron bioavailability on plants assessed with a whole-cell GFP-based bacterial biosensor. *Microbiology* **2000**, *146 (Pt 10)*, 2435–2445.

79 Selifonova, O., Burlage, R., Barkay, T. Bioluminescent sensors for detection of bioavailable Hg(II) in the environment. *Appl. Environ. Microbiol.* **1993**, *59*, 3083–3090.

80 Virta, M., Lampinen, J., Karp, M. A luminescence-based mercury biosensor. *Anal. Chem.* **1995**, *67*, 667–669.

81 Lyngberg, O. K., Stemke, D. J., Schottel, J. L., Flickinger, M. C. A single-use luciferase-based mercury biosensor using *Escherichia coli* HB101 immobilized in a latex copolymer film. *J. Ind. Microbiol. Biotechnol.* **1999**, *23*, 668–676.

82 Hansen, L. H., Sørensen, S. J. Versatile biosensor vectors for detection and quantification of mercury. *FEMS Microbiol. Lett.* **2000**, *193*, 123–127.

83 Rasmussen, L. D., Sorensen, S. J., Turner, R. R., Barkay, T. Application of a mer-lux biosensor for estimating bioavailable mercury in soil. *Soil Biol. Biochem.* **2000**, *32*, 639–646.

84 Ivask, A., Hakkila, K., Virta, M. Detection of organomercurials with sensor bacteria. *Anal. Chem.* **2001**, *73*, 5168–5171.

85 Tibazarwa, C., Corbisier, P., Mench, M., Bossus, A., Solda, P., Mergeay, M., Wyns, L., van der Lelie, D. A Microbial biosensor to predict bioavailable nickel in soil and its transfer to plants. *Environ. Pollut.* **2001**, *113*, 19–26.

86 Sticher, P., Jaspers, M. C. M., Stemmler, K., Harms, H., Zehnder, A. J. B., van der Meer, J. R. Development and characterization of a whole-cell bioluminescent sensor for bioavailable middle-chain alkanes in contaminated ground water samples. *Appl. Environ. Microbiol.* **1997**, *63*, 4053–4060.

87 Phoenix, P., Keane, A., Patel, A., Bergeron, H., Ghoshal, S., Lau, P. C. K. Characterization of a new solvent-responsive gene locus in *Pseudomonas putida* F1 and its functionalization as a versatile biosensor. *Environ. Microbiol.* **2003**, *5*, 1309–1327.

88 Kobatake, E., Niimi, T., Haruyama, T., Ikariyama, Y., Aizawa, M. Biosensing of benzene derivatives in the environment by luminescent *Escherichia coli*. *Biosens. Bioelectron.* **1995**, *10*, 601–605.

89 Applegate, B. M., Kehrmeyer, S. R., Sayler, G. S. A chromosomally based *tod-luxCDABE* whole-cell reporter for benzene, toluene, ethybenzene, and xylene (BTEX) sensing. *Appl. Environ. Microbiol.* **1998**, *64*, 2730–2735.

90 Stiner, L., Halverson, L. J. Development and characterization of a green fluorescent protein-based bacterial biosensor for bioavailable toluene and related compounds. *Appl. Environ. Microbiol.* **2002**, *68*, 1962–1971.

91 Heitzer, A., Malachowsky, K., Thonnard, J. E., Bienkowski, P. R., White, D. C., Sayler, G. S. Optical biosensor for environmental on-line monitoring of naphthalene and salicylate bioavailability with an immobilized bioluminescent catabolic reporter bacterium. *Appl. Environ. Microbiol.* **1994**, *60*, 1487–1494.

92 Abd-El-Haleem, D., Ripp, S., Scott, C., Sayler, G. S. A *luxCDABE*-based bioluminescent bioreporter for the detection of phenol. *J. Ind. Microbiol. Biotechnol.* **2002**, *29*, 233–237.

93 Layton, A. C., Muccini, M., Ghosh, M. M., Sayler, G. S. Construction of a bioluminescent reporter strain to detect polychlorinated biphenyls. *Appl. Environ. Microbiol.* **1998**, *64*, 5023–5026.

94 Mbeunkui, F., Richaud, C., Etienne, A. L., Schmid, R. D., Bachmann, T. T. Bioavailable nitrate detection in water by an immobilized luminescent cyanobacterial reporter strain. *Appl. Microbiol. Biotechnol.* **2002**, *60*, 306–312.

95 Taylor, C. J., Bain, L. A., Richardson, D. J., Spiro, S., Russell, D. A. Construction of a whole-cell gene reporter for the fluorescent bioassay of nitrate. *Anal. Biochem.* **2004**, *328*, 60–66.

96 Dollard, M. A., Billard, P. Whole-cell bacterial sensors for the monitoring of phosphate bioavailability. *J. Microbiol. Methods* **2003**, *55*, 221–229.

97 Shetty, R. S., Ramanathan, S., Badr, I. H. A., Wolford, J. L., Daunert, S. Green fluorescent protein in the design of a living biosensing system for L-arabinose. *Anal. Chem.* **1999**, *71*, 763–768.

98 Miller, W. G., Brandl, M. T., Quiñones, B., Lindow, S. E. Biological sensor for sucrose availability: relative sensitivities of various reporter genes. *Appl. Environ. Microbiol.* **2001**, *67*, 1308–1317.

99 Miller, M. B., Bassler, B. L. Quorum sensing in bacteria. *Annu. Rev. Microbiol.* **2001**, *55*, 165–199.

100 Fuqua, C., Parsek, M. R., Greenberg, E. P. Regulation of gene expression by cell-to-cell communication: acyl-homoserine lactone quorum sensing. *Annu. Rev. Genet.* **2001**, *35*, 439–468.

101 Winson, M. K., Swift, S., Fish, L., Throup, J. P., Jørgensen, F., Chhabra, S. R., Bycroft, B. W., Williams, P., Stewart, G. S. A. B. Construction and analysis of *luxCDABE*-based plasmid sensors for investigating *N*-acyl homoserine lactone-mediated quorum sensing. *FEMS Microbiol. Lett.* **1998**, *163*, 185–192.

102 Middleton, B., Rodgers, H. C., Cámara, M., Knox, A. J., Williams, P., Hardman, A. Direct detection of *N*-acylhomoserine lactones in cystic fibrosis sputum. *FEMS Microbiol. Lett.* **2002**, *207*, 1–7.

103 Burgess, N. A., Kirke, D. F., Williams, P., Winzer, K.,

Hardie, K. R., Meyers, N. L., Aduse-Opoku, J., Curtis, M. A., Cámara, M. LuxS-dependent quorum sensing in *Porphyromonas gingivalis* modulates protease and haemagglutinin activities but is not essential for virulence. *Microbiology* **2002**, *148*, 763–772.

104 Degrassi, G., Aguilar, C., Bosco, M., Zahariev, S., Pongor, S., Venturi, V. Plant growth-promoting *Pseudomonas putida* WCS358 produces and secretes four cyclic dipeptides: cross-talk with quorum sensing bacterial sensors. *Curr. Microbiol.* **2002**, *45*, 250–254.

105 Hentzerm, M., Givskov, M. Pharmacological inhibition of quorum sensing for the treatment of chronic bacterial infections. *J. Clin. Invest.* **2003**, *112*, 1300–1307.

106 Morin, D., Grasland, B., Vallée-Réhel, K., Dufau, C., Haras, D. On-line high-performance liquid chromatography-mass spectrometric detection and quantification of *N*-acylhomoserine lactones, quorum sensing signal molecules, in the presence of biological matrices. *J. Chromatogr. A* **2003**, *1002*, 79–92.

107 Andersen, J. B., Heydorn, A., Hentzer, M., Eberl, L., Geisenberger, O., Christensen, B. B., Molin, S., Givskov, M. *gfp*-Based *N*-acyl homoserine-lactone sensor systems for detection of bacterial communication. *Appl. Environ. Microbiol.* **2001**, *67*, 575–585.

108 Wu, H., Song, Z., Hentzer, M., Andersen, J. B., Heydorn, A., Mathee, K., Moser, C., Eberl, L., Molin, S., Høiby, N., Givskov, M. Detection of *N*-acylhomoserine lactones in lung tissues of mice infected with *Pseudomonas aeruginosa*. *Microbiology* **2000**, *146*, 2481–2493.

109 Hentzer, M., Riedel, K., Rasmussen, T. B., Heydorn, A., Andersen, J. B., Parsek, M. R., Rice, S. A., Eberl, L., Molin, S., Høiby, N., Kjelleberg, S., Givskov, M. Inhibition of quorum sensing in *Pseudomonas aeruginosa* biofilm bacteria by a halogenated furanone compound. *Microbiology* **2002**, *148*, 87–102.

110 Burmølle, M., Hansen, L. H., Oregaard, G., Sørensen, S. J. Presence of *N*-acyl homoserine lactones in soil detected by a whole-cell biosensor and flow cytometry. *Microb. Ecol.* **2003**, *45*, 226–236.

111 Hansen, L. H., Sørensen, S. J. Detection and quantification of tetracyclines by whole cell biosensors. *FEMS Microbiol. Lett.* **2000**, *190*, 273–278.

112 Korpela, M. T., Kurittu, J. S., Karvinen, J. T., Karp, M. T. A recombinant *Escherichia coli* sensor strain for the detection of tetracyclines. *Anal. Chem.* **1998**, *70*, 4457–4462.

113 Kurittu, J., Karp, M., Korpela, M. Detection of tetracyclines with luminescent bacterial strains. *Luminescence* **2000**, *15*, 291–297.

114 Kurittu, J., Lönnberg, S., Virta, M., Karp, M. A group-specific microbiological test for the detection of tetracycline residues in raw milk. *J. Agric. Food Chem.* **2000**, *48*, 3372–3377.

115 Pellinen, T., Bylund, G., Virta, M., Niemi, A., Karp, M. Detection of traces of tetracyclines from fish with a bioluminescent sensor strain incorporating bacterial luciferase reporter genes. *J. Agric. Food Chem.* **2002**, *50*, 4812–4815.

116 Hansen, L. H., Ferrari, B., Sørensen, A. H., Veal, D., Sørensen, S. J. Detection of oxytetracycline production by *Streptomyces rimosus* in soil microcosms by combining whole-cell biosensors and flow cytometry. *Appl. Environ. Microbiol.* **2001**, *67*, 239–244.

117 Bahl, M. I., Hansen, L. H., Licht, T. R., Sørensen, S. J. In vivo detection and quantification of tetracycline by use of a whole-cell biosensor in the rat intestine. *Antimicrob. Agents Chemother.* **2004**, *48*, 1112–1117.

118 Ikariyama, Y., Nishiguchi, S., Koyama, T., Kobatake, E., Aizawa, M., Tsuda, M., Nakazawa, T. Fiber-optic-based biomonitoring of benzene derivatives by recombinant *E. coli* bearing luciferase gene-fused TOL-plasmid immobilized on the fiber-optic end. *Anal. Chem.* **1997**, *69*, 2600–2605.

119 Polyak, B., Bassis, E., Novodvorets, A., Belkin, S., Marks, R. S. Optical fiber bioluminescent whole-cell microbial biosensors to genotoxicants. *Water Sci. Technol.* **2000**, *42*, 305–311.

120 Gil, G. C., Kim, Y. J., Gu, M. B. Enhancement in the sensitivity of a gas biosensor by using an advanced immobilization of a recombinant bioluminescent bacterium. *Biosens. Bioelectron.* **2002**, *17*, 427–432.

121 Sun, Y., Zhou, T., Guo, J., Li, Y. Dark variants of luminous bacteria whole cell bioluminescent optical fiber sensor to genotoxicants. *J. Huazhong Univ. Sci. Technolog. Med. Sci.* **2004**, *24*, 507–509.

122 Hakkila, K., Green, T., Leskinen, P., Ivask, A., Marks, R., Virta, M. Detection of bioavailable heavy metals in EILATox-Oregon samples using whole-cell luminescent bacterial sensors in suspension or immobilized onto fibre-optic tips. *J. Appl. Toxicol.* **2004**, *24*, 333–342.

123 Biran, I., Walt, D. R. Optical imaging fiber-based single live cell arrays: a high-density cell assay platform. *Anal. Chem.* **2002**, *74*, 3046–3054.

124 Simpson, M. L., Sayler, G. S., Patterson, G., Nivens, D. E., Bolton, E. K., Rochelle, J. M., Arnott, J. C., Applegate, B. M., Ripp, S., Guillorn, M. A. An integrated CMOS microluminometer for low-level luminescence sensing in the bioluminescent bioreporter integrated circuit. *Sens. Actuators B Chem.* **2001**, *72*, 134–140.

125 Bolton, E. K., Sayler, G. S., Nivens, D. E., Rochelle, J. M., Ripp, S., Simpson, M. L. Integrated CMOS photodetectors and signal processing for very low-level chemical sensing with the bioluminescent bioreporter integrated circuit. *Sens. Actuators B Chem.* **2002**, *85*, 179–185.

126 Roda, A., Pasini, P., Mirasoli, M., Guardigli, M., Russo, C., Musiani, M., Baraldini, M. Sensitive determination of urinary mercury(II) by a bioluminescent transgenic bacteria-based biosensor. *Anal. Letters* **2001**, *34*, 29–41.

127 Kim, B. C., Gu, M. B. A bioluminescent sensor for high throughput toxicity classification. *Biosens. Bioelectron.* **2003**, *18*, 1015–1021.

128 Premkumar, J. R., Rosen, R., Belkin, S., Lev, O. Sol-gel luminescence biosensors: encapsulation of recombinant *E. coli* reporters in thick silicate films. *Anal. Chim. Acta* **2002**, *462*, 11–23.

129 Barron, J. A., Rosen, R., Jones-Meehan, J., Spargo, B. J., Belkin, S., Ringeisen, B. R. Biological laser printing of genetically modified *Escherichia coli* for biosensor applications. *Biosens. Bioelectron.* **2004**, *20*, 246–252.

130 Van Dyk, T. K., DeRose, E. J., Gonye, G. E. LuxArray, a high-density, genomewide transcription analysis of *Escherichia coli* using bioluminescent reporter strains. *J. Bacteriol.* **2001**, *183*, 5496–5505.

9
Luminescent Proteins in Binding Assays

Aldo Roda, Massimo Guardigli, Elisa Michelini, Mara Mirasoli, and Patrizia Pasini

9.1
Introduction

Unraveling the mechanisms of natural phenomena occurring in various marine and terrestrial organisms and resulting in light production has provided exciting new tools in biosciences. The study of such mechanisms has moved from curiosity towards such fascinating phenomena to their effective application in analytical chemistry. Various bioluminescent or fluorescent proteins have been isolated from luminescent organisms, and the corresponding genes have been cloned. In addition, mutational studies have provided a number of luminescent proteins with improved spectral characteristics or emission at different colors.

Bioluminescent and fluorescent proteins allow for the development of ultra-sensitive binding assays, thanks to their high detectability and to the availability of highly sensitive light detection instruments. Bioluminescent proteins are referred to with the general term "luciferases". They are the key enzymes that catalyze bioluminescent reactions, which involve oxidation of a substrate generically known as luciferin. The reaction yields a singlet-excited intermediate that decays with photon emission. Because the enzyme active site provides a reaction microenvironment that is favorable to radiative decay of the excited intermediate product, the bioluminescent reaction is characterized by a very high quantum yield (0.88 for the reaction of firefly luciferase, which is the highest reported for a biological reaction).

Among luciferases, calcium-activated photoproteins, such as aequorin from *Aequorea victoria* or obelin from *Obelia longissima*, can be defined as proteins that emit light without turnover. Indeed, the product of the reaction is not released from the active site, where it remains tightly bound, preventing turnover of the reaction. Fluorescent proteins emit light as a result of photoexcitation. At present, a number of fluorescent proteins are available for applications in biosciences. Along with the green fluorescent protein (GFP), originally isolated from *A. victoria*, natural or mutated variants are available that display interesting features such as improved photophysical properties, emission at different colors, or photoactivation

Photoproteins in Bioanalysis. Edited by Sylvia Daunert and Sapna K. Deo

ISBN: 3-527-31016-9

Table 9.1 Emission properties of luminescent proteins.

	Origin	*Emission wavelength (nm)*
Fluorescent proteins		
GFP	*Aequorea victoria*	509 (absorption at 395 nm)
EB(lue)FP	GFP mutant	440 (absorption at 380 nm)
EG(reen)FP	GFP mutant	508 (absorption at 489 nm)
EC(yan)FP	GFP mutant	477 (absorption at 434 nm)
EY(ellow)FP	GFP mutant	527 (absorption at 514 nm)
RFP	*Discosoma striata*	583 (absorption at 558 nm)
Photoproteins		
Aequorin	*Aequorea victoria*	465 (Ca^{2+}-triggered emission)
Obelin	*Obelia longissima*	485 (Ca^{2+}-triggered emission)
Luciferases		
Bacterial luciferases	*Vibrio* spp., *Photorabdus* spp.	489
Firefly luciferase	*Photinus pyralis*	562
Renilla luciferase	*Renilla reniformis*	470

properties. Relevant properties of common luminescent proteins are gathered in Table 9.1.

Luminescent proteins have been employed as labels in different types of binding assays, including protein–protein and protein–ligand interaction assays, immunoassays, biotin–avidin-based assays, and nucleic acid hybridization assays. The principles of binding assays involving luminescent proteins are depicted in Fig. 9.1 and 9.2.

Besides their fluorescent or bioluminescent properties, a further reason for the increasing use of luminescent proteins in binding assays derives from the recent advancements in biotechnology. The availability of cDNA encoding these proteins and the possibility of expressing them in bacterial systems has allowed for their production in virtually unlimited amounts. Furthermore, fusion proteins obtained by genetic in-frame fusion of a luminescent protein to a target protein have enabled a number of revolutionary applications. Such bifunctional molecules have the interesting feature of conjugating luminescence with the target protein properties. Interestingly, the same approach can be used to obtain *in vivo* biotinylated luminescence proteins by fusion with a suitable biotin acceptor peptide. Gene fusion presents several advantages when compared with conventional chemical conjugation approaches. Indeed, it provides bifunctional conjugates uniform in their stoichiometry and in the position of luminescent protein attachment. In addition, such conjugates are readily available in virtually unlimited amounts after gene expression of the fusion gene, usually in bacteria, and protein purification. In addition, the luminescence activity of the label, which is often partially lost upon chemical conjugation, is usually retained upon protein fusion.

In the following, the most recent and innovative applications of luminescent proteins in binding assays will be described.

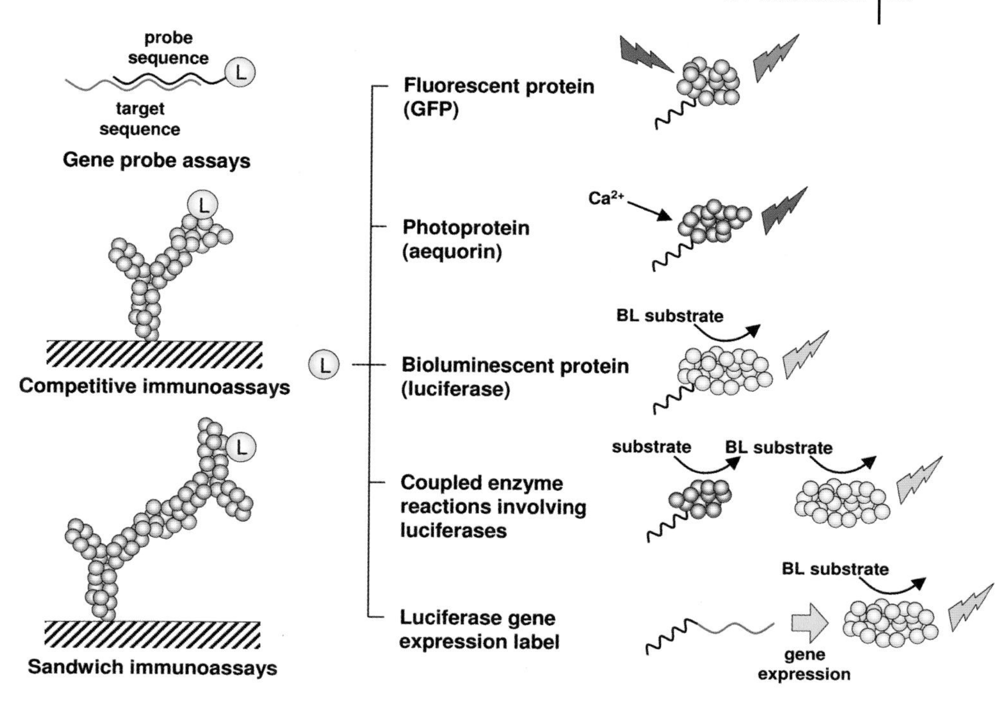

Fig. 9.1 Use of luminescent proteins as labels in binding assays.

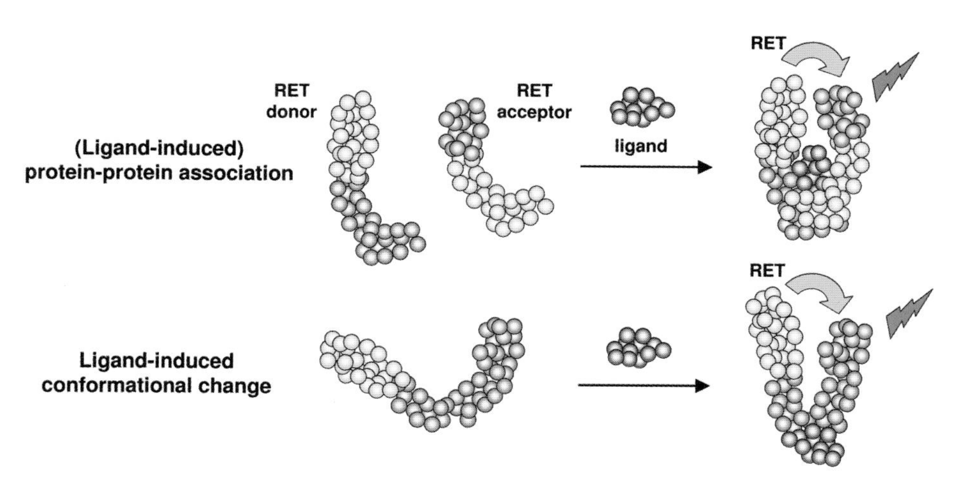

Fig. 9.2 Resonance energy transfer (RET)-based binding assays involving luminescent proteins. In FRET assays both the donor and the acceptor are fluorescent proteins, while in BRET assays the donor and the acceptor are bioluminescent and fluorescent proteins, respectively.

9.2
Protein–Protein and Protein–Ligand Interaction Assays

Luminescent proteins have been used extensively in the development of binding assays involving proteins, which have allowed us to obtain insights into many important biological processes.

9.2.1
FRET and BRET Techniques

Most of these assays involve resonance energy transfer (RET) techniques. RET is a short-range, nonradiative energy transfer between donor and acceptor molecules that takes place only if the two species are in close proximity (< 10 nm) to each other. RET systems are therefore "spectroscopic rulers" that are suitable for monitoring interactions between two partners, provided that each component of the donor–acceptor couple is linked to one partner. The same strategy can be used to detect protein conformation changes if donor and acceptor are fused at the target protein terminals.

In the case of fluorescence RET (FRET), both the donor and the acceptor are fluorescent molecules, whereas in bioluminescence RET (BRET) a bioluminescent molecule acts as the energy donor. FRET and, more recently, BRET are the methods of choice for imaging protein association inside living cells. In addition, RET processes can be evaluated using microtiter plate readers, thus allowing for the development of FRET- and BRET-based quantitative assays.

FRET methods can be broadly classified as intensity-based and decay kinetics-based methods. Intensity-based methods rely on the measurement of either the acceptor fluorescence or the acceptor:donor fluorescence intensity ratio. However, they are not suitable for detecting low-level protein–protein interactions because their sensitivity is reduced by several factors (e.g., sample autofluorescence and overlap between donor and acceptor emission). Such problems have been partially overcome by other FRET technologies, such as time-resolved FRET (TR-FRET), in which the donor is a lanthanide chelate with long-lived fluorescence. Decay kinetics-based FRET methods rely on the measurement of the kinetics of the light emission. In fluorescence lifetime imaging microscopy (FLIM), the decrease in the donor fluorescence lifetime that is due to FRET is measured, providing a reliable, donor concentration-independent estimation of FRET efficiency. Because the fluorescence lifetimes of donors are in the pico- or nanosecond range, even extremely rapid interactions can be monitored. In addition, unlike intensity-based FRET, FLIM measures can be also performed with spectrally similar fluorophores.

The BRET process occurs naturally in some marine organisms, such as the sea pansy *Renilla reniformis* and the jellyfish *Aequorea victoria*. In both cases the green fluorescent protein (GFP) is the acceptor, while *Renilla* luciferase (Rluc) or aequorin is the donor, respectively. BRET possesses some advantages in comparison with FRET because it avoids the problems associated with illumination. Therefore, it can be used even in photoresponsive cells or in cell types with significant autofluorescence. A further advantage of BRET is its higher sensitivity, which allows protein–protein interactions to be measured at low protein concentration, thus reducing nonspecific association phenomena of the target proteins. Because BRET assays are essentially intensity-based assays, spectral separation of donor and acceptor emissions is required.

9.2.2
FRET and BRET Applications

The great variety of available fluorescent and bioluminescent proteins emitting at different colors and the possibility to produce genetic in-frame fusions have had a great impact on FRET applications (which were originally based on organic fluorophores) and made BRET possible. Further advancements came as a result of the availability of improved substrates for bioluminescent proteins. For example, Packard has developed a BRET technology ($BRET^2$) based on a proprietary coelenterazine substrate and a GFP mutant that provides a large spectral shift between Rluc and GFP emission, thus allowing for more reliable BRET measurement (Dionne et al. 2002). The performance of FRET assays has also been significantly improved by new FRET techniques, such as two-photon excitation FRET and time-correlated single photon-counting FRET, which was made possible by the availability of more sensitive imaging devices and ultrafast light detection systems.

In recent years, an increasing number of studies have employed FRET and BRET techniques based on luminescent proteins, both *in vitro* and *in vivo* in intact cells (Eidne et al. 2002; Boute et al. 2002). In addition to those described in the following paragraphs, selected recent FRET and BRET binding assays involving luminescent proteins are reported in Table 9.2.

FRET assays can involve either a donor–acceptor couple of fluorescent proteins or a fluorescent protein, usually acting as the energy donor, and an organic fluorophore. FRET between enhanced yellow fluorescent protein (EYFP) and enhanced cyan fluorescent protein (ECFP) was used to study receptor–arrestin interactions (Kraft et al. 2001). FRET allowed detection of a rapid, time-dependent interaction between EYFP-tagged β-arrestin 2 and ECFP-tagged receptor 5, while co-immunoprecipitation techniques showed interaction only after 30 min of ligand stimulation. This result clearly demonstrates the superiority of RET methods for the assessment of protein–protein interactions.

FRET has been used to study homodimerization of the hY(1), hY(2), and hY(5) neuropeptide Y (NPY) receptors, which belong to the large family of G protein-coupled receptors, in living cells. Fusion proteins of NPY receptors and GFP or its spectral cyan, yellow, and red variants (CFP, YFP, and RFP, respectively) were generated and used as FRET pairs in either FRET fluorescence microscopy or fluorescence spectroscopy. Both techniques clearly showed that all the receptor subtypes are able to form homodimers and that dimers constitute a significant fraction of the receptors (Dinger et al. 2003).

Table 9.2 Additional selected BRET and FRET binding assays reported in the literature.

Type of binding interaction	*RET process*	*Reference*
Association of neuropeptide Y Y4 receptor	BRET ($BRET^2$)	Berglund et al. 2003
Heterodimerization of μ-opioid receptor (MOR1) and substance P receptor (NK1)	BRET (Rluc-GFP)	Pfeiffer et al. 2003
Homo-oligomerization of leukotriene C4 synthase	BRET ($BRET^2$)	Svartz et al. 2003
Oligomerization of adenosine A2A and dopamine D2 receptors	BRET ($BRET^2$)	Kamiya et al. 2003
Conformational change of protein B (PKT)/Akt resulting from lipid binding	FRET (GFP-YFP)	Calleja et al. 2003
Dimerization of CXCR4 receptor	BRET ($BRET^2$)	Babcock et al. 2003
Homo- and heterodimerization of oxytocin and vasopressin V1a and V2 receptors	BRET (Rluc-EYFP)	Terrillon et al. 2003
Interaction between insulin receptor and protein tyrosine-phosphatase 1B	BRET (Rluc-YFP)	Boute et al. 2003
Oligomerization of transcriptional intermediary factor 1 regulators	BRET (Rluc-GFP)	Germain-Desprez et al. 2003
Homodimerization of herpes simplex virus thymidine kinase (TK) monomers	FRET (CFP-YFP)	Tramier et al. 2002
Oligomerization of human thyrotropin receptor (TSHR)	FRET (EGFP-Cy3)	Latif et al. 2002
Interaction between the heart-specific FHL2 protein and the DNA-binding nuclear protein hNP220	FRET (BFP-GFP)	Ng et al. 2002
Dimerization and ligand-induced conformational changes of melatonin receptors	BRET (Rluc-YFP)	Ayoub et al. 2002
Homo- and heterodimerization of G protein-coupled receptors	BRET (Rluc-EYFP, $BRET^2$)	Ramsay et al. 2002
Oligomerization between adenosine A(1) and P2Y(1) receptors	BRET ($BRET^2$)	Yoshioka et al. 2002

The YFP-RFP donor–acceptor FRET couple has been employed to study the dimerization of the human thyrotropin (TSH) receptor (Latif et al. 2001). Observation of FRET between receptors differently tagged with the GFP variants provided evidence for the close proximity of individual receptor molecules, according to previous studies demonstrating the presence of receptor dimers and oligomers in thyroid tissue.

The mechanisms underlying the human immunodeficiency virus's (HIV) inhibitory activity of the chemokine stromal cell-derived factor 1 (SDF-1) were investigated. This chemokine, which binds to the CXC chemokine receptor 4 (CXCR4), is able to inhibit infection of $CD4^+$ cells by HIV strains through binding to cell-surface heparan sulfate proteoglycans (HSPGs). Using FRET between an enhanced green fluorescent protein (EGFP)-tagged CXCR4 and a Texas Red-labeled SDF-1 measured by laser confocal microscopy, the concomitant binding of SDF-1 to CXCR4 and cell-surface HSPGs was demonstrated. This binding may increase the local concentration of the chemokine in the surrounding environment of CXCR4, thus promoting its HIV-inhibitory activity (Valenzuela-Fernandez et al. 2001).

Signal transducer and activator of transcription 3 (Stat3) dimerization was studied by live-cell fluorescence spectroscopy and imaging microscopy combined with FRET. Stat3 fusion proteins with CFP and YFP were constructed and expressed in live cells. Fusion proteins showed functionality and intracellular distribution similar to wild-type Stat3, and FRET studies demonstrated the existence of Stat3 dimers. Visualization of the FRET signals also allowed insights into the spatiotemporal dynamics of Stat3 signal transduction (Kretzschmar et al. 2004).

The association and disassociation of heterotrimeric G proteins were monitored in living cells by tagging the $G\alpha_2$ and $G\beta\gamma$ proteins with CFP and YFP. Data from emission spectra were used to detect the FRET fluorescence and to determine kinetics and dose–response curves of bound ligand and analogues. Extending G-protein FRET to other G proteins should enable direct *in situ* mechanistic studies and applications, such as drug screening and identifying ligands of G protein-coupled receptors (Janetopoulos and Devreotes 2002).

An improved decay kinetics-based FRET method has been employed to study the association of the α and β_1 subunits of the human cardiac sodium channel (Biskup et al. 2004). FRET measurements performed using a mode-locked laser, a confocal microscope, and a streak camera demonstrated that the subunits, which were labeled with YFP and CFP, associate in the endoplasmic reticulum before they reach the plasma membrane. This approach has been described as superior to other decay kinetics-based methods because the fluorescence spectra and lifetimes of both donor and acceptor can be measured simultaneously, allowing distinction between FRET and other donor emission-quenching processes.

FRET has also been applied to flow cytometry for the development of cell-sorting assays (He et al. 2003). CFP and YFP were fused to the FUSE-binding protein (FBP)-interacting repressor and to the binding domain of FBP, respectively, and these tagged proteins were used to detect FRET in viable cells by flow cytometry. FRET between the interacting species was easily detected, suggesting that FRET could represent a generally available flow cytometry technique.

BRET was applied for the first time in 1999 for the study of the interactions between the circadian clock genes KaiA and KaiB of a cyanobacterium (Xu et al. 1999), showing the usefulness of this methodology over standard techniques for identifying protein interactions. From that time, the number of BRET applications reported in the literature, which mainly employed Rluc as the bioluminescent donor, has rapidly increased.

BRET and TR-FRET were used to study the oligomerization of the human δ-opioid receptor. Both approaches allowed the detection of homo-oligomers, whose formation was influenced by neither agonist nor inverse agonist ligands.

Hetero-oligomers between co-expressed human δ-opioid receptor and human β_2-adrenoreceptor were also studied. Interestingly, heterodimerization was detected by BRET, using the Rluc-EYFP BRET couple, but not by TR-FRET, and the extent of heterodimerization was increased in the presence of receptor agonists. The lack of observation of heterodimerization by TR-FRET has been attributed to the lower sensitivity of the technique or, alternatively, to the intracellular localization of the heterodimers, which in this case could not be detected by TR-FRET (McVey et al. 2001).

BRET has been extensively applied for the study of G protein-coupled receptor (GPCR) oligomerization in living cells, which is emerging as crucial aspect of GPCR function. BRET was shown to be a more powerful tool for studying GPCR oligomerization than standard co-immunoprecipitation methods, which are often limited by the formation of artifactual aggregates (Cheng and Miller 2001), and demonstrated that that GPCRs exist as either homo- or hetero-oligomeric complexes (Angers et al. 2001). It was also used to study GPCR receptor–β-arrestin interactions involved in receptor desensitization and trafficking in mammalian cells. Because of the highly hydrophobic nature and cellular localization of GPCRs, BRET presented significant advantages in comparison with other techniques (e.g., co-immunoprecipitation and yeast two-hybrid screening) for measuring protein–protein interactions (Kroeger and Eidne 2004).

BRET techniques also allow the real-time monitoring of ligand–receptor interactions. BRET from Rluc to YFP was used to study the interaction between a G protein-coupled receptor kinase (GRK2) and the human oxytocin receptor (OTR) in living cells (Hasbi et al. 2004). While previous attempts to demonstrate involvement of GRK have failed, the real-time BRET assay clearly showed the GRK2–OTR interactions and also offered insights into the dynamics of the process.

BRET- and FRET-based binding assays represent a powerful tool in drug discovery because they could permit homogeneous high-throughput screening assays. BRET is particularly suitable for these assays, because the FRET background signal that is due to sample autofluorescence is difficult to eliminate in non-imaging methods, such as in microtiter plate format assays.

BRET was used to monitor *in vitro* the activation state of the insulin receptor. Fusion proteins in which the human insulin receptor is fused to either Rluc or EYFP were expressed in HEK293 cells. Spontaneous *in vivo* dimer formation through disulfide bonds led to functional insulin receptors that, upon insulin binding, underwent conformational changes detectable by BRET. The procedure allowed for rapid determination of the activation state of the receptor, and it could be used in the search for new molecules with insulin-like activity (Boute et al. 2001).

9.2.3
Other Detection Principles

Fluorescent proteins can be directly used to monitor protein–protein association processes or conformation changes that are due to ligand binding. Fluorescent cellular sensors expressing YFP fused to the ligand-binding domains of estrogen, androgen, and glucocorticoid receptors were developed by Muddana and Peterson (2003). The fusion proteins were tethered through a short linker and expressed in *Saccharomyces cerevisiae* yeast cells. A dose-dependent fluorescence enhancement was observed in the presence of steroid receptor ligands. The correlation with the known relative receptor-binding affinity values of the compounds suggested that the fluorescence enhancement was due to ligand-induced receptor dimerization, perhaps through stabilization of YFP protein folding. These fluorescent cellular sensors could represent a novel, high-throughput method to identify and analyze ligands of nuclear hormone receptors.

Fluorescence anisotropy decay microscopy was used to determine, in individual living cells, the spatial and temporal monomer–dimer distribution of GFP-tagged herpes simplex virus thymidine kinase (TK). Measurement of the rotational time of the proteins by confocal microscopy and a time-correlated single photon-counting technique allowed the determination of their oligomeric state in both the cytoplasm and the nucleus. It was demonstrated that tagged TK was initially produced in a monomeric state and then formed dimers that grew into aggregates. Picosecond time-resolved fluorescence anisotropy microscopy was thus proposed as a promising technique for obtaining structural information on proteins in living cells, even in the case of low protein expression levels (Gautier et al. 2001).

9.3
Antibody-based Binding Assays

Several binding assays take advantage of the high specificity and avidity of antigen–antibody binding. Among these, immunoassays are the method of choice for the direct, sensitive, and specific quantification in complex matrices of analytes of clinical, pharmacological, environmental, or alimentary interest. What makes immunoassays very appealing is their rapidity, as well as the possibility of automation and high-throughput analyses. In addition, immunohistochemical methods are widely employed for the sensitive and specific localization of analytes in tissue sections or cells.

Various highly detectable labels have been used in the development of immunoassays, radioisotopes being the first example. However, safety and disposal issues, as well as limited shelf life, have prompted the search for alternative labels. Nowadays, enzymes are the most commonly used labels, thanks to signal amplification deriving from substrate turnover. This, together with the availability of chemiluminescent substrates, allowed reaching the highest detectability, which is similar or even higher than that obtained with radioisotopes. Among enzymes that can be detected using a chemiluminescent substrate, horseradish peroxidase (HRP) is extensively used, and the recent cloning and heterologous expression of its gene is opening new exciting applications in biosciences (Grigorenko et al. 2001). The most popular substrate for HRP, based on the H_2O_2–luminol–enhancer system, even if highly optimized, still presents a relatively low quantum yield, on the order of 0.01. Bioluminescence and fluorescence should provide superior detectability because they are characterized by much higher luminescence quantum yields. The various approaches used to employ luminescent proteins in antibody-based binding assays will be discussed in the following section.

9.3.1
Chemical Conjugation

Chemical conjugation of the protein label with either the analyte or the binder (e.g., antibody or protein A) is traditionally used in order to obtain tracers suitable for developing bioluminescent or fluorescent immunoassays. This rather simple approach takes advantage of the presence of reactive groups (usually primary amino or sulfhydryl groups) on the protein molecule. Various ultrasensitive assays have been developed with this strategy, using photoproteins (Erikaku et al. 1991; Stults et al. 1992; Zatta 1996; Mirasoli et al. 2002; Desai et al. 2002) and GFP (Deo and Daunert 2001a) as labels. In order to provide signal amplification, and thus a potential increase in assay sensitivity, streptavidin-biotinylated luciferase complexes were used (Valdivieso-Garcia et al. 2003). Alternatively, the efficiency of the analyte chemical conjugation to obelin was increased by introducing additional sulfhydryl groups into the obelin sequence, with the use of Traut's reagent (2-iminothiolane). This approach allowed development of a competitive immunoassay for thyroxin and a sandwich immunoassay for thyrotropin, both characterized by a sensitivity comparable with that obtained using radioisotope labels. However, it required careful experimental optimization to avoid excessive obelin conjugation and, consequently, a decrease in its bioluminescent activity (Frank et al. 2004).

To circumvent low reproducibility and label inactivation drawbacks, site-directed chemical conjugation has been proposed. In particular, aequorin mutants containing a single cysteine residue have been produced and used to obtain one-to-one thyroxine–aequorin homogeneous conjugates that allowed development of immunoassays for thyroxine with detection limits on the order of 10^{-12} M, which is three orders of magnitude lower than that reported for commercially available immunoassays (Lewis and Daunert 2001).

Because luciferase is most often inactivated by the chemical reactions involved in direct labeling of antigens or binding molecules, gene-fusion strategies have been used to exploit its high detectability in binding assay development. Alternatively, the luciferase reaction was coupled with a second enzymatic reaction. This allowed the use of the coupled enzyme as a conjugated label, followed by firefly luciferase addition in solution. For example, luciferin derivatives available as a substrate for firefly luciferase only upon activation by a specific enzymatic reaction have been synthesized (Miska and Geiger 1988). Alternatively, the firefly luciferase reaction was coupled with an enzymatic reaction producing ATP, such as those catalyzed by acetate kinase or pyruvate phosphate dikinase (Maeda 2003).

9.3.2 Gene Fusion

A new approach for producing tracers suitable for the development of fluorescent or bioluminescent immunoassays takes advantage of the possibility to produce genetic in-frame fusions of a target protein (either the analyte or a binding protein) with the bioluminescent label.

Several fusions of binding and luminescent proteins have been produced and proposed as universal reagents for immunoassays or immunohistochemistry techniques. In particular, a fusion protein between protein A and GFP has been proposed as a reagent for Western blotting (Aoki et al. 1996). Alternatively, the IgG-binding domain of protein A (ZZ) was fused to obelin and utilized for developing quantitative assays of rabbit, mouse, and human IgGs, with the same sensitivity obtained with commercially available protein A–HRP conjugate (Frank et al. 1996). With a similar strategy, GFP was fused to the single-chain antibody variable fragment of antibodies of interest, such as the antibody specific for hepatitis B surface antigen (Casey et al. 2000), for the E6 protein of human papillomavirus type 16 (Schwalbach et al. 2000), or for the herbicide picloram (Kim et al. 2002). In all cases the fusion protein retained both binding and luminescence properties and was suitable for developing sensitive and rapid antibody-based assays. An alternative approach consists of engineering the GFP amino acid sequence by inserting antibody-binding loops into surface-exposed loops at one end of the GFP molecule. The resulting so-called fluorobody works as a binding protein with intrinsic fluorescence characteristics (Zeytun et al. 2003). A fusion protein between streptavidin and luciferase from *Pyrophorus plagiophthalamus* was produced in an insect cell line using the baculovirus system and was used in both sandwich- and competitive-type immunoassays (Karp et al. 1996). An identical cloning strategy was used to obtain a fusion protein between streptavidin and GFP, which was successfully used for identification of biotinylated antibodies in living or fixed cells by fluorescence microscopy, proving that GFP is an excellent marker for histochemical applications. However, in the microtiter plate format the GFP–streptavidin fusion exhibited much lower detectability than the luciferase–streptavidin fusion. The binding domain from a human hyaluronan-binding protein was fused to *Renilla reniformis* luciferase, and the fusion protein was used to develop an assay for hyaluronan (Chang et al. 2003). An emerging number of luciferases from marine and terrestrial organisms are finding their way to biotechnological applications. Different types of fusion proteins between eukaryotic luciferases and binding molecules, along with their possible applications, have been discussed (Karp and Oker-Blom 1999). As an example, an *in vivo* biotinylated firefly luciferase was constructed by gene fusion of a thermostable mutant of the *Luciola lateralis* luciferase with a biotin acceptor peptide (Tatsumi et al. 1996). This construct, which retains luciferase activity and is able to bind to streptavidin, is an extremely versatile bioluminescent label, and it has been applied to the development of various immunoassays (Seto et al. 2001a, 2001b).

The gene-fusion approach has also been used to obtain bioluminescent tracers, where a bioluminescent protein is fused to a protein analyte. Using this approach, ultrasensitive competitive immunoassays have been developed for various analytes, such as a model octapeptide (Ramanathan et al. 1998), the mammalian opioid leucine–enkephalin peptide (Deo and Daunert 2001b), and protein C (Desai et al. 2001). These works demonstrate the versatility of aequorin in such a strategy, since the target analyte can be fused at either the aequorin N-terminal or the C-terminal. In addition, the antibody-binding region of a large protein can be fused to aequorin, rather than the whole protein, thus avoiding problems usually encountered when expressing large eukaryotic proteins in bacteria (Desai et al. 2001). The methods developed using such aequorin-based tracers were shown to be more sensitive than immunoassays based on conventional enzyme- or fluorescence-based assays. With a similar strategy, an analyte–obelin fusion protein was used to develop an immunoassay for a model octapeptide (Matveev et al. 1999).

9.3.3 Dual-analyte Assays

The possibility of assaying more than one analyte in one assay is particularly appealing in analytical chemistry, though still not very often reported. A dual-analyte competitive binding assay using tandem flash luminescence from the photoprotein aequorin and an acridinium-9-carboxamide label has been described. The assay allowed quantification of two analytes in one tube, by sequential triggering and measuring of the two luminescent reactions. The assay, which requires very short measurement times, is amenable for high-throughput screening (Adamczyk et al. 2002). A competitive fluorescence immunoassay that allowed quantification of two model peptide analytes in one well was developed by using two mutants of GFP emitting at different wavelengths as labels. Each peptide was genetically fused to either red-shifted GFP (rsGFP) or blue fluorescent protein (BFP), and the conjugates were used to develop an assay in which the peptides could be independently measured in one well, with detection limits on the order of 10^{-9} M to 10^{-10} M (Lewis and Daunert 1999). Because of the great variety of available fluorescent and bioluminescent proteins, further advances in this field are expected.

9.3.4 Expression Immunoassays

Use of enzyme labels is successful in the development of immunoassays because they allow signal amplification as a result of their high substrate turnover. Expression immunoassays take advantage of commercially available cell-free transcription–translation systems to introduce a further amplification system in the assay. In particular, an enzyme-coding DNA fragment, which also contains control elements for the transcription–translation process, is used as a label, instead of the protein itself. The amplification factor obtained by using this approach is potentially very high because, theoretically, several mRNA molecules are generated from each DNA molecule during the transcription–translation process, and, similarly, several protein molecules are generated from each mRNA molecule. Subsequently, the enzyme is detected by adding the suitable substrate. Various expression immunoassays have been developed using a firefly luciferase-coding DNA sequence that, in addition to luciferase high detectability, offers the advantage of being composed of a single and relatively short (550-amino-acid) polypeptide chain, which enhances its *in vitro* expression efficiency. As an example, an immunoassay for anti-thyrotropin immunoglobulins characterized by a limit of detection of 5×10^4 molecules per well (White et al. 2000) and a sandwich immunoassay for prostate-specific antigen characterized by a limit of detection of 30 ng L^{-1} (Chiu and Christopoulos 1999) have been developed. In order to circumvent difficulties encountered when linking the antibody to the reporter gene, a bifunctional cross-linking molecule was recently produced. The linker molecule consists of (strept)avidin bound to a poly(dA) oligonucleotide, which is able to bind both to a biotinylated antibody and to a poly(dT) sequence properly enzymatically added to the reporter gene (Tannous et al. 2002).

9.3.5 BRET-based Immunoassays

The possibility of fusing a binding with a luminescent protein has been exploited for developing FRET- and BRET-based immunoassays. An example is the development of open sandwich fluorescent immunoassays, where the antibody heavy-chain and light-chain fragments are fused to FRET or BRET partners. The assay is based on the RET phenomenon observed upon antigen-dependent re-association of antibody variable domains. Using this strategy, a FRET-based assay (Arai et al. 2000) and a BRET-based assay for egg lysozyme (Arai et al. 2001) were developed. However, the authors observed that such an interesting and innovative approach is not always guaranteed to work. For example, when the epitopes recognized by the couple of antibody fragments are quite far from one another on the analyte molecule, a situation in which FRET or BRET partners are not close enough for energy transfer, even after antibody variable domain association, may occur. For this reason, the authors further developed the method by fusing an artificial dimerization motif of leucine zippers at the C-terminal of FRET partners. These additional dimerization motifs were able to bring FRET partners in close proximity upon antigen–antibody reactions, thus enhancing FRET efficiency. Using this strategy, a homogeneous sandwich immunoassay for a rather large molecule, such as human serum albumin, was developed (Ohiro et al. 2002).

9.4 Biotin–Avidin Binding Assays

Biotin (vitamin H) is a vitamin found in tissue and blood that binds to the glycoprotein avidin, which contains four identical binding sites for biotin. Biotin and its binding protein avidin have been used extensively in the development of a variety of bioanalytical techniques that take advantage of the extremely high affinity ($k_a \approx 10^{15}$ M^{-1}) between the two biomolecules. Biotin has also been selected as a model analyte to develop binding assays in order to explore the possibility of applying luminescent proteins as labels for the determination of given analytes in competitive binding assay formats. These assays are based upon the competition between free biotin in a sample and the biotin–label conjugate for the binding sites on avidin (Lewis and Daunert 2000).

A biotin–aequorin conjugate was produced by conventional chemical conjugation methods and used to develop both a homogeneous and a heterogeneous assay for biotin. The homogeneous assay relied on the inhibition of the signal from the biotin–aequorin conjugate upon binding with avidin. The degree of inhibition was inversely proportional to the free biotin concentration, and a detection limit for biotin of 1×10^{-14} M was achieved (Witkowski et al. 1994). Alternatively, the heterogeneous assay employed a solid phase of avidin-coated microspheres to separate the bound and free biotin-aequorin conjugate, which permitted lowering the detection limit to 1×10^{-15} M (Feltus et al. 1997). Such a low detection limit, corresponding to 100 zmol of biotin in the sample, allowed for miniaturization of the assay. A homogeneous competitive binding assay for the detection of biotin in microfabricated picoliter vials using biotinylated recombinant aequorin was developed.

The results showed that detection limits on the order of 10^{-14} mol of biotin were possible (Grosvenor et al. 2000). A binding assay for avidin in picoliter-volume vials in which avidin could be detected at femtomole levels was also described (Crofcheck et al. 1997). These binding assays based on picoliter volumes have potential applications in different fields, such as microanalysis and single-cell analysis, in which the amount of sample is limited. In addition, the small volumes allow for instantaneous mixing of the reagents, thus rendering these miniaturized assays suitable for high-throughput screening of biopharmaceuticals and potential drugs synthesized by combinatorial methods. The detection of biotin in individual cells using a BL competitive binding assay was reported. An aequorin–biotin conjugate, free biotin, and avidin were microinjected into sea urchin oocytes, and then the resulting bioluminescence within the oocyte upon triggering of aequorin was measured by a photomultiplier tube-based microscope photometry system. The method allowed for the detection of approximately 20 amol of biotin inside individual oocytes (Feltus et al. 2001). Recombinant GFP and its mutant enhanced GFP (EGFP) have been used as labels in heterogeneous and homogeneous competitive binding assays for biotin, respectively. Detection limits on the order of 10^{-8} M and 10^{-9} M were obtained (Hernandez and Daunert 1998; Deo and Daunert 2001a).

Homogeneous binding assays for biotin were recently developed using luminescent proteins as labels and BRET as the measurement technique. Biotinylated aequorin and a quencher dye conjugated to avidin were used as the two partners of the BRET pair. When biotin bound to avidin, bioluminescence from aequorin was transferred to the chemical dye, which was selected on the basis of its ability to quench the BL emission. A water-soluble biotin derivative or a biotinylated protein was used as a model analyte to develop homogeneous competitive assays. In both cases dose–response curves showing an inverse relationship between percent bioluminescence quenching and analyte concentration were generated, thus demonstrating the feasibility of assays based on this BRET format. Furthermore, the approach is quite general because biotin–avidin could be replaced with a different pair, e.g., hapten–antibody, and other quencher dyes could be used (Adamczyk et al. 2001). Fusions of aequorin with streptavidin (SAV) and enhanced green fluorescent protein (EGFP) with biotin carboxyl carrier protein (BCCP) were obtained after expression of the corresponding genes in *Escherichia coli* cells. Association of SAV–aequorin and BCCP–EGFP was followed by BRET between aequorin (donor) and EGFP (acceptor). It was shown that free biotin inhibited BRET in a dose-dependent manner because of its competition with BCCP-EGFP for binding to SAV–aequorin. The distinctive feature of this assay is the use of protein fusions that provide bifunctionally active conjugates that are uniform in their stoichiometry and in the point of reporter protein attachment and that are readily available after expression of the corresponding genes in *E. coli* and purification. This represents an advantage with respect to methods that use the chemical conjugation of biotin and avidin with donor and acceptor, since chemical modifications are often difficult to reproduce; in addition, in the case of aequorin particular precautions should be taken while biotinylating and storing the conjugate to avoid loss of BL activity (Gorokhovatsky et al. 2003).

9.5
Nucleic Acid Hybridization Assays

The analysis of specific nucleic acid sequences by hybridization is a fast-growing area of laboratory medicine. These methods are based on the specific hybridization of a suitably labeled oligonucleotide probe with a target nucleic acid sequence, followed by the hybrid detection and quantification. Luminescent proteins are becoming more and more popular as labels because luminescence-based hybridization assays are amenable to automation and offer higher detectability over conventional spectrophotometric ones.

A microtiter well-based hybridization assay using aequorin as a detection molecule has been developed. The target DNA was hybridized simultaneously with a digoxigenin (dig)-labeled capture probe and a biotinylated detection probe. The hybrids were immobilized onto the wells through dig–anti-dig interaction and were detected by aequorin covalently attached to streptavidin or by complexes of biotinylated aequorin with streptavidin. Linearity of response in the range from

5 amol to 10 fmol of target DNA was shown. The method was combined with reverse transcriptase polymerase chain reaction (RT-PCR) and applied to the determination of the mRNA for prostate-specific antigen (PSA) that was detected from a single cell in the presence of 10^6 cells not expressing PSA. The method proved to be very fast, since the detection of aequorin is practically instantaneous as compared to the long incubation times required for the detection of enzyme-labeled probes (Galvan and Christopoulos 1996).

An aequorin-based immunoassay combined with RT-PCR for quantitating cytokine mRNA has been developed. In this system the hybridized duplex was captured onto a streptavidin-coated microtiter plate and quantitated by anti-dig antibody conjugated with aequorin. The method detected as low as 40 amol of amplified cytokine products, corresponding to 500 copies of template when 27 PCR cycles were used (Xiao et al. 1996). The same capture and detection format, in combination with RT-PCR, was exploited for investigating the induction of human cytokine expression in peripheral blood mononuclear cells. A statistically significant increase in cDNA for several cytokines in stimulated cells compared to unstimulated cells was demonstrated. The BL method exhibited significant advantages when compared with radioactive methods, such as the ability to quantitate amplicons in a PCR cycle range where linear detection is more robust and to analyze products in an automated, open-architecture microtiter plate format (Actor et al. 1998).

Hybridization assays that use luminescent proteins as labels have been successfully coupled to PCR, in order to quantify the amplification product in a more accurate manner. A dual-analyte luminescence hybridization assay for quantitative PCR has been developed. This method allowed the simultaneous determination of both amplified target DNA and internal standard (IS) in the same reaction vessel. Biotinylated PCR products from target DNA and IS were captured on a single microtiter well coated with streptavidin. The amplified target DNA was hybridized with a dig-labeled specific probe, and the hybrids were detected by anti-dig antibody labeled with aequorin. The amplified IS DNA was hybridized in the same well with a fluorescein-labeled specific probe, and the hybrids were detected by anti-fluorescein antibody labeled with AP. The ratio of the luminescence values obtained from the target DNA and IS amplification products was linearly related to the number of target DNA molecules present in the sample prior to amplification, over a range of 430 to 315 000 target DNA molecules (Verhaegen and Christopoulos 1998).

Novel hybridization assay configurations based on *in vitro* expression of DNA reporter molecules have been proposed. *In vitro* expression of DNA consists of the cell-free transcription of DNA to mRNA followed by translation of the mRNA into protein. In one study a system was described that involved simultaneous hybridization of the single-stranded target DNA with a biotinylated capture probe, immobilized on streptavidin-coated microtiter wells, and a dATP-tailed detection probe. The hybrids were reacted with dTTP-tailed firefly luciferase-coding DNA fragments followed by *in vitro* expression of the DNA on the solid phase. Several enzyme molecules per molecule of enzyme-coding DNA label were generated in solution, thus providing signal amplification (Laios et al. 1998). Another study described a similar configuration that used a DNA label encoding apoaequorin. After *in vitro* expression, apoaequorin was converted to active aequorin and each DNA label was estimated to produce 156 aequorin molecules. A detection limit as low as 0.25 amol of target DNA was achieved, with a linear range over four orders of magnitude (White and Christopoulos 1999). The significance of this approach lies in the use of DNA in hybridization assays as a signal-generating molecule (reporter), rather than solely as a recognition molecule, which forms the basis of a highly sensitive analytical system for binding assays.

A different strategy to enhance the sensitivity of aequorin-based hybridization assays relied on the introduction of multiple aequorin labels per DNA hybrid, through enzyme amplification. After simultaneous hybridization of the target DNA with a biotinylated capture probe and a dig-labeled detection probe, the hybrids were reacted with anti-dig antibody labeled with HRP. Peroxidase catalyzed the oxidation of digoxigenin–tyramine by H_2O_2, with the production of tyrosyl radicals that reacted with the tyrosine residues of the streptavidin immobilized on the surface of the microtiter wells, forming protein-bound dityrosine. This resulted in the covalent attachment of multiple dig moieties to the solid phase that were detected by aequorin-labeled anti-dig antibody. An eightfold improvement in detectability was observed with the enzyme amplification as compared to an assay that used only aequorin-labeled anti-dig antibody, resulting in the detection of as low as 1 amol per well of target DNA (Laios et al. 2001).

Biotinylated aequorin complexed with streptavidin can be used as a reporter molecule in hybridization assays to detect target DNA either labeled with biotin through PCR or hybridized with biotin-labeled probe, as well as in immunoassays in combination with biotinylated antibodies. In order to avoid the inconvenience of chemical cross-linking, bacterial expression of *in vivo*-biotinylated aequorin was carried out. The entire process was completed in less than two days, and it was calculated that 1 L of bacterial culture provided enough biotinylated aequorin for 300 000 hybridization assays (Verhaegen and Christopoulos 2002a).

A new conjugation strategy based on the use of recombinant aequorin fused to a hexahistidine tag was recently proposed to obtain aequorin–oligonucleotide conjugates. Affinity capture-facilitated purification of the conjugates enabled the rapid and effective removal of the unreacted oligonucleotide. The conjugates were applied to the development of rapid bioluminometric hybridization assays that exhibited a dynamic range of 2–2000 pmol L^{-1} of target DNA (Glynou et al. 2003).

Recently, the potential of *Gaussia princeps* luciferase (GLuc) as a new label for DNA hybridization has been examined. Because luciferases are inactivated upon conjugation to other biomolecules, such as DNA probes or antibodies, their use as labels is limited. To overcome inactivation problems, a bacterial overexpression system that produces *in vivo*-biotinylated GLuc has been designed. Purified GLuc activity was maintained, allowing enzyme detection down to 1 amol. Biotinylated GLuc was complexed with streptavidin and used as a detection reagent in a microtiter well-based DNA hybridization assay, in which an analytical range of 1.6–800 pmol L^{-1} of target DNA was shown (Verhaegen and Christopoulos 2002b).

The difference in light emission kinetics between the Ca^{2+}-triggered BL reaction of aequorin and the alkaline phosphatase (AP)-catalyzed CL hydrolysis of dioxetane-based substrates has been exploited for the analysis of both alleles of biallelic polymorphisms in a single microtiter well. Genomic DNA isolated from blood was first subjected to PCR. Then a single-oligonucleotide ligation reaction employing two allele-specific probes, labeled with biotin and digoxigenin, and a common probe carrying a characteristic tail was performed. The ligation product was captured in the microtiter well by hybridization of the tail with an immobilized complementary oligonucleotide. The products were detected by adding a mixture of streptavidin–aequorin complex and anti-dig–alkaline phosphatase conjugate. The ratio of the luminescence signals obtained from AP and aequorin gave the genotype of each sample (Tannous et al. 2003).

9.6 Other Binding Assays

A different type of binding assay that exploits mutants of the EGFP was developed for the detection of bacterial endotoxin, or lipopolysaccharide (LPS). EGFP variants containing binding sites for LPS and lipid A (LA), the bioactive component of LPS, were obtained by virtual mutagenesis. DNA mutant constructs were expressed in COS-1 cells. LPS or LA binding to the EGFP mutant proteins caused concentration-dependent fluorescence quenching. Thus, the EGFP mutant can represent the basis of a novel fluorescent biosensor for bacterial endotoxin (Goh et al. 2002a). In another study it was demonstrated that the EGFP mutant can specifically tag gram-negative bacteria such as *Escherichia coli* and *Pseudomonas aeruginosa* in contaminated environmental water samples. This dual function in detecting both free endotoxin and live gram-negative bacteria extends the potential of this novel fluorescent biosensor (Goh et al. 2002b).

GFP mutants were designed, created, and characterized after identifying potential metal-binding sites on the surface of GFP. These metal-binding mutants of GFP exhibit fluorescence quenching at lower transition metal ion concentrations than those of the wild-type protein, thus representing a new class of luminescent protein-based metal sensors (Richmond et al. 2000).

9.7 Concluding Remarks

The naturally occurring luminescent proteins aequorin, obelin, GFP, luciferases, and their mutants have proved to be highly effective labels in the development of different types of binding assays, including protein–protein and protein–ligand interaction assays, immunoassays, biotin–avidin-based assays, and nucleic acid hybridization assays. These proteins can be detected down to very low levels, thus allowing ultrasensitive detection of the target analytes. This also enables the analysis of small-volume samples, which leads to the development of miniaturized and high-throughput assays. Further advantages include the opportunities to produce fusion proteins and to genetically engineer the native proteins to emit luminescence signals at different wavelengths or with higher efficiency. These features, along with the discovery of new luminescent proteins, allow for their use as labels in the simultaneous array detection of several biomolecules in a given sample, as well as for the development of improved FRET- and BRET-based assays.

References

Actor, J. K., Kuffner, T., Dezzutti, C. S., Hunter, R. L., McNicholl, J. M. A flash-type bioluminescent immunoassay that is more sensitive than radioimaging: quantitative detection of cytokine cDNA in activated and resting human cells. *J. Immunol. Methods* **1998**, *211*, 65–77.

Adamczyk, M., Moore, J. A., Shreder, K. Quenching of biotinylated aequorin bioluminescence by dye-labeled avidin conjugates: application to homogeneous bioluminescence resonance energy transfer assays. *Org. Lett.* **2001**, *3*, 1797–1800.

Adamczyk, M., Moore, J. A., Shreder, K. Dual analyte detection using tandem flash luminescence. *Bioorg. Med. Chem. Lett.* **2002**, *12*, 395–398.

Angers, S., Salahpour, A., Bouvier, M. Biochemical and biophysical demonstration of GPCR oligomerization in mammalian cells. *Life Sci.* **2001**, *68*, 2243–2250.

Aoki, T., Takahashi, Y., Koch, K. S., Leffert, H. L., Watabe, H. Construction of a fusion protein between protein A and green fluorescent protein and its application to western blotting. *FEBS Lett.* **1996**, *384*, 193–197.

Arai, R., Ueda, H., Tsumoto, K., Mahoney, W. C., Kumagai, I., Nagamune, T. Fluorolabeling of antibody variable domains with green fluorescent protein variants: application to an energy transfer-based homogeneous immunoassay. *Protein Eng.* **2000**, *13*, 369–376.

Arai, R., Nakagawa, H., Tsumoto, K., Mahoney, W., Kumagai, I., Ueda, H., Nagamune, T. Demonstration of a homogeneous noncompetitive immunoassay based on bioluminescence resonance energy transfer. *Anal. Biochem.* **2001**, *289*, 77–81.

Ayoub, M. A., Conturier, C., Lucas-Meunier, E., Angers, S., Foisser, P., Bouvier, M., Jockers, R. Monitoring of ligand-independent dimerization and ligand-induced conformational changes of melatonin receptors in living cells by bioluminescence resonance energy transfer. *J. Biol. Chem.* **2002**, *277*, 21522–21528.

Babcock, G. J., Farzan, M., Sodroski, J. Ligand-independent dimerization of CXCR4, a principal HIV-1 coreceptor. *J. Biol. Chem.* **2003**, *278*, 3378–3385.

Berglund, M. M., Schober, D. A., Esterman, M. A., Gehlert, D. R. Neuropeptide Y Y4 receptor homodimers dissociate upon agonist stimulation. *J. Pharmacol. Exp. Ther.* **2003**, *307*, 1120–1126.

Biskup, C., Zimmer, T., Benndorf, K. FRET between cardiac Na^+ channel subunits measured with a confocal microscope and a streak camera. *Nat. Biotechnol.* **2004**, *22*, 220–224.

Boute, N., Pernet, K., Issad, T. Monitoring the activation state of the insulin receptor using bioluminescence resonance energy transfer. *Mol. Pharmacol.* **2001**, *60*, 640–645.

Boute, N., Jockers, R., Issad, T. The use of resonance energy transfer in high-througput screening: BRET versus FRET. *Trends Pharmacol. Sci.* **2002**, *23*, 351–354.

Boute, N., Boubekeur, S., Lacasa, D., Issad, T. Dynamics of the interaction between the insulin receptor and protein tyrosine-phosphatase 1B in living cells. *EMBO Rep.* **2003**, *4*, 313–319.

Calleja, V., Ameer-Beg, S. M., Vojnovic, B., Woscholski, R., Downward, J., Larijani, B. Monitoring conformational changes of proteins in cells by fluorescence lifetime imaging microscopy. *Biochem. J.* **2003**, *372*, 33–40.

Casey, J. L., Coley, A. M., Tilley, L. M., Foley, M. Green fluorescent antibodies: novel *in vitro* tools. *Protein Eng.* **2000**, *13*, 445–452.

Chang, T. S., Wan, H. M., Chen, C. C., Giridhar, R., Wu, W. T. Fusion protein of the hyaluronan binding domain from human TSG-6 with luciferase for assay of hyaluronan. *Biotechnol. Lett.* **2003**, *25*, 1037–1040.

Cheng, Z. J., Miller, L. J. Agonist-dependent dissociation of oligomeric complexes of G protein-coupled cholecystokinin receptors demonstrated in living cells using bioluminescence resonance energy transfer. *J. Biol. Chem.* **2001**, *276*, 48040–48047.

Chiu, N. H., Christopoulos, T. K. Two-site expression immunoassay using a firefly luciferase-coding DNA label. *Clin. Chem.* **1999**, *45*, 1954–1959.

Crofcheck, C. L., Grosvenor, A. L., Anderson, K. W., Lumpp, J. K., Scott, D. L., Daunert, S. Detecting biomolecules in picoliter vials using aequorin bioluminescence. *Anal. Chem.* **1997**, *69*, 4768–4772.

Deo, S. K., Daunert, S. Green fluorescent protein mutant as label in homogeneous assays for biomolecules. *Anal. Biochem.* **2001a**, *289*, 52–59.

Deo, S. K., Daunert, S. An immunoassay for Leu-enkephalin based on a C-terminal aequorin-peptide fusion. *Anal. Chem.* **2001b**, *73*, 1903–1908.

Desai, U. A., Wininger, J. A., Lewis, J. C., Ramanathan, S., Daunert, S. Using epitope-aequorin conjugate recognition in immunoassays for complex proteins. *Anal. Biochem.* **2001**, *294*, 132–140.

Desai, U. A., Deo, S. K., Hyland, K. V., Poon, M., Daunert, S. Determination of prostacyclin in plasma through a bioluminescent immunoassay for 6-keto-prostaglandin F1alpha: implication of dosage in patients with primary pulmonary hypertension. *Anal. Chem.* **2002**, *74*, 3892–3898.

Dinger, M. C., Bader, J. E., Kobor, A. D., Kretzschmar, A. K., Beck-Sickinger, A. G. Homodimerization of neuropeptide y receptors investigated by fluorescence resonance energy transfer in living cells. *J. Biol. Chem.* **2003**, *278*, 10562–10571.

Dionne, P., Caron, M., Labontè, A., Carter-Allen, K., Houle, B., Joly, E., Taylor, S. C., Menard, L. BRET[2]: Efficient energy transfer from Renilla luciferase to GFP[2] to measure protein-protein interactions and intracellular signaling events in living cells. In *Luminescence biotechnology. Instruments and applications* (Van Dyke, K., Van Dyke, C., Woodfork, K., Eds.), CRC Press, Boca Raton, pp. 539–555, **2002**.

Eidne, K. A., Kroeger, K. M., Hanyaloglu, A. C. Applications of novel resonance energy transfer techniques to study dynamic hormone receptor interactions in living cells. *Trends Endocrinol. Metab.* **2002**, *13*, 415–421.

Erikaku, T., Zenno, S., Inouye, S. Bioluminescent immunoassay using a monomeric Fab′-photoprotein aequorin conjugate. *Biochem. Biophys. Res. Commun.* **1991**, *174*, 1331–1336.

Feltus, A., Ramanathan, S., Daunert, S. Interaction of immobilized avidin with an aequorin-biotin conjugate: an aequorin-linked assay for biotin. *Anal. Biochem.* **1997**, *254*, 62–68.

Feltus, A., Grosvenor, A. L., Conover, R. C., Anderson, K. W., Daunert, S. Detection of biotin in individual sea urchin oocytes using a bioluminescence binding assay. *Anal. Chem.* **2001**, *73*, 1403–1407.

Frank, L. A., Illarionova, V. A., Vysotski, E. S. Use of proZZ-obelin fusion protein in bioluminescent immunoassay. *Biochem. Biophys. Res. Commun.* **1996**, *219*, 475–479.

Frank, L. A., Petunin, A. I., Vysotski, E. S. Bioluminescent immunoassay of thyrotropin and thyroxine using obelin as a label. *Anal. Biochem.* **2004**, *325*, 240–246.

Galvan, B., Christopoulos, T. K. Bioluminescence hybridization assays using recombinant aequorin. Application to the detection of prostate-specific antigen mRNA. *Anal. Chem.* **1996**, *68*, 3545–3550.

Gautier, I., Tramier, M., Durieux, C., Coppey, J., Pansu, R. B., Nicolas, J. C., Kemnitz, K., Coppey-Moisan, M. Homo-FRET microscopy in living cells to measure monomer-dimer transition of GFP-tagged proteins. *Biophys. J.* **2001**, *80*, 3000–3008.

Germain-Desprez, D., Bazinet, M., Bouvier, M., Aubry, M. Oligomerization of transcriptional intermediary factor 1 regulators and interaction with ZNF74 nuclear matrix protein revealed by bioluminescence resonance energy transfer in living cells. *J. Biol. Chem.* **2003**, *278*, 22367–22373.

Glynou, K., Ioannou, P. C., Christopoulos, T. K. Affinity capture-facilitated preparation of aequorin-oligonucleotide conjugates for rapid hybridization assays. *Bioconjugate Chem.* **2003**, *14*, 1024–1029.

Goh, Y. Y., Frecer, V., Ho, B., Ding, J. L. Rational design of green fluorescent protein mutants as biosensor for bacterial endotoxin. *Protein Eng.* **2002a**, *15*, 493–502.

Goh, Y. Y., Ho, B., Ding, J. L. A novel fluorescent protein-based biosensor for gram-negative bacteria. *Appl. Environ. Microbiol.* **2002b**, *68*, 6343–6352.

Gorokhovatsky, A. Y., Rudenko, N. V., Marchenkov, V. V., Skosyrev, V. S., Arzhanov, M. A., Burkhardt, N., Zakharov, M. V., Semisotnov, G. V., Vinokurov, L. M., Alakhov, Y. B. Homogeneous assay for biotin based on *Aequorea victoria* bioluminescence resonance energy transfer system. *Anal. Biochem.* **2003**, *313*, 68–75.

Grigorenko, V., Andreeva, I., Borchers, T., Spener, F., Egorov, A. A genetically engineered fusion protein with horseradish peroxidase as a marker enzyme for use in competitive immunoassays. *Anal. Chem.* **2001**, *73*, 1134–1139.

Grosvenor, A. L., Feltus, A., Conover, R. C., Daunert, S., Anderson, K. W. Development of binding assays in microfabricated picoliter vials: an assay for biotin. *Anal. Chem.* **2000**, *72*, 2590–2594.

Hasbi, A., Devost, D., Laporte, S. A., Zingg, H. H. Real time detection of interactions between the human oxytocin receptor and G protein-coupled receptor kinase-2. *Mol. Endocrinol* **2004**, *18*, 1277–1286.

He, L., Bradrick, T. D., Karpova, T. S., Wu, X., Fox, M. H., Fischer, R., McNally, J. G., Knutson, J. R., Grammer, A. C., Lipsky, P. E. Flow cytometric measurement of fluorescence (Forster) resonance energy transfer from cyan fluorescent protein to yellow fluorescent protein using single-laser excitation at 458 nm. *Cytometry* **2003**, *53*, 39–54.

Hernandez, E. C., Daunert, S. Recombinant green fluorescent protein as a label in binding assays. *Anal. Biochem.* **1998**, *261*, 113–115.

Janetopoulos, C., Devreotes, P. Monitoring receptor-mediated activation of heterotrimeric G-proteins by fluorescence resonance energy transfer. *Methods* **2002**, *27*, 366–373.

Kamiya, T., Saitoh, O., Yoshioka, K., Nakata, H. Oligomerization of adenosine A2A and dopamine D2 receptors in living cells. *Biochem. Biophys. Res. Commun.* **2003**, *306*, 544–549.

Karp, M., Lindqvist, C., Nissinen, R., Wahlbeck, S., Akerman, K., Oker-Blom, C. Identification of biotinylated molecules using a baculovirus-expressed luciferase-streptavidin fusion protein. *Biotechniques* **1996**, *20*, 452–459.

Karp, M., Oker-Blom, C. A streptavidin-luciferase fusion protein: comparisons and applications. *Biomol. Eng.* **1999**, *16*, 101–104.

Kim, I. S., Shim, J. H., Suh, Y. T., Yau, K. Y., Hall, J. C., Trevors, J. T., Lee, H. Green fluorescent protein-labeled recombinant fluobody for detecting the picloram herbicide. *Biosci. Biotechnol. Biochem.* **2002**, *66*, 1148–1151.

Kraft, K., Olbrich, H., Majoul, I., Mack, M., Proudfoot, A., Oppermann, M. Characterization of sequence determinants within the carboxyl-terminal domain of chemokine receptor CCR5 that regulate signaling and receptor internalization. *J. Biol. Chem.* **2001**, *276*, 34408–34418.

Kretzschmar, A. K., Dinger, M. C., Henze, C., Brocke-Heidrich, K., Horn, F. Analysis of Stat3 (signal transducer and activator of transcription 3) dimerization by fluorescence resonance energy transfer in living cells. *Biochem. J.* **2004**, *377*, 289–297.

Kroeger, K. M., Eidne, K. A. Study of g-protein-coupled receptor-protein interactions by bioluminescence resonance energy transfer. *Methods Mol. Biol.* **2004**, *259*, 323–334.

Laios, E., Ioannou, P. C., Christopoulos, T. K. Novel hybridization assay configurations based on *in vitro* expression of DNA reporter molecules. *Clin. Biochem.* **1998**, *31*, 151–158.

Laios, E., Ioannou, P. C., Christopoulos, T. K. Enzyme-amplified aequorin-based bioluminometric hybridization assays. *Anal. Chem.* **2001**, *73*, 689–692.

Latif, R., Graves, P., Davies, T. F. Oligomerization of the human thyrotropin receptor: fluorescent protein-tagged hTSHR reveals post-translational complexes. *J. Biol. Chem.* **2001**, *276*, 45217–45224.

Latif, R., Graves, P., Davies, T. F. Ligand-dependent inhibition of oligomerization at the human tyrotropin receptor. *J. Biol. Chem.* **2002**, *277*, 45059–45067.

Lewis, J. C., Daunert, S. Dual detection of peptides in a fluorescence binding assay by employing genetically fused GFP and BFP mutants. *Anal. Chem.* **1999**, *71*, 4321–4327.

Lewis, J. C., Daunert, S. Photoproteins as luminescent labels in binding assays. *Fresenius J. Anal. Chem.* **2000**, *366*, 760–768.

Lewis, J. C., Daunert, S. Bioluminescence immunoassay for thyroxine employing genetically engineered mutant aequorins containing unique cysteine residues. *Anal. Chem.* **2001**, *73*, 3227–3233.

Maeda, M. New label enzymes for bioluminescent enzyme immunoassay. *J. Pharm. Biomed. Anal.* **2003**, *30*, 1725–1734.

Matveev, S. V., Lewis, J. C., Daunert, S. Genetically engineered obelin as a bioluminescent label in an assay for a peptide. *Anal. Biochem.* **1999**, *270*, 69–74.

McVey, M., Ramsay, D., Kellett, E., Rees, S., Wilson, S., Pope, A. J., Milligan, G. Monitoring receptor oligomerization using time-resolved fluorescence resonance energy transfer and bioluminescence resonance energy transfer. The human delta-opioid receptor displays constitutive oligomerization at the cell surface, which is not regulated by receptor occupancy. *J. Biol. Chem.* **2001**, *276*, 14092–10499.

Mirasoli, M., Deo, S. K., Lewis, J. C., Roda, A., Daunert, S. Bioluminescence immunoassay for cortisol using recombinant aequorin as a label. *Anal. Biochem.* **2002**, *306*, 204–211.

Miska, W., Geiger, R. A new type of ultrasensitive bioluminogenic enzyme substrates. I. Enzyme substrates with D-luciferin as leaving group. *Biol. Chem. Hoppe. Seyler.* **1988**, *369*, 407–411.

Muddana, S. S., Peterson, B. R. Fluorescent cellular sensors of steroid receptor ligands. *Chembiochem.* **2003**, *4*, 848–855.

Ng, E. K., Chan, K. K., Wong, C. H., Tsui, S. K., Ngai, S. M., Lee, S. M., Kotaka, M., Lee, C. Y., Waye, M. M., Fung, K. P. Interaction of the heart-specifc LIM domain protein, FHL2, with DNA-binding nuclear protein, hNP220. *J. Cell Biochem* **2002**, *84*, 556–566.

Ohiro, Y., Arai, R., Ueda, H., Nagamune, T. A homogeneous and noncompetitive immunoassay based on the enhanced fluorescence resonance energy transfer by leucine zipper interaction. *Anal. Chem.* **2002**, *74*, 5786–5792.

PFEIFFER, M., KIRSCHT, S., STUMM, R., KOCH, T., WU, D., LAUGSCH, M., SCHRODER, H., HOLLT, V., SCHULTZ, S. Heterodimerization of substance P and μ-opioid repeptors regulates receptor trafficking and resensitization. *J. Biol. Chem.* **2003**, *278*, 51630–51637.

RAMANATHAN, S., LEWIS, J. C., KINDY, M. S., DAUNERT, S. Heterogeneous bioluminescence binding assay for an octapeptide using recombinant aequorin. *Anal. Chim. Acta* **1998**, *369*, 181–188.

RAMSAY, D., KELLETT, E., MCVEY, M., REES, S., MILLIGAN, G. Homo- and hetero-oligomeric interactions between G-protein coupled receptors in living cells monitored by two variants of bioluminescence resonance energy transfer (BRET): hetero-oligomers between receptor subtypes form more efficiently than between less closely related sequences. *Biochem. J.* **2002**, *365*, 429–440.

RICHMOND, T. A., TAKAHASHI, T. T., SHIMKHADA, R., BERNSDORF, J. Engineered metal binding sites on green fluorescence protein. *Biochem. Biophys. Res. Commun.* **2000**, *268*, 462–465.

SCHWALBACH, G., SIBLER, A. P., CHOULIER, L., DERYCKERE, F., WEISS, E. Production of fluorescent single-chain antibody fragments in Escherichia coli. *Protein Expr. Purif.* **2000**, *18*, 121–132.

SETO, Y., IBA, T., ABE, K. Development of ultra-high sensitivity bioluminescent enzyme immunoassay for prostate-specific antigen (PSA) using firefly luciferase. *Luminescence* **2001a**, *16*, 285–290.

SETO, Y., OHKUMA, H., TAKAYASU, S., IBA, T., UMEDA, A., ABE, K. Development of highly sensitive bioluminescent enzyme immunoassay with ultra-wide measurable range for thyroid-stimulating hormone using firefly luciferase. *Anal. Chim. Acta* **2001b**, *429*, 19–26.

STULTS, N. L., STOCKS, N. F., RIVERA, H., GRAY, J., MCCANN, R. O., O'KANE, D., CUMMINGS, R. D., CORMIER, M. J., SMITH, D. F. Use of recombinant biotinylated aequorin in microtiter and membrane-based assays: purification of recombinant apoaequorin from Escherichia coli. *Biochemistry.* **1992**, *31*, 1433–1442.

SVARTZ, J., BLOMGRAN, R., HAMMARSTROM, S., SODERSTROM, M. Leukotriene C4 synthase homo-oligomers detected in living cells by bioluminescence resonance energy transfer. *Biochim. Biophys. Acta* **2003**, *21*, 90–95.

TANNOUS, B. A., CHIU, N. H., CHRISTOPOULOS, T. K. Heterobifunctional linker between antibodies and reporter genes for immunoassay development. *Anal. Chim. Acta* **2002**, *459*, 169–176.

TANNOUS, B. A., VERHAEGEN, M., CHRISTOPOULOS, T. K., KOURAKLI, A. Combined flash- and glow-type chemiluminescent reactions for high-throughput genotyping of biallelic polymorphisms. *Anal. Biochem.* **2003**, *320*, 266–272.

TATSUMI, H., FUKUDA, S., KIKUCHI, M., KOYAMA, Y. Construction of biotinylated firefly luciferases using biotin acceptor peptides. *Anal. Biochem.* **1996**, *243*, 176–180.

TERRILLON, S., DURROUX, T., MOUILLAC, B., BREIT, A., AYOUB, M. A., TAULAN, M., JOCKERS, R., BARBERIS, C., BOUVIER, M. Oxytocin and vasopressin V1a and V2 receptors form constitutive homo- and heterodimers during biosynthesis. *Mol. Endocrinol.* **2003**, *17*, 677–691.

TRAMIER, M., GAUTIER, I., PIOLOT, T., RAVALET, S., KEMNITZ, K., COPPEY, J., DURIEUX, C., MIGNOTTE, V., COPPEY-MOISAN, M. Picosecond-hetero-FRET microscopy to probe protein-protein interactions in live cells. *Biophys. J.* **2002**, *83*, 3570–3577.

VALDIVIESO-GARCIA, A., DESRUISSEAU, A., RICHE, E., FUKUDA, S., TATSUMI, H. Evaluation of a 24-hour bioluminescent enzyme immunoassay for the rapid detection of Salmonella in chicken carcass rinses. *J. Food Prot.* **2003**, *66*, 1996–2004.

VALENZUELA-FERNANDEZ, A., PALANCHE, T., AMARA, A., MAGERUS, A., ALTMEYER, R., DELAUNAY, T., VIRELIZIER, J. L., BALEUX, F., GALZI, J. L., ARENZANA-SEISDEDOS, F. Optimal inhibition of X4 HIV isolates by the CXC chemokine stromal cell-derived factor

1 alpha requires interaction with cell surface heparan sulfate proteoglycans. *J. Biol. Chem.* **2001**, *276*, 26550–26558.

Verhaegen, M., Christopoulos, T. K. Quantitative polymerase chain reaction based on a dual-analyte chemiluminescence hybridization assay for target DNA and internal standard. *Anal. Chem.* **1998**, *70*, 4120–4125.

Verhaegen, M., Christopoulos, T. K. Bacterial expression of *in vivo*-biotinylated aequorin for direct application to bioluminometric hybridization assays. *Anal. Biochem.* **2002a**, *306*, 314–322.

Verhaegen, M., Christopoulos, T. K. Recombinant *Gaussia* luciferase. Overexpression, purification, and analytical application of a bioluminescent reporter for DNA hybridization. *Anal. Chem.* **2002b**, *74*, 4378–4385.

White, S. R., Christopoulos, T. K. Signal amplification system for DNA hybridization assays based on *in vitro* expression of a DNA label encoding apoaequorin. *Nucleic Acids Res.* **1999**, *27*, e25 i–viii.

White, S. R., Chiu, N. H., Christopoulos, T. K. Expression immunoassay. *Methods* **2000**, *22*, 24–32.

Witkowski, A., Ramanathan, S., Daunert, S. Bioluminescence binding assay for biotin with attomole detection based on recombinant aequorin. *Anal. Chem.* **1994**, *66*, 1837–1840.

Xiao, L., Yang, C., Nelson, C. O., Holloway, B. P., Udhayakumar, V., Lal, A. A. Quantitation of RT-PCR amplified cytokine mRNA by aequorin-based bioluminescence immunoassay. J. Immunol. *Methods* **1996**, *199*, 139–147.

Xu, Y., Piston, D. W., Johnson, C. H. A bioluminescence resonance energy transfer (BRET) system: application to interacting circadian clock proteins. *Proc. Natl. Acad. Sci. USA* **1999**, *96*, 151–156.

Yoshioka, K., Saitoh, O., Nakata, H. Agonist-promoted heteromeric oligomerization between adenosine A(1) and P2Y(1) receptors in living cells. *FEBS Lett.* **2002**, *523*, 147–151.

Zatta, P. F. A new bioluminescent assay for studies of protein G and protein A binding to IgG and IgM. *J. Biochem. Biophys. Methods.* **1996**, *32*, 7–13.

Zeytun, A., Jeromin, A., Scalettar, B. A., Waldo, G. S., Bradbury, A. R. Fluorobodies combine GFP fluorescence with the binding characteristics of antibodies. *Nat. Biotechnol.* **2003**, *21*, 1473–1479.

10
Luminescent Proteins: Applications in Microfluidics and Miniaturized Analytical Systems

Emre Dikici, Laura Rowe, Elizabeth A. Moschou, Anna Rothert, Sapna K. Deo, and Sylvia Daunert

10.1
Miniaturization and Microfluidics

The need for fast, accurate, and low-cost analytical methods for the high-throughput analysis of complex and low-volume samples is fueling up advancements in the area of microfluidic devices. Great efforts have been made toward the miniaturization of conventional bench-top analytical techniques into microfluidic "chip"-based formats [1]. These microfluidic chips are based on architectures of microreservoirs and microchannels (Fig. 10.1) with very small dimensions, which range from tens to hundreds of micrometers, approximating the thickness of a human hair [2]. Small aliquots of sample, on the order of microliters or even smaller, are forced to flow through the microfluidic architecture and participate in analytical processes carried out in the microscale [3]. The fluid physics at the microscale offers some unique advantages as compared to that at the macroscale. These advantages include faster mass transport and much higher surface-to-volume ratios [4]. Microfluidic devices utilize these unique characteristics, which arise from performing sample analysis in miniature, to offer the additional advantages of low reagent consumption, low waste production, a decrease in the cost and time of the analysis, and the capability of increasing the throughput and automating the processes of sample analysis [5, 6].

Along with the miniaturization of various analytical techniques, each separate into a microfluidic format, significant efforts are also currently being made toward the integration of multiple analytical procedures into a single chip [7]. The microfluidic architecture, which is the basic element of microfluidic devices, can be carefully designed to combine microfluidic features for various analytical methods so that the analytical processes can be performed sequentially on the same microfluidic platform [8]. These microfluidic devices, the so-called micro-total-analysis systems (μTAS), or lab-on-a-chip [9], are able to perform multiple analytical procedures, such as sample pretreatment, reaction, separation, and analyte detection, sequentially on a single chip. Ideally, these sophisticated

Photoproteins in Bioanalysis. Edited by Sylvia Daunert and Sapna K. Deo

ISBN: 3-527-31016-9

Fig. 10.1 Image of microfluidic architectures composed of miniature microreservoirs and microchannels.

devices should be capable of automation, thereby increasing the throughput of routine sample analysis, as well as specially designed to meet the needs of specific applications. Additionally, the small physical dimensions of the microfluidic devices offer the advantage of portability, which makes them suitable for use in field analysis of complex real samples.

Microfluidic devices can be made out of a large variety of materials, including glass, silicon, or polymeric materials such as poly(dimethylsiloxane) (PDMS), or they can be hybrids composed of different materials [10, 11]. A variety of methods can be used for the fabrication of microfluidic devices, including wet and dry etching, photolithography, soft and hot embossing, injection molding, laser ablation, and CNC machining [12, 13]. The development of the microfluidic devices can be based on either the direct microfabrication of the chip, such as with laser ablation of the microfluidic architecture on the glass or poly(methylmethacrylate) (PMMA) chip, or the fabrication of a master molding device. In the latter case, the negative of the microfluidic pattern is fabricated on the master. For instance, the pattern can be made on a photoresist on a silicon substrate, and the multiple chips can then be manufactured by replica molding, i.e., by casting the polymer precursor (such as PDMS) onto the master and removing the polymerized chip [14]. The ability of the latter technique to fabricate microfluidic chips with high speed, along with the selection of appropriate low-cost materials such as polymeric materials, reduces the cost of fabrication of the chips and allows the production of disposable microfluidic devices.

The principle of operation of microfluidic devices is based on the manipulation of fluids through a microfluidic architecture. The microfluidic architecture is composed of microfluidic features arranged on a desired pattern. These microfluidic features may be simple, such as microreservoirs and microchannels, or more complex. More complex features include microvalves for stopping and starting the flow of the fluid; mixing devices for the efficient mixing of the fluids; filtration, separation, and fractionation devices; and many others, depending on the need of each specific application (Fig. 10.2) [15].

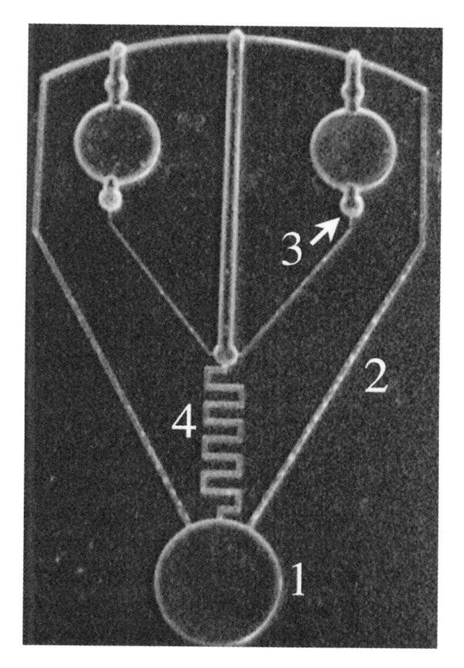

Fig. 10.2 Microfluidic architecture incorporating various microfluidic features, such as microreservoirs (1), microchannels (2), microvalves (3), and a mixing device (4).

The propulsion of the fluids through the microfluidic architectures can be accomplished by utilizing various methods based on different physical parameters. One of the most popular methods of controlling the flow of fluids in microfluidics is based on electrokinetic control generated by the application of an electric field [16]. In this case, the flow of fluid is controlled by switching the voltages on and off, thus preventing the need for valves and allowing the use of simple microfluidic structures. The limitations associated with this method include the need for multiple power supplies or multiple switches to accommodate frequent voltage changes. A more important limitation to this technique is the dependence of the electrokinetic flow on the ionic strength and pH of the fluids. Another commonly used method for fluid propulsion is pressure-driven flow generated by pressure differences [17]. This is a generic method of fluid propulsion that is compatible with a wide range of solvents and a series of microfluidic materials. A disadvantage of pressure-driven flow is that it generates parabolic flow profiles, which can result in sample dispersion [18]. Other less-utilized methods of fluid manipulation include thermal- [19], magnetic- [20], and capillary action-driven [21] methods.

Another promising method of fluid propulsion utilizes the centrifugal force that arises by spinning the microfluidic platform with the help of a rotating motor. The microfluidic devices utilized with this fluid propulsion method have the form of a compact disk (CD) [22]. Each CD can incorporate multiple identical

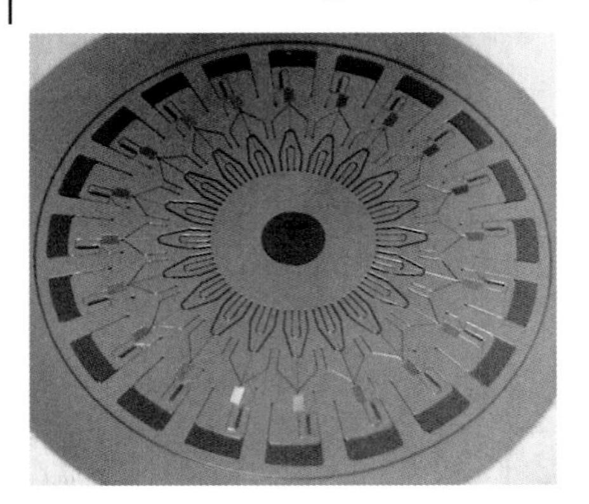

Fig. 10.3 Image of CD microfluidic architectures, for the simultaneous analysis of 20 samples by a single spin of the CD.

microfluidic networks or an assortment of different structures, offering the ability for high-throughput analysis of multiple samples with the simple spinning of the CD (Fig. 10.3) [23, 24]. Additionally, the fluid flow is not affected by matrix effects of the samples when centrifugal microfluidics pumping is used [25]. All the samples experience the same conditions throughout the identical microfluidic architectures on the CD, resulting in reproducible results for multiple sample analysis [26, 27]. Furthermore, the small weight and size dimensions of the CDs offer the advantage of portability, and they can be made out of low-cost polymeric materials for the fabrication of disposable microfluidic devices. Another advantage of the CDs is the fact that, besides offering the ability for integrating multiple analytical processes, they are also amenable to on-chip product isolation and analyte detection [28], allowing the development of complete and automated analytical microdevices [29].

The detection methods that are more commonly used with the microfluidic platforms include electrochemical detection [30], which has the capability of being compact and entirely integrated on the chip; mass spectrometry [31], a powerful technique for the detection of large biological molecules or fragments thereof; and optical detection [32], with fluorescence being the most popular detection method used in microfluidics. The wide range of fluorescence applications results from the high sensitivity of the method, which is desirable for the analysis of miniature samples, and its on-chip analyte detection ability [33, 34].

The area of microfluidics is continuously advancing, and microfluidic chips are finding a wide range of applications in many fields, including clinical [35] and forensic analysis [36], high-throughput screening [37], drug discovery [38], and genomic [39] and proteomic analysis [40, 41]. Many bench-top analytical techniques have been successfully miniaturized and integrated on microfluidic chips, such as various separation techniques including chromatography [42, 43], electrophoresis [44–46], and isoelectric focusing [47]; polymerase chain reaction [48]; sequencing [49]; cell handling [50]; and chemical [51], enzymatic [52], or immunoassay reactions [53–55]. In addition, great efforts have been made recently regarding the adaptation of various sample pretreatment methods on microfluidic devices, such as sample filtration, dialysis, pre-concentration, and derivatization [56]. The aspiration is to integrate the sample pretreatment processes along with the sample analysis methods on a single microfluidic chip. Ideally, these lab-on-a-chip devices will be sophisticated enough to perform all the steps required for the complete analysis of a complex real sample autonomously.

One of the most promising and significant fields of sample analysis integrated onto microfluidic devices is the field of microfluidic assays. A wide variety of microfluidic detection schemes have been employed so far, including enzymes [57, 58], ion-selective membranes [59], supramolecules [60], nanoparticles [61], binding proteins [62] and whole cells [63–65]. In this chapter, we will mainly focus on photoprotein-based assays integrated on the miniaturized microfluidic platforms.

10.2 Photoproteins and Applications in Miniaturized Detection Systems

The term "photoprotein" refers to a protein that participates directly in the light-emitting reaction of a living organism. Additionally, a photoprotein is capable of emitting light of an intensity proportional to the amount of the protein present and is not an unstable, transient intermediate of an enzyme-substrate reaction. Worms, insects, bacteria, crustaceans, fireflies, fungi, and fish are among the luminescent organisms that incorporate approximately 30 different bioluminescence systems [66]. In some of these organisms light is emitted in a glow-type kinetics, whereas in others light is emitted in the form of a rapid flash, lasting for approximately 10 s. These flash-type light emissions can occur from a change in pH, the addition of molecular oxygen, or the binding to calcium ions. Seven Ca^{2+}-binding photoproteins have been isolated: thalassicolin, aequorin, mitrocomin, clytin, obelin, mnemiopsin, and berovin. Four of these proteins have been sequenced and cloned, and obelin and aequorin have been used extensively as labels in the development of various *in vitro* binding assays and *in vivo* cell studies [67].

Luciferase is an enzyme that emits a glow-type bioluminescence when the substrate luciferin is added. Luciferase has been widely used in a variety of bioanalytical applications. Several luciferases that exhibit different luminescent properties have been isolated from various organisms and are attractive tools for the development of multi-analyte detection systems. Another interesting luminescent protein, the green fluorescent protein (GFP), has become one of the most popular biochemical tools in analysis today. The isolation and large-scale purification of these proteins have been milestones in analytical and biological chemistry. Because of their high sensitivity, ease of use, adaptability to miniaturization, and biological safety, they have been gaining popularity in many aspects of today's research. Applications of these photoproteins in miniaturized analytical systems are discussed in the following sections of this chapter.

10.2.1
Green Fluorescent Protein

The green fluorescent protein (GFP) is a naturally occurring fluorescent protein that is isolated from the jellyfish *Aequorea victoria* and some other marine organisms [68]. The role of GFP in nature is to shift the blue bioluminescence emitted by aequorin into the green region. Such a shift improves the quantum efficiency and penetration of light through water and reduces light scattering. GFP from *Aequorea victoria* is a 238-amino-acid protein, with a molecular weight of approximately 27 000 Da. The X-ray crystal structure of the protein reveals an 11-stranded β barrel with a central α helix (Fig. 10.4). GFP consists of an internal chromophore, 4-(*p*-hydroxybenzylidene)-imidazolidone-5-one, which is post-translationally formed from the amino acids Ser 65-Tyr 66-Gly 67 [69]. Several studies involving mutations of the wild-type GFP produced GFP mutants with improved brightness, stability, efficiency of folding, and chromophore formation. These studies also produced GFPs with different emission spectra, so that there are now several GFP colors to choose from [70]. The general bioluminescent properties of these mutants are listed in Table 10.1.

As can be seen from Table 10.1, the blue light of BFP can excite EGFP; therefore, this pair has been used in fluorescence resonance energy transfer (FRET) assays. The major advantage of GFP is that it does not require any additional cofactors or substrates for fluorescence emission. GFP is relatively nontoxic to the cells and is easily imaged and quantified. It is stable in environments of pH 5–12, in the presence of proteases and oxidants, and at temperatures as high as 65 °C [71]. Additionally, it can be used to genetically engineer fusion proteins, without affecting the function and the localization of its fusion partner, because of its small size. For these reasons, GFP has been extensively used as a reporter in studies of cellular dynamics and cell trafficking, in intact tissue studies, and as a reporter in both homogenous and cell-based assays [72–93]. Moreover, GFP has

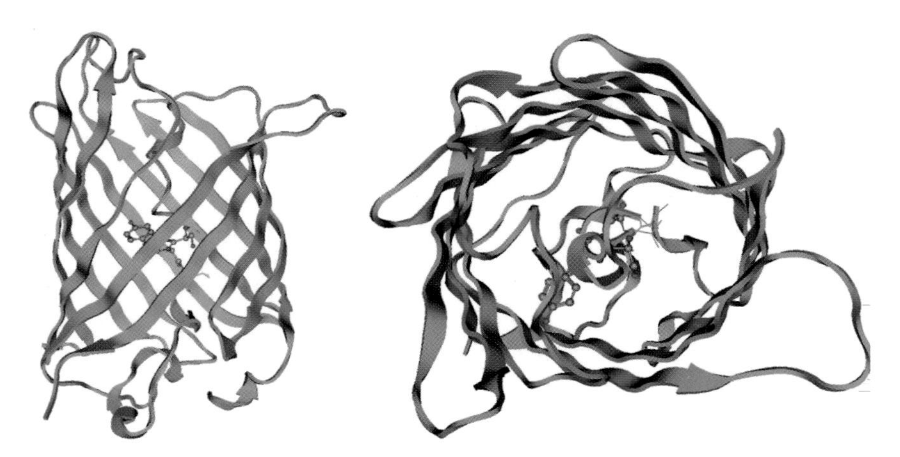

Fig. 10.4 X-ray crystal structure of GFP.

Table 10.1 Bioluminescent properties of various GFP mutants.

Mutant	*Mutations*	*Excitation wavelength (nm)*	*Emission wavelength (nm)*	*Quantum yield*
Wild type	–	395 (470)	509	0.77
EGFP (enhanced green fluorescent protein)	S65T	484	510	0.70
BFP (blue fluorescent protein)	Y66H, Y145F	380	440	0.20
YFP (yellow fluorescent protein)	S65G, V68L, S72A, T203Y	512 (498)	529	0.63
CFP (cyan fluorescent protein)	T66W	439 (453)	476 (501)	0.24

been used in high-throughput drug screening, for the evaluation of viral vectors in gene therapy, for biological pest control, and for monitoring genetically altered microorganisms in bioremediation processes [73].

10.2.1.1 GFP in Miniaturized Microfluidic-based Assays

Because of the inherent advantages of GFP, this fluorescent protein has been successfully employed in numerous small-volume assays, including microassays integrated onto microfluidic platforms. One such example includes the development of a microfluidic assay employing the GFP mutant GFP_{uv} for the detection of arsenite and antimonite [74]. In this work, a whole-cell assay was developed incorporating the gene for the reporter protein GFP_{uv}. For that, a plasmid that encoded the gene for GFP_{uv} was constructed and transformed into the bacteria *E. coli*. The GFP_{uv} gene was placed under the control of the promoter for ArsR. In the absence of arsenic or antimonite, the ArsR protein binds to the *ars* promoter sequence and blocks the transcription of GFP_{uv}. However, in the presence of arsenite or antimonite, arsenite or antimonite will bind to the ArsR protein and cause a conformational change, which dissociates the ArsR protein from the *ars* promoter sequence. Transcription and translation can then occur, which leads to the production of GFP_{uv} and the generation of a fluorescence signal. The intensity of the GFP_{uv} fluorescence signal was correlated to the concentration of antimonite or arsenite present in the sample. This assay was adapted to a CD microfluidic platform, which employed the microfluidic architecture presented in Fig. 10.2. The experimental procedure included the loading of 15 μL of cell aliquot, which contained the plasmid pSD10, on the CD, along with 15 μL of arsenite and antimonite samples in varying concentrations. The simple spinning of the CD at 1000 rpm resulted in the efficient mixing of the cell reagent with the arsenite/antimonite samples and the emission of the GFP_{uv} fluorescence signal. The CD microfluidic platform was optically transparent, which allowed on-chip detection of the fluorescence signal using a fiber optic fluorometer. This approach provided a fast, simple, sensitive, and selective method for the detection of arsenite and antimonite, with detection limits on the order of 1×10^{-6} M. Given the fast assay time, low reagent consumption, and ability to do parallel analysis, this method may potentially be employed for high-throughput detection of important pollutants. Furthermore, the CD microfluidic platforms offer the additional advantage of portability, which is desirable for on-site monitoring of target pollutants.

Another mutant of GFP, the enhanced GFP (EGFP), has been employed for the high-throughput screening of calmodulin antagonists in a microtiter plate and a CD microfluidic platform [70, 75]. A class-selective homogeneous assay for the detection of phenothiazine-type antidepressants has been developed. Phenothiazine drugs are known to bind the Ca^{2+}-binding protein calmodulin effectively, inhibiting the interaction of calmodulin with other peptides or proteins in various biological pathways. A fusion protein of calmodulin (CaM) and EGFP was first genetically constructed. The conformational change of calmodulin, upon binding to the phenothiazine class of drugs, alters the fluorescence properties of the fused EGFP, and these changes can be correlated to the concentration of the drug present in the sample. The comparison of the assay performance using the two different measurement platforms, the microtiter plate, and the CD microfluidic platform is presented in Table 10.2.

The microassay integrated on the CD microfluidic platform was proven to be an efficient method for the screening of phenothiazine drugs, offering a significant decrease in the sample volume used and in the analysis time of the microassay. The utilization of a CD, incorporating multiple microfluidic architectures, makes this platform suitable for high-throughput drug screening. Additionally, the authors successfully established the ability to dry biological reagents directly on the CD and later reconstitute them, without deterioration of the performance of the assay. For this reconstitution assay, 12 μL of the protein solution was loaded in a microreservoir of the CD and allowed to dry for 3.5 h under humidity-controlled conditions. An aliquot of 12 μL of water was later added to the reservoir containing the dried protein, allowing the protein to reconstitute within the timeframe of 5 min. The assay utilizing the dried and later reconstituted protein solution exhibited comparable performance with the assay employing the liquid protein reagent. This work demonstrated the ability to develop a microfluidic platform that can be stored with the required reagent incorporated into its microstructures. Such reagent storage advances can be utilized for the development of microfluidic systems that can be stored and be ready to use as dictated by the needs of the specific application.

Table 10.2 Performance of phenothiazine drug detection assays using two measurement platforms.

	Detection limit	*Assay volume*	*Analysis time*
Microtiter plate	4.0×10^{-5} M	100–300 μL	15 min
CD microfluidic platform	7.4×10^{-4} M	25 μL	5 min

The ability to image the fluorescence of GFP in assays *in vitro* and *in vivo* has further expanded the applications of this protein as a reporter. Yu et al. employed GFP to demonstrate solid-phase extraction by using a highly porous monolithic column within the microchannel of a microchip [76]. The monolith, composed of poly(butyl methacrylate-*co*-ethylene dimethacrylate), was used to extract the protein from a 18.5-nmol L^{-1} GFP solution, which was pumped through the monolith with a flow rate of 3 μL min^{-1}. The protein that was loaded on the monolith was eluted with a 1 : 1 water:acetonitrile mixture, resulting in a 1000-fold more concentrated GFP solution. Similar experiments for the preconcentration of proteins, which employed different polymers, were also successfully performed employing GFP as the model protein [77]. Lee et al. demonstrated the isoelectric focusing (IEF) of EGFP and BSA in a 1.2-cm-long microchannel (100 μm in width and 40 μm in depth). The estimated pI values of both EGFP and BSA were in good agreement with the known values of the two proteins [78, 79]. GFP was also used as a visualization protein on a microchip that was designed to detect protein–protein interactions. In this work, Gibbons and coworkers employed a 6×His-tagged GFP in order to visualize the interaction of 6×His-GFP with an anti-6×His antibody. Because the anti-6×His antibody is not fluorescent, it cannot be detected with a laser-induced fluorescence (LIF) detector. But the interaction of the anti-6×His antibody with 6×His-GFP on a microfluidic device allows the measurement of the former with LIF [80].

The fluorescent protein EGFP has been used as a model protein for the evaluation of the performance of chromatographic separation media integrated on a CD microfluidic platform [81]. The separation media developed were methacrylate-based monolithic columns with selected physical characteristics and chemical functionalities that generated ion-exchange [78] and affinity [82] interactions for the successful binding and elution of the target protein. The CD employed was a closed-channel device with a microfluidic architecture that allowed the chromatographic separation on a monolith and the subsequent isolation and detection of the purified protein in a detection chamber. The CD microfluidic device (Fig. 10.5) was made of the optically transparent PDMS (poly(dimethylsiloxane)), which allowed for the fluorometric determination of the isolated protein directly on the CD platform. The ion-exchange interaction and elution of 4 μL of 3×10^{-6} M EGFP, the concentration of 40 μL of 3×10^{-7} M EGFP, and the affinity purification of 4 μL of 3×10^{-6} M of fused protein FLAG–EGFP from lysed cells have been performed efficiently on the CD, each in a timeframe of less than 5 min.

GFP has also been used as a label in a microfluidic cell sorter. In this device, Quake and coworkers sorted *E. coli* HB101 cells that expressed GFP [83]. The cell sorting was carried out by the introduction of different ratios of wild-type GFP-expressing *E. coli* HB101 cells into the microfluidic device, and the mobility of the cells in the microchannel using electrokinetic propulsion. Laser-induced fluorescence emission was used to monitor the flow direction from the waste reservoir to the collection reservoir. After sorting, the cells were collected and streaked onto a LB-agar plate for colony counting. By using this method, an

Fig. 10.5 CD microfluidic platform containing five microfluidic architectures for the preconcentration and purification of EGFP. Each microfluidic architecture contains a microreservoir (L) for loading the sample solution, a microreservoir (C) incorporating the microcolumn, a fractionation device (F), the microreservoir for storage of the waste solutions (W), and the detection chamber (D) for the isolation and fluorometric detection of the purified protein EGFP on the CD.

enrichment of 30-fold was achieved. The throughput of the system is only 20 cells per second, which is a considerably slower throughput than that of conventional fluorescence-activated cell sorter (FACS) instruments. However, it is expected that with the ability to manufacture parallel systems, the throughput of this GFP microfluidic cell sorter could be increased considerably.

Li et al. demonstrated real-time fluorescence measurements that indicated cellular changes in single Jurkat T yeast cells. Yeast cells were transported and retained in special U-shaped microstructures on a glass microfluidic chip, which could retain the cells while allowing the liquid to pass. The inhibitory protein IκB, which is involved in the NF-κB signaling pathway in yeast cells, was fused to EGFP and expressed in Jurkat T yeast cells. The cellular degradation of the inhibitory IκB protein was subsequently monitored within a single cell by measuring the decrease of the EGFP fluorescence signal. This work demonstrated that microfluidic chips can successfully facilitate the monitoring of cellular changes of biological cells at the single-cell level [84].

Fluorescence resonance energy transfer (FRET) has been widely used to study protein–protein and protein–ligand interactions [85–87]. Interactions at the subcellular level can also be monitored by employing FRET analysis using a fluorescence microscope [88, 89]. Only a few FRET-based assays adapted to microfluidic platforms have been reported [90–94]. One example of a FRET-based microfluidic assay is the DNA mutation assay developed by Wabuyele and colleagues [111]. They used FRET detection integrated on a microfluidic platform for the development of a molecular beacon assay, which was used to detect a point mutation in unamplified genomic DNA. The point mutation is located in the kRAS oncogene, and this mutation is highly correlated to colorectal cancer.

Allele-specific primers carrying donor/acceptor fluorescent dyes (Cy5/Cy5.5) would link together only when the primers flanked the kRAS oncogene mutation. This ligation reaction occurred in a microfluidic chip and resulted in a detectable FRET signal that was monitored with a near-IR single-molecule fluorescence detection system. Although GFP and its red-shifted mutants have not yet been used in FRET microfluidics platforms, they have been extensively used as FRET pairs, and it is therefore likely that they will be used in various FRET microfluidic protocols in the future [95–99]. The FRET-based approach coupled with miniaturized systems will have direct applications in the HTS of pharmaceuticals and other biologically active compounds.

As can be seen from the examples above, because of its unique properties, GFP has become a versatile tool in biological, analytical, and microfluidic applications. Mutants of GFP with different emission characteristics should further expand the range of applications of GFP in miniaturized analytical platforms and in array detection methodologies.

10.2.2
Luciferase

Luciferase is a bioluminescent protein that exists in diverse organisms such as bacteria, insects, and marine coelenterates. Because these systems have evolved independently, homology is not typically observed among the luciferase genes from different groups. Therefore, the most commonly used luciferase genes are distinguishable based on the bioluminescent organism from which they have been isolated, using the abbreviations *lux* (bacterial), *luc* (firefly), and *Rluc* (sea pansy *Renilla*) [100].

As a result of their inherent differences, different luciferases have advantages in different types of reporter gene-based sensing and screening applications. The bacterial luciferases are useful in measuring and analyzing prokaryotic gene transcription but have limited use in mammalian cells because they are generally heat-labile, dimeric proteins. Firefly luciferase has high sensitivity and a broad dynamic range, making it a popular reporter gene for use in mammalian cell systems. Because the substrate for firefly luciferase is not naturally membrane permeable, the cells used in these assays originally have to be lysed before the substrate, luciferin, can be added. This complication has been resolved through the use of firefly luciferin esters that are membrane permeable and photolysable. The luciferase from the bioluminescent sea pansy *Renilla reniformis* has recently found more applications in reporter gene assays. Much like the photoproteins aequorin and obelin, the *Renilla* luciferase catalyzes the oxidation of a coelenterazine substrate. Unlike the firefly luciferin, coelenterazine is membrane permeable, and therefore, this system has inherent advantages in whole-cell reporter gene applications [101].

10.2.2.1 **Luciferase in Miniaturized Microfluidic-based Assays**

Several of the luciferase reporter gene assays developed so far demonstrate characteristics that make them suitable for miniaturization and incorporation onto high-throughput screening microfluidic devices. However, there have been only a few reports of the successful integration of luciferase-based assays onto miniaturized analytical systems. Such examples include the utilization of luciferase in miniaturized 384- and 1536-well plate formats for the screening of drugs. Luciferase reporter gene assays on 384-well plate platforms have been developed for the measurement of Tp53 response to anticancer drugs [102] and the screening of compounds that inhibit the replication of the West Nile virus [103]. Additionally, successful small-volume, high-throughput analyses have been achieved with 1536-well plates for the identification of inhibitors of bacterial protein synthesis [104], as well as in the detection of compounds that inhibit luciferase expression in a mammalian cell-based reporter gene assay [105]. These screening systems utilize volumes of 2–3 μL that are dispensed using an automated system, thereby reducing the consumption of often costly reagents. The increased well density of the 1536-well plate microtiter platform allows for more rapid screening of a larger number of compounds. The luciferase reporter is ideal for these miniaturized screens because its high sensitivity enables detection even when it is present in very minute amounts.

ATP and ATP-conjugated metabolites have been detected with a microfluidics-based luciferase system using 10 μL aliquots of sample [106]. The microfluidic device developed was based on a closed-channel PDMS substrate that contained a Y-shaped microfluidic architecture. The mixing of the sample with the luciferase was accomplished at the end of the stem of the Y microfluidic network. The resulting bioluminescent signal was able to determine the ATP levels down to a 0.2-μM concentration. This signal was observed within 30 s after the introduction of the sample onto the microfluidic platform. The microfluidic chip developed was used for the determination of ATP levels in cell lysates, as well as for the determination of the concentration of galactose, an ATP-conjugated metabolite, in urine samples.

Tani's group has developed a microfluidic system for the detection of mutagenicity based on a whole-cell bioassay employing the bioluminescent reporter protein luciferase [107]. This system employed mutagen-sensitive *E. coli* sensor strains. In the presence of mutagens these sensor strains expressed luciferase, thereby producing a bioluminescent signal. These cells were immobilized on a microfluidic chip by entrapment in an agarose gel, which was formed *in situ* on the device. The microfluidic chip developed was based on a silicon substrate that contained multiple 700-μm^2 microwells with a volume of 0.25 μL each. These microwells were used for the immobilization of the cells. The microfluidic device was completed by sandwiching the silicone substrate between two PDMS layers. Each PDMS layer contained a series of microchannels (700 μm wide and 200 μm deep, volume of 3 μL), which, after assembling the device, interconnected the microwells of the chip. The microchannels were filled with various concentrations of the genotoxic substances tested, and after 1 h incubation the bioluminescence triggered by the introduction of luciferin and ATP was used for the determination of the levels of these mutagens. The ability to immobize a series of sensor bacteria on different microwells on the same microfluidic chip can allow for the high-throughput screening of various substances, such as environmental pollutants and drugs.

Emneus and coworkers developed another microfluidic system that utilizes the expression of luciferase for the real-time and high-throughput screening of ligands for transmembrane receptors. In this work, genetically modified HeLa cells, incorporating the highly efficient reporter system HFF11, were immobilized on a silicon microfluidic chip. The design of the chip contained 28 parallel V grooves, with each groove being 10 mm long, 100 μm-wide, and 71 μm-deep and having a total volume of 1.9 μL [108]. The introduction of an aliquot of 2 μL of a transmembrane receptor ligand on the chip with the immobilized HeLa cells initiates a series of intracellular reactions, leading to the expression of luciferase. Following the injection of luciferin and ATP, the bioluminescence signal emitted from luciferase was used for the determination of the transmembrane receptor ligands. The expression of luciferase on the microfluidic chip was detected 30 min after stimulation with the ligand, which is a much shorter analysis time than the 6 h required for the analogous microtiter plate assay. Additionally, the microfluidic system developed was able to perform real-time and continuous monitoring of dynamic cellular events for a period up to 30 h.

These early successes using various luciferases as reporters in highly miniaturized systems will likely serve as a model for future applications of bioluminescent proteins. The ability to incorporate these proteins into miniaturized assays using microfluidics systems provides the possibility of automating several steps in the process and therefore creating assays and screening systems that are rapid and highly efficient.

10.2.3
Aequorin

Aequorin is a bioluminescent photoprotein isolated from the same jellyfish as GFP, *Aequorea victoria*. Aequorin is composed of apoaequorin, a chromophoric unit, coelenterazine, and molecular oxygen. The binding of Ca^{2+} to aequorin results in a conformational change that causes the protein to behave as an oxygenase towards coelenterazine. Coelenterazine is oxidized to an excited coelenteramide, and the relaxation of this coelenteramide to its ground state is accomplished by the emission of 469-nm light and CO_2 [109, 110]. The quantum yield of the photoprotein in the light-emitting reaction is between 0.15 and 0.20 [111].

Aequorin has been used extensively as a label in the past years, most notably for the detection of calcium concentrations *in vivo*, and as a label in immunoassays and DNA hybridization assays. Aequorin has been utilized in these analytical applications because of the inherent advantages it has as a label. Some of these advantages include that aequorin is active at physiological pH, is nontoxic to cells, does not require an excitation light source, and can be successfully cloned and expressed in orthogonal species and cell types [112]. Moreover, aequorin is an extremely sensitive reporter. As opposed to fluorescence, there is a low background level of bioluminescence in most biological fluids [113]. This low background noise makes aequorin more sensitive than fluorescent labels, such as GFP, when biological fluids are analyzed. Such sensitivity makes aequorin an attractive reporter for miniaturized microfluidics platforms, whose small sample volumes require labels with very low detection limits. The implementation of aequorin-based immunoassays and DNA hybridization assays in miniaturized and microfluidic platforms will be discussed in the next section.

10.2.3.1 Aequorin in Miniaturized Microfluidic-based Assays

Homogeneous assays are often preferred for miniaturization, since their protocols are simpler and faster than those of heterogeneous assays. In homogeneous assays all the necessary reagents are present in solution, and the concentration of an analyte can be determined in one step. Homogeneous assays eliminate time-consuming and error-introducing incubation and washing steps. They also provide a directly measurable signal without the need for separation procedures. Moreover, because homogeneous assays are simpler, they are preferred in clinical use because there is no need for highly trained personnel to perform them [114].

Homogeneous assays have been developed for biotin [115, 116], digoxin [117], G protein-coupled receptors [118], and protease activity [119] using aequorin. Moreover, Grosevnor et al. reported the development of a biotin assay that can be performed using picoliter volumes [115], with the small volume of sample still retaining a low detection limit (10^{-14} mol biotin). For that purpose, picoliter vials were created by laser-ablating glass cover slips until a vial volume of 350 pL per well was formed. Figure 10.6 shows a scanning electron microscopic image of the picoliter vials.

In this assay a biotin–aequorin conjugate was employed. It was determined that the biotin component of the biotin–aequorin conjugate binds tightly to the protein avidin and that increasing amounts of avidin bound to the conjugate result in a dose-dependent decrease in the conjugate's bioluminescent signal.

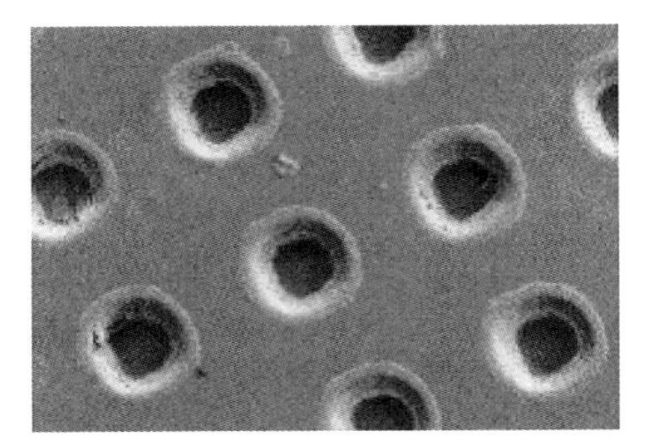

Fig. 10.6 SEM image of picoliter vials. The volume of each vial is approximately 350 pL.

For this homogenous biotin assay, varying amounts of biotin were incubated with a constant amount of free avidin in the picoliter vials 5 min prior to injecting a constant amount of the biotin–aequorin conjugate. The free avidin would bind with the biotin in the sample, reducing the amount of avidin available to bind with the conjugate. Therefore, increasing concentrations of free biotin in solution corresponded to an increase in bioluminescence following the addition of a Ca^{2+}-containing buffer. Given the small sample volume employed in the picoliter vials, this homogeneous biotin assay could potentially be adapted to a miniaturized microfluidic platform.

The DNA hybridization assay is an additional bench-top analytical technique that is being incorporated onto miniaturized microfluidic platforms [120]. DNA hybridization assays utilize, and are used to detect, specific DNA sequences that are unique to individual pathogens or genetic diseases. Hybridization assays take advantage of the specific interactions and bonding between one string of nucleotides and a complementary nucleotide sequence. The complementary nucleotide sequence competes with an identical complementary sequence that has been conjugated to a detectable label. Enzymes, fluorophores, and bioluminescent proteins have all been incorporated into various DNA hybridization assay designs.

Recently, Jia and colleagues employed aequorin in one such DNA hybridization assay using a miniaturized microfluidics platform [121]. In this assay single-stranded DNA probes were first immobilized in the hybridization column of a microfluidics flow cell (Fig. 10.6). The disposable, polydimethylsiloxane (PDMS), flow cells were replicated from a master mold designed using lithography techniques. Four individual flow cells were aligned on a single compact disc platform. This CD platform utilizes centrifugal force to manipulate the reagents in the flow cells. Changing the angular velocity of the spinning CD allows sequential release of reagents from subsequent chambers (Fig. 10.7) and allows for the fine-tuning of flow rates.

Biotinylated DNA target strands that were complementary to the immobilized DNA probe strands were located in the sample chamber (Fig. 10.7). This bio-

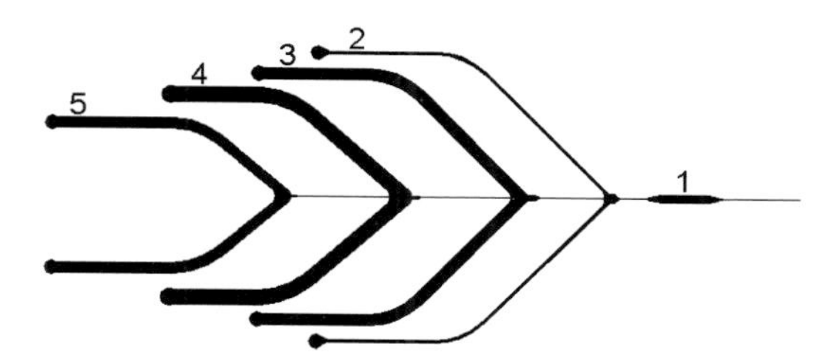

Fig. 10.7 Design of the flow cell:
1. Hybridization column.
2. Sample chamber.
3. Blocking buffer chamber.
4. Wash buffer chamber.
5. Streptavidin–aequorin conjugate chamber.

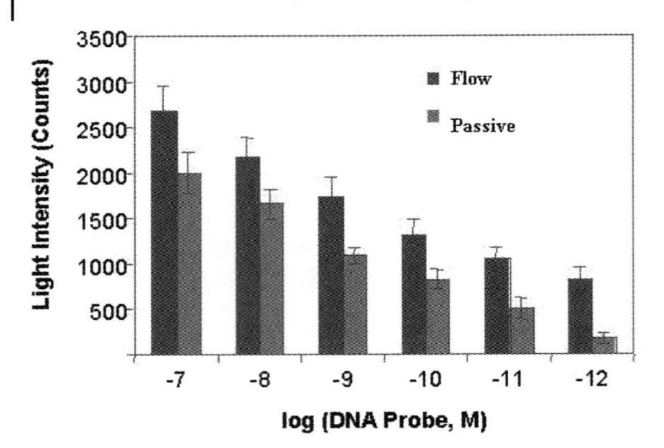

Fig. 10.8 Dose–response curve for the microfluidic, aequorin-based DNA hybridization assay. Flow and passive indicates the assay performed with fluids flowing via centrifugal force (flow) and fluid being allowed to simply remain in the hybridization column for 3 min without flow conditions (passive).

tinylated DNA was pushed into the hybridization column by spinning the CD at a specific velocity (500 rpm). The complementary biotinylated DNA became bound to the immobilized DNA in the hybridization column, and excess, unbound DNA was removed with subsequent washing steps.

Next, streptavidin–aequorin conjugates were released into the hybridization column (Fig. 10.7) by spinning the CD at 1000 rpm. The streptavidin component of the streptavidin–aequorin conjugates binds tightly to the biotinylated target DNA, and the aequorin produces a bioluminescent signal following the addition of a Ca^{2+}-containing buffer. The bioluminescent light signal detected was found to be proportional to the concentration of biotinylated target DNA present in a sample, as shown in the following dose–response curve in Fig. 10.8.

This microfluidics aequorin-based hybridization assay was able to detect the presence of the complementary target DNA down to 1-pM concentrations. This detection limit is two orders of magnitude lower than the corresponding fluorescence hybridization assay performed with a fluorogenic substrate.

The implementation of aequorin onto miniaturized microfluidic platforms has to date been very limited. To our knowledge, the aforementioned DNA hybridization assay is the only case in which aequorin has been employed in a miniaturized microfluidics device. However, the sensitivity, versatility, and widespread use of aequorin in assay development suggest that it may be utilized more and more as miniaturization and microfluidics technologies advance.

10.3 Future Perspectives

There are a number of photoproteins in addition to those described in this chapter that have been discovered and characterized. The examples include obelin, mitrocomin, clytin, red fluorescent proteins, etc. Among these, obelin, mitrocomin, and clytin are calcium-binding photoproteins similar to aequorin in that they consist of apoprotein, coelenterazine, and molecular oxygen. Applications of obelin as a label in bioanalysis have been established previously, but it has not yet been employed in miniaturized systems. Red fluorescent proteins isolated from a variety of organisms are structurally very similar to GFP but have red-shifted excitation and emission maxima when compared to GFP. This property should allow easy analysis in biological samples because of the low background signal in the red region of the spectrum. The proteins described here and the new proteins that are continually being discovered should provide a large arsenal for the development of bioanalysis methods and their applications in miniaturized microfluidic systems.

Research into miniaturized, microfluidic technology amenable to automation has grown rapidly during the past decade and shows no indication of plateauing. High-throughput screening during drug development and portable on-site analytical devices are two important applications of miniaturized microfluidic platforms. This trend towards miniaturization is driven both by the increased speed and reproducibility automation confers to experiments and by the considerable expense of biomaterials and reagents that is reduced with small volumes. Miniaturization is not without drawbacks, however. One of those drawbacks is the necessity of incorporating extremely sensitive labels with detection limits low enough to be discriminated from noise in such small sample volumes. Luminescent proteins, such as GFP, luciferase, and aequorin, are sufficiently sensitive to have been successfully employed as reporters in multiple miniaturized microfluidic systems, as hence reviewed. The versatility, biological safety, and sensitivity of these luminescent proteins give them the potential to become an even more prominent label in future miniaturized microfluidic technologies.

Ackowledgments

This work was funded by the National Institute of Health (NIH), the National Science Foundation (NSF), the National Aeronautics and Space Administration (NASA), and the National Institute of Environmental Health Sciences (NIEHS).

References

1 Ismagilov, R. F. *Angew. Chem. Int. Ed.* **2003**, *42*, 4130.

2 Meldrum, D. R., Holl, M. R. *Science* **2002**, *297*, 1197.

3 Erickson, D., Li, D. *Anal. Chim. Acta* **2004**, *507*, 11.

4 Hansen, C., Quake, S. R. *Current Opinion in Structural Biology* **2003**, *13*, 538.

5 De Mello, A. J. *Anal. Bioanal. Chem.* **2002**, *372*, 12.

6 Felton, M. J. *Anal. Chem.* **2003**, 505A.

7 Bilitewski, U., Genrich, M., Kadow, S., Mersal, G. *Anal. Bioanal. Chem.* **2003**, *377*, 556.

8 Polson, N. A., Hayes, M. A. *Anal. Chem. 2001, 312A.*

9 Gould, P. *Materials Today* **2004**, 48.

10 Fiorini, G. S., Chiu, D. T. *BioTechniques* **2005**, *38*, 429.

11 Jakeway, S. C., de Mello, A. J., Russel, E. L. *Fresenius Anal. Chem.* **2000**, *366*, 525.

12 Mijatovic, D., Eijkel, J. C. T., van den Berg, A. *Lab Chip* **2005**, *5*, 492.

13 Madou, M. J. *Fundamentals of Microfabrication-The Science of Miniaturization, Second Edition* **2002**, CRC Press LLC, Boca Raton, FL.

14 Beebe, D. J., Mensing, G. A., Walker, G. M. *Annu. Rev. Biomed. Eng.* **2002**, *4*, 261.

15 Ouellette, J. *The Industrial Physicist* **2003**, 14.

16 Bruin, G. M. *Electrophoresis* **2000**, *21*, 3931.

17 Sato, K., Hibara, A., Tokeshi, M., Hisamoto, H., Kitamori, T. *Adv. Drug. Deliv. Rev.* **2003**, *55*, 379.

18 Paul, P. H., Garguilo, M. G., Rakestraw, D. J. *Anal. Chem.* **1998**, *70*, 2459.

19 Sammarco, T. S., Burns, M. A. *AIChE J.* **1999**, *45*, 350.

20 Barbic, M., Mock, J. J., Gray, A. P., Schultz, S. *Applied Physics Letters* **2001**, *79*, 1399.

21 Juncker, D.,Schmid, H., Drechsler, U., Wolf, H., Wolf, M., Michel, B., de Rooij, N., Delamarche, E. *Anal. Chem.* **2002**, *74*, 6139.

22 Madou, M. J., Lu, Y., Lai, S., Lee, L. J., Daunert, S. *Micro Total Analysis Systems* **2000**, 565.

23 Duffy, D. C., Gillis, H. L., Lin, J., Sheppard, N. F., Jr., Kellogg, G. J. *Anal. Chem.* **1999**, *71*, 4669.

24 Badr, I. H. A., Johnson, R. D., Madou, M. J., Bachas, L. G. *Anal. Chem.* **2002**, *74*, 5569.

25 Felton, M. J. *Anal. Chem.* **2003**, 302A.

26 Puckett, L. G.,Dikici, E., Lai, S., Madou, M., Bachas, L. G., Daunert, S. *Anal. Chem.* **2004**, *76*, 7263.

27 Johnson, R. D., Badr, I. H. A., Barrett, G., Lai, S., Lu, Y., Madou, Marc, J., Bachas, L. G. *Anal. Chem.* **2001**, *73*, 3940.

28 Moschou, E. A., Nicholson, A. D., Jia, G., Zoval, J. V., Madou, M. J., Bachas, L. G., Daunert, S. Submitted in *Anal. Chem.*

29 Madou, M. J., Lee, L. J., Daunert, S., Lai, S., Shih, C.-H. *Biomed. Microdev.* **2001**, *3*, 245.

30 Wang, J. *Talanta* **2002**, *56*, 223.

31 Lion, N., Reymond, F., Girault, H. H., Rossier, J. S. *Current Opinion in Biotechnology* **2004**, *15*, 31.

32 Mogensen, K. B., Klank, H., Kutter, J. P. *Electrophoresis* **2004**, *25*, 3498.

33 Verpoorte, E. *Lab Chip* **2003**, *3*, 42N.

34 Johnson, R. D., Badr, I. H., Barrett, B., Lai, S., Lu, Y., Madou, M. J., Bachas, L. G. *Anal. Chem.* **2001**, *73*, 3940.

35 Tudos, A. J., Besselink, G. A. J., Schasfoort, R. B. M. *Lab on a Chip* **2001**, *1*, 83.

36 Verpoorte, E. *Electrophoresis* **2002**, *23*, 677.

37 Wolcke, J., Ullmann, D. *Drug Discovery Today* **2001**, *6*, 637.

38 Pang, H. M., Kenseth, J., Coldiron, S. *Drug Discovery Today* **2004**, *9*, 1072.

39 Walter, G., Bussow, K., Lueking, A., Glokler, J. *Trends Molec. Med.* **2002**, *8*, 250.

40 Lion, N., Rohner, T. C., Dayon, L., Arnaud, I. L., Damoc, E., Youhnovski, N., Wu, Z. Y., Roussel, C., Josserand, J., Jensen, H., Rossier, J. S.,

Przybylski, M., Girault, H. H. *Electrophoresis* **2003**, *24*, 3533.
41 Cooper, J. W., Wang, Y., Lee, C. S. *Electrophoresis* **2004**, *25*, 3913.
42 Fintschenko, Y., Ngola, S. M., Shepodd, T. J., Arnold, D. W. *Proceedings of Micro Total Analysis Systems 2000*, Kluwer Academic Publishers: Dordrecht, The Netherlands **2000**, 411.
43 Moschou, E. A., Nicholson, A. D., Jia, G., Zoval, J. V., Madou, M. J., Bachas, L. G., Daunert, S. *Anal. Bioanal. Chem.* **2006**, in press.
44 Chiem, N., Harrison, D. J. *Anal. Chem.* **1997**, *69*, 373.
45 Hutt, D. L., Glavin, D. P., Bada, J. L., Mathies, R. A. *Anal. Chem.* **1999**, *71*, 4000.
46 Hong, J. W., Hosokawa, K., Fujii, T., Seki, M., Endo, I. *Biotechnol. Prog.* **2001**, *17*, 958.
47 Cabrera, C. R., Finlayson, B., Yager, P. *Anal. Chem.* **2001**, *73*, 658.
48 Wook Hong, J., Qauke, S. R. *Nature Biotechnology* **2003**, *21*, 1179.
49 Khandurina, J., Guttman, A. *J. Chromatogr. A* **2002**, *943*, 159.
50 Sia, S. K., Whitesides, G. M. *Electrophoresis* **2003**, *24*, 3563.
51 Eijkel, J. C., Prak, A., Cowen, S., Craston, D. H., Manz, A. *J. Chromatogr. A* **1998**, *815*, 265.
52 Tanaka, Y., Slyadnev, M. N., Hibara, A., Tokeshi, M., Kitamori, T. *J. Chromatogr. A* **2000**, *894*, 2514.
53 Rossier, J. S., Girault, H. H. *Lab on a Chip* **2001**, *1*, 153.
54 Eteshola, E., Balberg, M. *Biomed. Microdev.* **2004**, *6*, 7.
55 Koutny, L. B., Schmalzing, D., Taylor, T. A., Fuchs, M. *Anal. Chem.* **1996**, *68*, 18.
56 Lichtenberg, J., de Rooij, N. F., Verpoorte, E. *Talanta* **2002**, *56*, 233.
57 Hadd, A. G., Jacobson, S. C., Ramsey, J. M. *Anal. Chem.* **1999**, *71*, 5206.
58 Hadd, A. G., Raymond, D. E., Halliwell, J. W., Jacobson, S. C., Ramsey, J. M. *Anal. Chem.* **1997**, *69*, 3407.
59 Badr, I. H. A., Johnson, J. D., Madou, M., Bachas, L. G. *Anal. Chem.* **2002**, *74*, 5569.
60 Rudzinski, C. M., Young, A. M., Nocera, D. G. *J. Am. Chem. Soc.* **2002**, *124*, 1723.
61 Perez, J. M., Josephson, L., O'Loughlin, T., Hogemann, D., Weissleder, R. *Nature Biotech.* **2002**, *20*, 816.
62 Galanina, O. E., Mecklenburg, M., Nifantiev, N. E., Pazynina, G. V., Bovin, N. V. *Lab on a Chip* **2003**, *3*, 260.
63 Walker, G. M., Ozers, M. S., Beebe, D. J. *Biomed. Microdev.* **2002**, *4*, 161.
64 Chang, W.-J., Akin, D., Sedlak, M., Ladish, M. R., Bashir, R. *Biomed. Microdev.* **2003**, *5*, 281.
65 Walker, G. M., Ozers, M. S., Beebe, D. J. *Sensors and Actuators B* **2004**, *98*, 347.
66 Hastings, J. W. *Gene* **1996**, *173*, 5.
67 Lewis, J. C., Daunert, S. *Fres. J. Anal. Chem.* **2000**, *366*, 760.
68 Wilson, T., Hastings, J. W. *Annu. Rev. Cell Dev. Biol.* **1998**, *14*, 197.
69 Cubitt, A. B., Heim, R., Adams, S. R., Boyd, A. E., Gross, L. A., Tsien, R. Y. *Trends Biochem. Sci.* **1995**, *20*, 448.
70 Pollok, B. A., Heim, R. *Trends Cell Biol.* **1999**, *9*, 57.
71 Tsien, R. Y. *Annu. Rev. Biochem.* **1998**, *67*, 509.
72 Periasamy, A., Day, R. N. *Meth. Cell Biol.* **1999**, *58*, 293.
73 Errampalli, D., Leung, K., Cassidy, M. B., Kostrzynska, M., Blears, M., Lee, H., Trevors, J. T. *J. Microbiol. Meth.* **1999**, *35*, 187.
74 Rothert, A., Deo, S. K., Puckett, L. G., Millner, L., Madou, M., Daunert, S. *Abstracts, 55th Southeast Regional Meeting of the American Chemical Society* **2004**, Atlanta, GA.
75 Puckett, L. G., Dikici, E., Lai, S., Madou, M., Bachas, L. G., Daunert, S. *Anal. Chem.* **2004**, *76*, 7263.
76 Yu, C., Xu, M., Svec, F., Frechet, J. M. J. *J. Polymer Sci.* **2002**, *40*, 755.
77 Yu, C., Davey, M. H., Svec, F., Frechet, J. M. J. *Anal. Chem.* **2001**, *73*, 5088.
78 Li, Y., DeVoe, D. L., Lee, C. S. *Electrophoresis* **2003**, *24*, 193.
79 Xu, J., Locascio, L., Gaitan, M., Lee, C. S. *Anal. Chem.* **2000**, *72*, 1930.

80 Tan, W., Fan, Z. H., Qiu, C. X., Ricco, A. J., Gibbons, I. *Eletrophoresis* **2002**, *23*, 3638.
81 Moschou, E., Nicholson, A., Guangyao, J., Zoval, J., Madou, M., Bachas, L., Daunert, S. *Anal. Chem.* **2006**, submitted.
82 Moschou, E., Nicholson, A., Guangyao, J., Zoval, J., Madou, M., Bachas, L., Daunert, S. In preparation.
83 Fu, A. Y., Spence, C., Scherer, A., Arnold, F. H., Quake, S. R. *Nature Biotech.* **1999**, *17*, 1109.
84 Li, P. C. H., de Camprieu, L., Cai, J. Sangar, M. *Lab on a Chip* **2004**, *3*, 174.
85 Yan, Y., Marriott, G. *Curr. Opin. Chem. Biol.* **2003**, *7*, 635.
86 Haas, E. *Chem. Phys. Chem.* **2005**, *6*, 858.
87 Hovius, R. *Trends Pharmacol. Sci.* **2000**, *21*, 266.
88 Wouters, F. S. *Trends Cell Biol.* **2001**, *11*, 203.
89 Yan, Y., Marriott, G. *Methods Enzymol.* **2003**, *48*, 187.
90 Abriola, L., Chin, M., Fuerst, P., Schweitzer, R., Sills, M. *J. Biomol. Screening* **1999**, *4*, 121.
91 Kakiuchi, N., Nishikawa, S., Hattori, M., Shimotohno, K. *J. Virol. Methods* **1999**, *80*, 77.
92 Wabuyele, M. B., Farquar, H., Stryjewski, W., Hammer, R. P., Soper, S. A., Cheng, Y. W., Barany, F. *J. Am. Chem. Soc.* **2003**, *125*, 6937.
93 Buranda, T., Huang, J., Perez-Luna, V. H., Schreyer, B., Sklar, L. A., Lopez, G. P. *Anal. Chem.* **2002**, *74*, 1149.
94 Lipman, E. A., Schuler, B., Bakajin, O., Eaton, W. A. *Science* **2003**, *301*, 1233.
95 Daly, C., McGrath, J. C. *Pharm. & Therap.* **2003**, *100*, 101.
96 Ellenberg, J., Lippincott-Schwartz, J., Presley, J. F. *Trends Cell Biol.* **1999**, *9*, 52.
97 Janetopoulos, C., Tang, L., Zhang, N., Devreotes, P. N. *Pharmacologist* **2002**, *44*, 49.
98 Sorkin, A., McClure, M., Huang, F., Carter, R. *Curr. Biol.* **2000**, *10*, 1395.
99 Wouters, F. S., Bastiaens, P. I. *Curr. Biol.* **1999**, *9*, 1127.
100 Wilson, T., Hastings, J. W. *Annu. Rev. Cell Dev. Biol.* **1998**, *14*, 197.
101 Naylor, L. H. *Biochem. Pharmacol.* **1999**, 58(5), 749.
102 Sohn, T. A., Bansal, R., Su, G. H., Murphy, K. M., Kern, S. E. *Carcinogenesis* **2002**, 23(6), 949.
103 Lo, M. K., Tilgner, M., Shi, P. *J. Virol.* **2003**, 77(23), 12901.
104 Kariv, I., Cao, H., Marvil, P. D., Bobkova, E. V., Bukhtiyaro, Y. E., Yan, Y. P., Patel, U., Coudurier, L., Chung, T. D. Y., Oldenburg, K. R. *J. Biomol. Screening* **2001**, 6(4), 233.
105 Maffia, A. M., Kariv, I., Oldenburg, K. R. *J. Biomol. Screening* **1999**, 4(3), 137.
106 Liu, B. F., Ozaki, M., Hisamoto, H., Luo, Q., Utsumi, Y., Hattori, T., Terabe, S. *Anal. Chem.* **2005**, *77*, 573.
107 Tani, H., Maehana, K., Kamidate, T. *Anal. Chem.* **2004**, *76*, 6693.
108 Davidsson, R., Boketoft, A., Bristulf, J., Kotarsky, K., Olde, B., Owman, C., Bengtsson, M., Laurell, T., Emneus, J. *Anal. Chem.* **2004**, *76*, 4715.
109 Prendergrast, F. *Nature* **2000**, *405*, 291–293.
110 Head, J., Inouye, S., Teranishi, K., Shimomura, O. *Nature* **2000**, *405*, 372–376.
111 Shimomura, O., Musicki, B., Kishi, Y. *Biochem. J.* **1988**, *251*, 405–410.
112 Chiesa, A. et al. *Biochem. J.* **2001**, *355*, 1–12.
113 Lewis, J. C., Daunert, S. *Fresenius J. Anal. Chem.* **2000**, *366*, 760–768.
114 Seethala, R. *Drugs and Pharmaceutical Sciences* **2001**, *114*, 69–128.
115 Witkowski, A., Ramanthan, S., Daunert, S. *Anal. Chem.* **1994**, *66*, 1837–1840.
116 Grosevnor, A., Feltus, A., Conover, R., Daunert, S., Anderson, K. *Anal. Chem.* **2000**, *72*, 2590–2594.
117 Shim, Y., Insook, R. *Analytical Sciences* **2001**, *17*, 41–44.
118 LePoul, E. et al. *J. of Biomolec. Screening* **2002**, *7*, 57–65.
119 Waud, J. et al. *Biochem. J.* **2001**, *357*, 687–697.
120 Haviv, I., Campbell, I. *Molec. and Cellular Endocrinology* **2002**, *191*, 121–126.
121 Jia, G. et al. *Proc. SPIE* **2004**, *5455*, 341–352.

11
Advances in Instrumentation for Detecting Low-level Bioluminescence and Fluorescence

Eric Karplus

11.1
Introduction

Many experimental methods in biological and biochemical research involve the use of bioluminescent and/or fluorescent reporter systems. There continue to be advances in high-sensitivity, low-noise light detection technology, and it can be a daunting task to determine which technology is best for a particular application.

When selecting a detection technology, there are important factors such as cost and quality of support that deserve careful consideration. However, it is also important to be able to consider as objectively as possible how well the different technologies will perform from a quantitative perspective. Some of the fundamental limits of detection will be considered, and some considerations for effectively detecting low-level fluorescent and bioluminescent signals will be discussed.

11.2
Low Light Levels

Visible light is a form of energy that can be observed by the human eye. It has been observed that visible light energy occurs in discrete units, and these units are called photons. Photons can be theoretically treated as particles having an energy E proportional to their wavelength λ as follows:

$$E = h\,c\,/\,\lambda,$$

where h is Planck's constant and c is the speed of light in a vacuum. The human eye perceives light of different wavelengths as different colors as shown in Table 11.1, which is provided as an approximate reference. Perception of a color by the human eye is not guaranteed to be a reliable indication of the wavelengths

Photoproteins in Bioanalysis. Edited by Sylvia Daunert and Sapna K. Deo

ISBN: 3-527-31016-9

Table 11.1 Approximate relationships between perceived color and wavelength.

Color	*Wavelength*
Violet	380–435 nm
Blue	435–500 nm
Cyan	500–520 nm
Green	520–565 nm
Yellow	565–590 nm
Orange	590–625 nm
Red	625–740 nm

present in the light observed for two main reasons. First, different people may perceive the same wavelength as a slightly different color. In addition, different combinations of different intensities of light at different wavelengths can be perceived by the human eye as equivalent colors. Such light combinations are called metamers, and these can be differentiated only by using wavelength-sensitive instrumentation.

Because light interacts with solid materials in wavelength-dependent ways, it is possible to construct wavelength-dependent optical systems that perform important functions in scientific research. One important characteristic of materials used for optical components is transmission efficiency, which will be discussed in more detail below. Careful engineering of the transmission characteristics of materials can be used to construct interference filters that allow only certain wavelengths of light to pass through.

Another important characteristic of materials used for optical components is wavelength-dependent index of refraction, which indicates how fast light travels inside the material. This characteristic can result in problems such as chromatic aberrations in lenses, but it can also be used to advantage in components such as prisms when wavelength-dependent behavior is of interest. Another method of separating light based on its wavelength is to use a diffraction grating consisting of small slits spaced very close together that reflect light at a wavelength-dependent angle. Diffraction gratings are commonly used in spectrographs and monochromators to process light based on its wavelength.

It has been observed that certain materials will absorb energy in the form of light at one wavelength and then reemit slightly less energy in the form of visible light at a longer (redder) wavelength. This phenomenon is known as fluorescence. A wide range of biological and biochemical methods take advantage of the fluorescence process to enable detection of biological and biochemical activity at a molecular level. The presence of inherent fluorescence, often called autofluorescence, in a number of biological and biochemical materials can limit the usefulness of fluorescence methods in certain situations.

In bioluminescent reactions, a chemical reaction spontaneously emits energy in the form of visible light. Some very useful reporters have been developed that cause energy in the form of visible light to be emitted when certain biological or biochemical reactions take place. Inherent luminescence, or autoluminescence, is extremely rare; therefore, bioluminescent methods are often a good choice when it is important to have a low background signal.

Both luminescent and fluorescent methods are finding increased applications in living cells as well as in *in vitro* environments. Proper use of fluorescent methods requires a careful consideration of how to properly excite the fluorescent molecules being investigated. Issues of fluorescence illumination are beyond the scope of this presentation, which focuses on optimization of the detection of the emitted light. Block diagrams of the components involved in detecting visible light emitted when using bioluminescent and fluorescent methods are shown in Fig. 11.1.

The lowest level of light that can be detected is a single photon. In practice, detection systems are less than 100% efficient, so a sample must produce more than one photon in order to guarantee that the emitted light can be reliably detected. For example, if a detection system is 5% efficient, then a sample must emit 20 photons in order for just one of them to be detected. Furthermore, practical detection systems are affected by various noise sources that cannot be distinguished from a true signal; thus, the output of a detection system always consists of an inseparable combination of signal and noise. In addition, when luminescent and/or fluorescent reporters are used to make quantitative measurements relating to biological or biochemical systems, it is necessary to have statistically significant measurements of the amount of detected light in order to make reliable statements about the amount of light emitted by the sample. Ultimately it is best to have a sample that emits as much light as possible and a detection system that has the highest detection efficiency and lowest noise possible. Issues affecting the design and implementation of such systems are considered in the balance of this chapter.

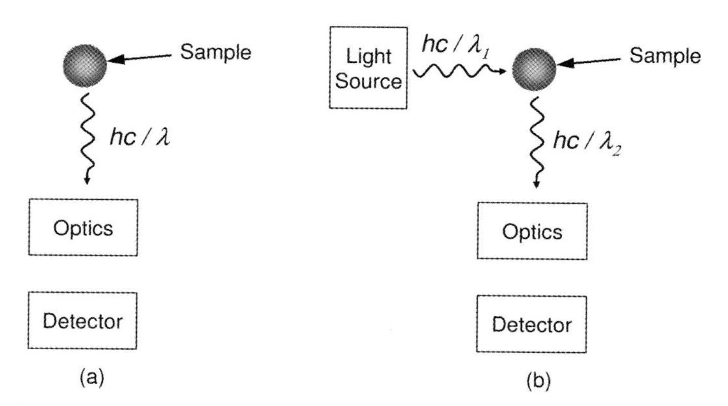

Fig. 11.1 Block diagrams of components involved in low-light-level (a) bioluminescence and (b) fluorescence detection.

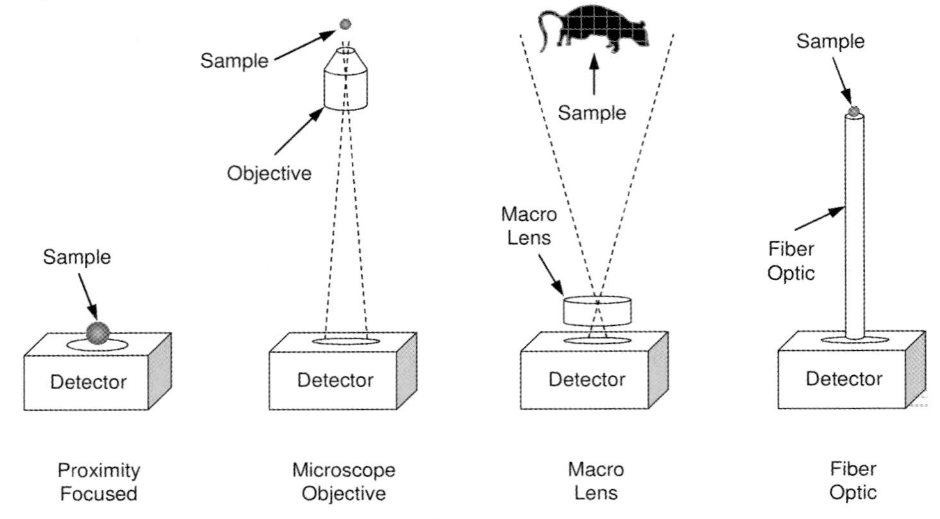

Fig. 11.2 Four general categories of optical systems used for low-light-level bioluminescence and fluorescence detection.

11.3
Methods of Coupling the Signal to the Detector

An important aspect of experimental design is the optical system used to collect light from the sample and direct it to the detector. The four general categories for optical systems shown in Fig. 11.2 will be considered: (1) proximity focused, (2) microscope objective, (3) macro lens, and (4) fiber optic. In some cases the physical constraints of the sample force selection of a certain configuration. Key aspects of each configuration will be introduced, and a discussion of how to evaluate the performance of an optical system will be considered next.

11.3.1
Proximity Focusing

In a proximity-focused system, the object is placed as close as possible to the detector's photosensitive layer. This system requires no, or only minimal, optical elements, but it precludes any magnification or reduction of the image and generally results in an out-of-focus image because of the thickness of the sample and the presence of a protective layer of glass or other material over the active area of most detectors, both vacuum based and solid state.

Proximity focusing is usually the most efficient method for coupling bioluminescent samples to a non-imaging detector such as a photomultiplier tube. It is usually not an effective method for coupling light into cooled detectors because of the large distance between the photosensitive layer of the detector and the outside surface of the input window of the detector's cooling chamber (see Fig. 11.11).

11.3.2 Microscope Objectives

Microscope objectives are generally used when the sample is smaller than the detector's photosensitive layer and/or there is a need to see features of the sample that are smaller than the spatial resolution of the detector would normally allow. For example, the smallest pixels on most scientific-grade charge-coupled devices (CCDs) are around 6 microns, which allow reliable imaging of features about 12 microns or larger. A 20× microscope objective will nominally project 0.3 microns of the sample onto one 6-micron pixel of the CCD, so the limiting resolution factor will no longer be the resolution of the detector but instead will be the diffraction limits of the light and the other optical elements being used in the image path.

Microscope objectives usually bear markings that indicate numerical aperture and magnification, and the impact of these specifications on the performance of an optical system will be discussed below. Microscope objectives are located close to the sample, and the physical working distance between the front surface of the objective and the focal plane where the sample must be placed can be extremely small, especially for high-magnification, high-numerical-aperture objectives. The physical thickness of a sample or the thickness of a glass element through which it must be viewed can preclude the use of certain microscope objectives because of working distance limitations.

11.3.3 Macro Lenses

Macro lenses are generally used when the sample is larger than the detector's photosensitive layer, e.g., with imaging gels or live animals such as rats or mice. Another situation when macro lenses are helpful is when a single detector is being used to detect light from regions that are spread out widely in space, e.g., when detecting simultaneous light emissions from multi-well plates.

Macro lenses usually bear markings that indicate F/stop, which is inversely proportional to numerical aperture. The impact of this specification on the performance of an optical system will be discussed below.

Macro lenses are located close to the detector and far from the sample. There are a variety of standard "lens mount" specifications such as C-mount, CS-mount, F-mount, etc. These specifications determine the diameter of the lens elements that can be used as well as the optical distance between the mounting flange surface and the photosensitive layer of the detector (see Fig. 11.11). Cooled detectors normally have a cooling chamber that requires an insulating window several millimeters in front of the photosensitive layer of the detector. This reduces the free distance behind the lens mount flange and can preclude the use of certain macro lenses that have lens elements that protrude too far behind the mounting flange.

11.3.4
Fiber Optics

Fiber optics are constructed by coating a core material having an index of refraction N_1 with a clad material having a lower index of refraction N_2. The interface between these two materials defines a surface that will reflect light incident at or below a critical angle defined as

$$A_{\text{crit}} = \sin^{-1} \{[(N_1)^2 - (N_2)^2]^{½}\} \tag{1}$$

Any light entering the core material with an angle of incidence that is smaller than the critical angle at the input surface will be reflected at the core–clad interface, resulting in the core–clad assembly functioning as a "light pipe". A schematic of a typical single fiber optic element is shown in Fig. 11.3. In situations where a sample does not need to be imaged, a proximity-focused fiber optic (i.e., placing the sample as close as possible to the input surface of a fiber optic and the detector as close as possible to the output surface of the fiber optic) is often the most efficient way to collect light from the sample and couple it to the detector.

Multiple fiber optic elements can be fused together to form an "image conduit". The cladding material occupies part of the surface area of such a fused bundle, so the overall transmission of the bundle is lower (typically 30–40% lower) than the transmission of an equivalent lens optic. However, the numerical aperture at the input surface of the bundle can approach 1.0. As discussed below, this high numerical aperture can provide a tremendous light-gathering benefit in applications where the sample can be proximity focused onto the input surface of the bundle.

In certain detector structures discussed below, a fused bundle can be directly integrated into the detector structure to achieve extremely high-numerical-aperture light collection. Fused fiber optic bundles can also be tapered to accomplish magnification or reduction, typically by a factor of up to three. However, this scale change occurs at the expense of a proportional reduction in numerical aperture, which is not always an acceptable tradeoff.

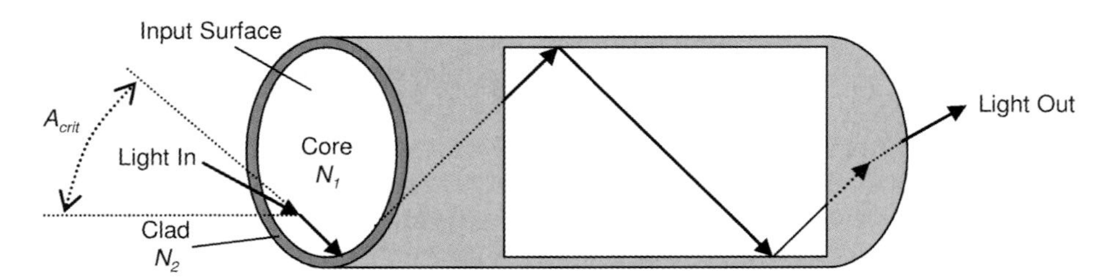

Fig. 11.3 Schematic of a fiber optic.

11.4
Evaluating the Performance of an Optical System

Important considerations regarding the performance of the optical system include numerical aperture, transmission efficiency, and magnification. Additional considerations such as spatial resolution and distortion correction will not be treated extensively here.

11.4.1
Numerical Aperture

The light collection efficiency of an optical system is largely determined by a quantity known as numerical aperture. The numerical aperture of an optic is described as

$$NA = n \sin \alpha \tag{2}$$

where n is the index of refraction of the media between the sample and the optic ($n_{air} = 1.0$, $n_{water} = 1.3$, $n_{oil} = 1.51$) and α is the angular aperture of the system, measured as the half-angle included by a cone with its apex at the sample and its base at the perimeter of the first surface of the collection optic when the sample is in the location at which it will be observed. A schematic of this is shown in Fig. 11.4.

The numerical aperture of an optic can be used to determine the percentage of total light that can be collected from a point source that is emitting uniformly in all directions. Considering the schematic in Fig. 11.4, the light emitted from a point

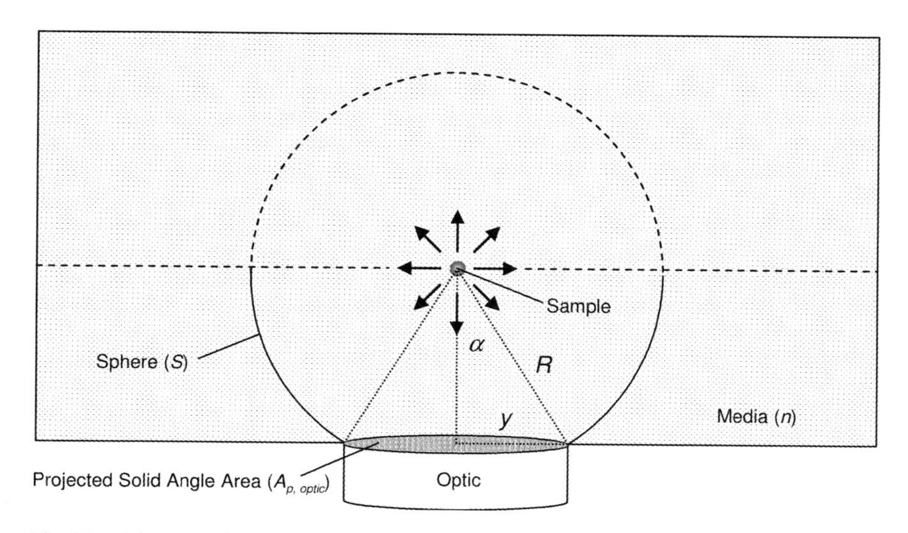

Fig. 11.4 Schematic showing the spatial interpretation of numerical aperture.

source at the center of the sphere S will be distributed evenly across the surface of the sphere. The area of the sphere that is intercepted by the optic is projected onto an area $A_{p,\,optic}$. The projected area can be computed as follows:

$$A_{p,optic} = R^2 \int_0^{2\pi} d\theta \int_0^{\alpha} (\cos\phi) \sin\phi \, d\phi = 2\pi R^2 \sin^2 \alpha \tag{3}$$

The maximum projected surface area on the optic side hemisphere of the spherical surface S occurs when $\alpha = 90$ degrees:

$$A_{hemisphere} = 2\pi R^2 \tag{4}$$

If we accept that the source is radiating uniformly in all directions, only half of the total light emitted propagates into the optic side hemisphere. Thus, the fraction of the total emitted light that is collected by the optic, Q_{optic}, is given by half of the ratio of these two areas:

$$Q_{optic} = (A_{p,\,optic} / A_{hemisphere}) / 2 = (2\pi R^2 \sin^2 \alpha / 2\pi R^2) / 2 = \sin^2 \alpha / 2 \tag{5}$$

From this it can be seen that an optic's collection efficiency for a luminous point source is proportional to the square of the optic's numerical aperture. Substituting in the definition for numerical aperture from Eq. (2) when the index of refraction is 1.0 (for air) gives

$$Q_{optic} = NA^2 / 2 \tag{6}$$

The impact of numerical aperture on overall collection efficiency is significant, and when very low levels of light are being detected, it is always important to seek the highest possible numerical aperture optics that will meet the other requirements of the system. The only exception to this is when the numerical aperture of the sample is fixed and known and can be oriented properly relative to the detector. In cases such as this, it is not helpful to have a collection system with a numerical aperture higher than the numerical aperture of the sample.

A graph of the percentage of light collected by an optic in air as a function of numerical aperture from a point source emitting uniformly in all directions is shown in Fig. 11.5.

Microscope objectives usually have the numerical aperture printed directly on the side after the magnification (e.g., 10× / 0.25 means a numerical aperture of 0.25). In practice, the highest angular aperture that can be readily achieved with a microscope objective is around 72 degrees, giving practical numerical aperture limits as shown in Table 11.2.

Most macro lenses are characterized by a quantity known as *f/number*, which is the ratio of the diameter of the lens d to the focal length of the lens f. This quantity can also be expressed as a function of numerical aperture as follows:

$$f/number = d / f = 1 / (2 \times NA) \tag{7}$$

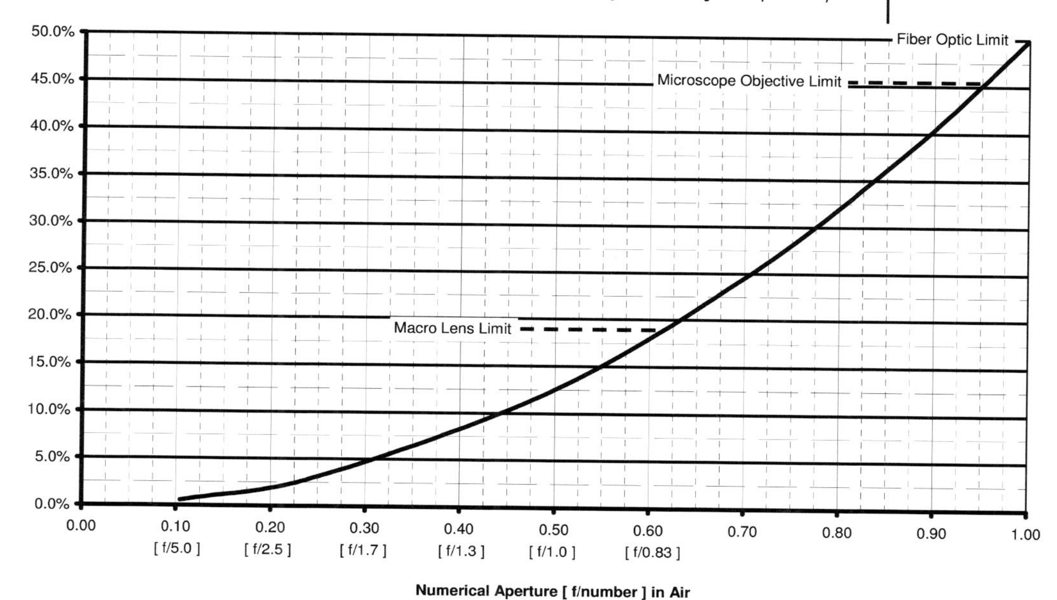

Fig. 11.5 Percentage of total light collected by an optic in air as a function of numerical aperture in the case of a uniformly emitting point source, showing practical limits for some common optical systems.

Table 11.2 Practical numerical aperture limits for microscope objectives.

Media	*NA limit*
Air	0.95
Water	1.26
Oil	1.44

Low *f*/*number* (high *NA*) lenses are called "fast" because they gather a lot of light, requiring a shorter exposure time to form a bright image. High *f*/*number* (low *NA*) lenses are called "slow" because they gather less light and require a longer exposure time to form a bright image. In practice, most of the fastest commercially available macro lenses have an *f*/*number* of 1.0. Some inexpensive aspheric macro lenses with an *f*/*number* as low as 0.8 are available, but these lenses may introduce geometric distortion that can cause trouble in certain imaging applications.

Numerical aperture also determines the depth of field, which describes the thickness of the section of the sample that appears to be in focus. High-numerical-aperture (low *f*/*number*, or fast) lenses have a smaller depth of field than low-numerical-aperture (high *f*/*number*, or slow) lenses. Depth of field can be defined in a variety of ways, but the following proportionality relationship is generally expressed:

$$DOF \sim 1 / NA \sim f/number \qquad (8)$$

Numerical aperture also determines the resolving power of a lens. There are a variety of methods for assessing resolving power, but all generally express the following proportionality relationship:

$$d \sim \lambda / NA \sim \lambda \times f/number \qquad (9)$$

where d represents the minimum size of the object that can be discerned, λ is the wavelength of the light being used to observe the sample, and NA is the numerical aperture of the optics being used. The general principle is that the best resolving power comes from using the shortest wavelength light and the highest numerical aperture lens.

From the above it appears that higher numerical apertures are always more desirable, but in practice there may be a tradeoff in signal-to-noise ratio. For example, in the case of microscope objectives, higher numerical apertures are normally available only with higher magnification lenses, so the increased signal is spread out over a larger area of the detector. As will be discussed below, this increases the amount of dark noise that is superimposed on the true signal, reducing the effective signal-to-noise ratio.

11.4.2 Transmission Efficiency

Transmission efficiency indicates how much light is lost because of reflections at optical surfaces and absorption within the lens elements. Most high-quality commercial lenses use antireflection coatings on the surfaces to minimize reflective losses. Most lens materials also transmit well in the 390–700 nm visible light spectrum. However, because the transmission efficiency of the total optical system is the product of the independent transmission efficiencies of each optical element, it is best to minimize the number of optical elements in the system. For example, if each surface of a lens loses 0.1% to reflection and absorbs 0.2% of the light passing through the interior of the lens, then the overall transmission efficiency of the single lens element is

$$T = 99.9\% \times 99.8\% \times 99.9\% = 99.6\%.$$

In a typical microscope objective, there may be eight individual lens elements. If we assign the above transmission to each element, the overall transmission of the objective will be

$$T = (99.6\%) \wedge 8 = 96.8\%.$$

This transmission percentage is quite high, but it also makes very optimistic assumptions about the transmission efficiency of each lens element. In practice, many objectives have overall transmission efficiencies that are better than 90% in the range of 400–700 nm. When comparing a variety of otherwise equivalent objectives for an application, transmission efficiency should be taken into consideration.

11.4.3
Magnification

Optical systems with higher magnification tend to have higher numerical apertures, while optical systems with greater reduction (i.e., increased field of view) tend to have lower numerical apertures. However, choosing a higher magnification optic to achieve a higher numerical aperture is not always a sound approach. Even though more light is collected, it is spread out over a larger area, affecting the apparent brightness of the image:

$$B_{\text{luminescence}} \sim NA^2 \,/\, M^2 \tag{10}$$

Consider the difference between a 5× / 0.25 NA lens and a 20× / 0.75 NA lens. From Eq. (6) this lens should collect $(0.75)^2 / (0.25)^2 = 9$ times more light; however, this light will be spread out over a 16-times larger area, resulting in an image that is 9/16 = 56% as "bright". Figure 11.6 shows brightness as a function of magnification in a luminescence detection configuration for some of the brightest commercially available microscope objectives.

In non-imaging applications, this reduction in brightness is of no consequence provided the detector is large enough to collect all of the light that the optics project from the sample. However, in imaging applications, the reduction in brightness causes a corresponding decrease in the signal-to-noise ratio. This decrease results because the dark noise that is superimposed on the signal is proportional to the area over which the signal occurs, so if the same signal is spread out over a larger area, it will have correspondingly more noise superimposed on it. An example of the impact of brightness on signal to noise is given below.

In fluorescence detection configurations, represented generally in Fig. 11.1b, the excitation light source usually includes optical elements. In cases such as epifluorescence microscopy, where the excitation and detection systems share some optical elements, it is necessary to optimize the optical elements from the perspective of the excitation source requirements as well as the detector requirements.

For example, if the fluorescence excitation light is in the ultraviolet range (e.g., < 390 nm), there can be a benefit from selecting lenses made from quartz or other materials that transmit well at short wavelengths. Because epi-illumination optics use the microscope as a condenser to collect light from the illuminator, the brightness of the image in fluorescence detection configurations is related to numerical aperture and magnification as follows:

$$B_{\text{fluorescence}} \sim NA^4 \,/\, M^2 \tag{11}$$

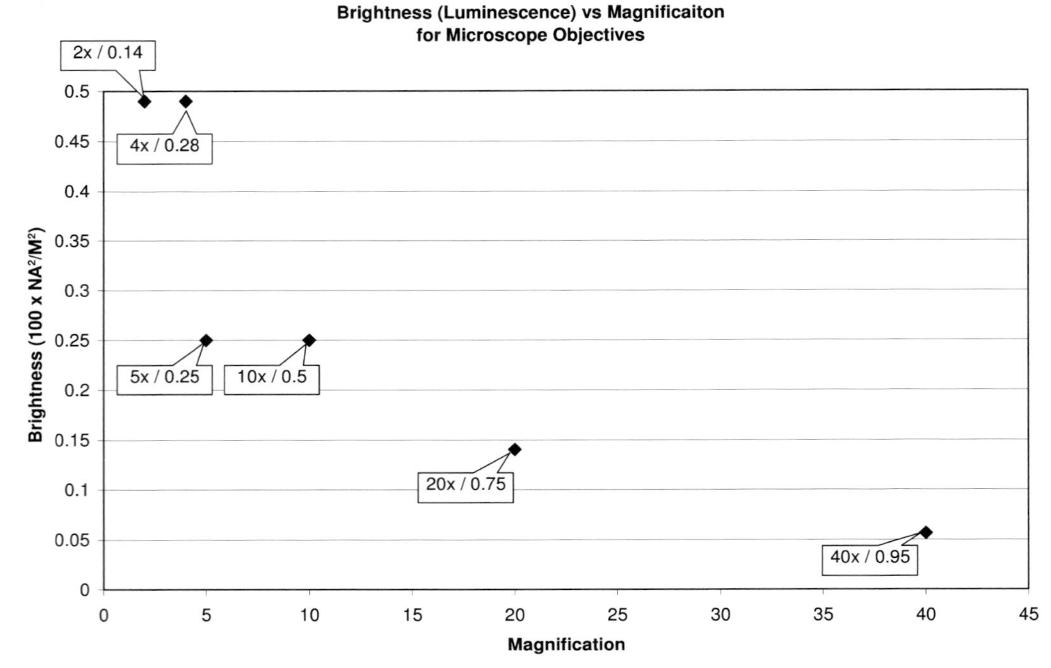

Fig. 11.6 Brightness as a function of magnification for luminescence detection with some of the brightest commercially available microscope objectives.

In all imaging applications, the distortion introduced by the optical elements should also be considered. As mentioned previously, aspherical macro lenses often have very high numerical apertures, but they also tend to introduce distortions in images. Sometimes this is an acceptable tradeoff because of the increase in signal collected.

11.5 Detector Technologies

Most light detector technologies can be grouped into two main categories: vacuum based and solid state. Schematics of these two main categories are presented in Fig. 11.7.

Vacuum-based light detectors utilize a photocathode, a thin layer of photosensitive material deposited on a solid surface internal to the vacuum, to convert incident photons into free photoelectrons in the photosensitive material. By applying an electric field across the vacuum enclosure, the free photoelectrons in the thin layer can be harvested into the vacuum and collected on an electrode at the other end of the field. The electrical signal collected on the electrode corresponds to the optical signal on the photosensitive material, thus forming an optical detector. Most

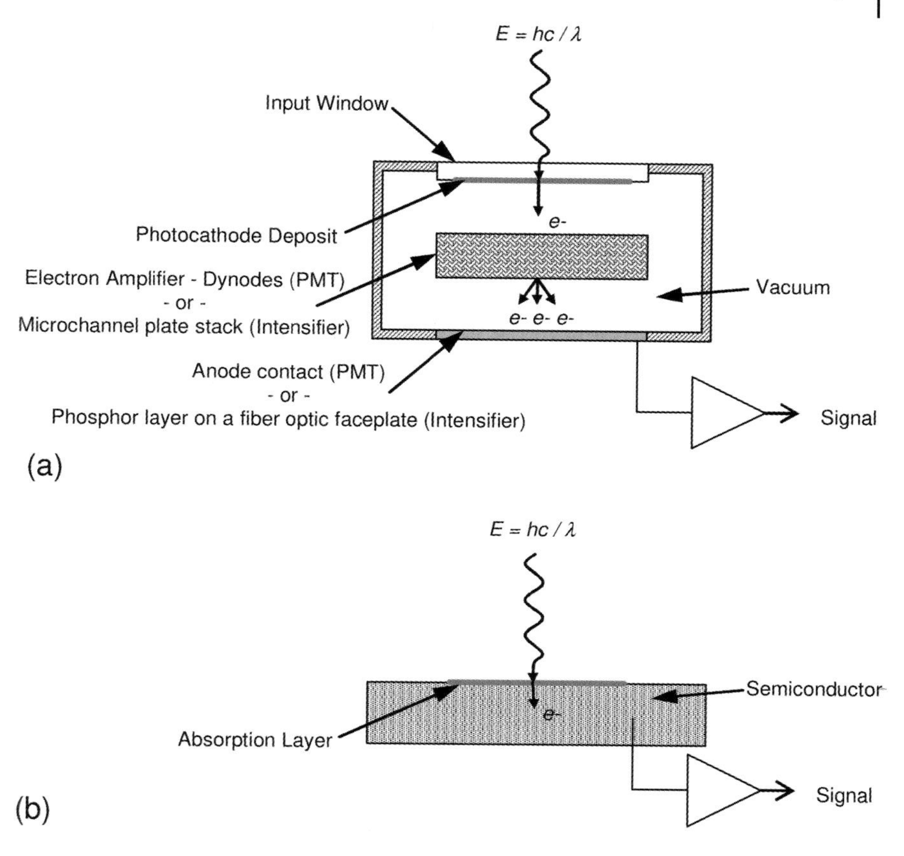

Fig. 11.7 (a) Vacuum-based detector structure (b) solid-state detector.

vacuum-based detectors have internal amplification mechanisms that multiply a single electron liberated from the photocathode deposit layer into a bundle of electrons that arrives on the output anode. Common examples of vacuum-based detectors are photomultiplier tubes and image intensifiers. In photomultipliers, the amplification mechanism is usually a series of metal plates or other structures called dynodes. In image intensifiers, the amplification mechanism is usually a stack of one or more microchannel plates. Image intensifiers normally have an anode consisting of a phosphor material coated on a fiber optic faceplate. When the bundle of electrons from the microchannel plate stack is accelerated into the phosphor, it produces a bright spot as the energy in the electrons is converted back into photons. This light can then be viewed with the human eye, as is common in night vision systems or with a solid-state detector such as a CCD camera, as is common for intensified CCD detectors.

In solid-state light detectors, the incident photons are converted into photoelectrons that move within the atomic structure of a solid material rather than within a vacuum enclosure. A common solid-state detector that has no internal gain is the photodiode, which provides direct current readout that is proportional to the number of electrons generated in the absorption layer of the device. A variation on this device called an avalanche photodiode has an internal gain mechanism that multiplies a single electron produced in the absorption layer into a bundle of electrons that is extracted from the output of the device.

A common solid-state detector used for imaging applications is called a charge-coupled device (CCD) camera. Two important versions of CCDs are illustrated in Fig. 11.8. These devices consist of an array of special electron storage areas called pixels fabricated on a monolithic substrate. Photoelectrons migrate from the absorption layer into the pixel wells during an integration period, and the array is periodically read out to determine how many photoelectrons were collected in each pixel. The readout process involves shifting charge laterally across the device to a readout channel and then serially down the readout channel to an output amplifier. For scientific grade CCDs, the efficiency of moving charge from one pixel to the next and then out the readout channel is high enough to make losses inconsequential.

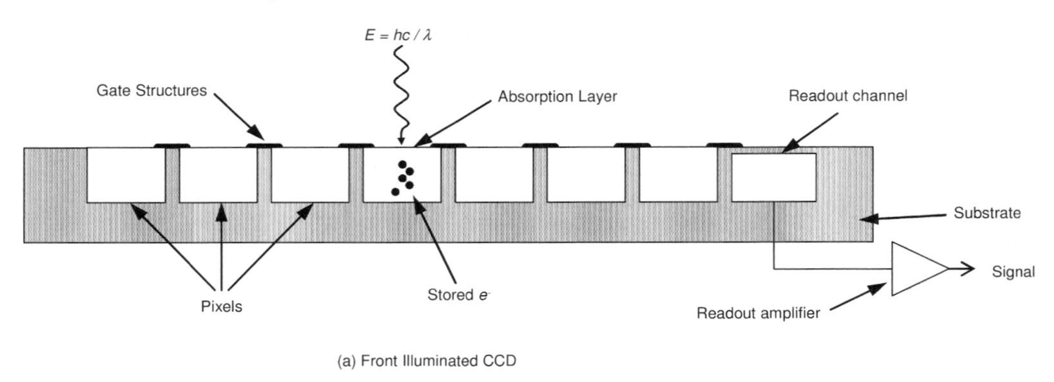

(a) Front Illuminated CCD

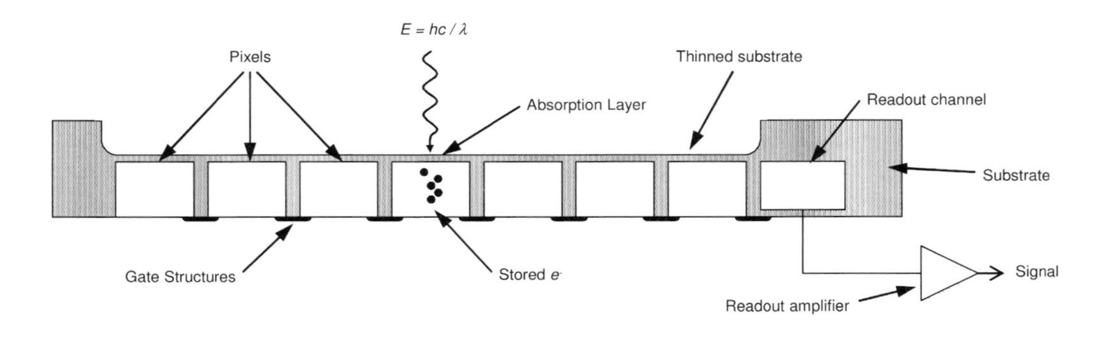

(b) Back Illuminated CCD

Fig. 11.8 Conceptual illustration of (a) a front-illuminated CCD and (b) a back-illuminated CCD.

The input signal must be blocked during the readout cycle so that it does not smear on top of the collected image during readout. This results in unavoidable dead time for the detector, which can be a significant problem in applications where dynamics of a low-light-level signal are being studied.

The main difference between the two CCD structures shown in Fig. 11.8 is which side of the substrate is used to convert photons into electrons. The front-illuminated structure is easier to fabricate and thus has a lower cost, but some of the light is inevitably blocked by the gate structures that are needed to accomplish the charge transfer process. This loss is eliminated by etching the back of the substrate so that there is only a thin layer behind the pixel wells. Then the unobstructed back side of the device can be used to convert photons into electrons that are stored in the pixel wells. The quantum efficiency of such devices can exceed 90% over a fairly broad range of the visible light spectrum. A significant economic drawback of these devices is that the fabrication process has a low yield, primarily because the etching process needs to be uniform and the resulting thinned device is extremely fragile. However, there have been marked improvements in performance and reductions in cost as the demand for back-illuminated CCDs has increased.

An important category of recently developed solid-state detectors is known as electron-multiplying CCDs (EM-CCDs). These devices have a special built-in electron-multiplying structure called a gain register inserted between the readout channel and the readout amplifier. The gain register performs very low noise amplification of the photoelectrons stored in each pixel. This feature can remove the limitation of read noise that prevents traditional CCDs from being used in some important low-light-level applications. The combination of a back illuminated CCD with an electron-multiplying readout amplifier can provide some of the best performance available for low-light-level detection; however, there are situations where vacuum-based detectors still provide a superior result, and not just because of cost.

A cousin of CCD detectors that deserves recognition is CMOS imaging sensors, which are array detectors similar to CCDs, but the fabrication and readout process is different. To date, CMOS imagers have had too much intrinsic dark noise for effective use in low-light-level detection.

A third category of light detector devices has been developing in recent years as well. It consists of a solid-state detector sealed inside a vacuum enclosure. In this configuration, a photocathode is used to convert photons into electrons, and the electrons are then accelerated towards the solid-state device. Some gain is realized by bombardment, when the electron strikes the solid-state device and produces many free electrons as its energy is absorbed. Additional gain can be realized either intrinsically in the solid-state structure (e.g., when the solid-state device is an avalanche photodiode) or more commonly in the readout amplifier, when the solid-state device is a CCD or CMOS imaging sensor. Some commercial products based on these structures are beginning to emerge. Their main benefit for low-light-level detection is the ability to use the low noise of a photocathode while achieving better spatial resolution than a comparable vacuum-only device would afford.

11.6
Selecting the Right Detector

Table 11.3 presents some broad generalizations about the differences between vacuum-based and solid-state detectors. Solid-state devices have earned the reputation of being superior to vacuum-based devices, but there are notable exceptions to this situation. Vacuum-based detectors usually have a much lower cost per unit of active area than comparable solid-state detectors. In addition, certain types of photocathodes can achieve a lower noise level per unit area than even the best-cooled solid-state devices. However, the high sensitivity, high spatial resolution, compact size, simple power requirements, and overall robust performance of solid-state detectors make them a better choice for many applications.

Table 11.3 General distinctions between vacuum-based and solid-state detectors.

	Vacuum based	*Solid state*
Cost per unit of active area	Lower	Higher
Noise per unit area	Lower	Low
Sensitivity	Lower	Higher
Spatial resolution	Lower	Higher
Detector volume	Larger	Smaller
Power consumption	Higher	Lower

Effective detector selection involves much more than a cursory consideration of generalizations. Within the two broad categories of detectors, there are many different products offered by several different vendors, often with a variety of claims about notable aspects of performance capabilities. It is important for a prospective customer to be able to assess the accuracy of such claims as well as their relevance to the application under consideration.

11.7
Detector Sensitivity

An important characteristic of a visible light detector is its wavelength-dependent response. There are many different ways that vendors report the wavelength-dependent response of the detectors they offer. Direct comparison of detectors can be done only after the methodologies have been examined carefully. Even if a published response curve is taken to be truly representative of a certain aspect of the detector's performance, it may not directly indicate the relationship between input and output. A notable example of this is with vacuum-based detectors such as photomultipliers and image intensifiers, where the sensitivity of the photocathode is often reported without indicating how much of the photocathode signal becomes real output signal. In addition, response curves published in product literature are usually only representative of typical performance and are not a guarantee of actual performance.

In order to compare different detector options, it is usually necessary to use each detector's published response curve to generate a "true response" curve that expresses the probability that a single photon with a given wavelength will be detected as a signal. Detection quantum efficiency (DQE) is a parameter that expresses the probability that an incident photon will be converted into a free electron inside the device, as shown in Fig. 11.7. Some vendors report a QE value in relative units, which is often a sign that the DQE is not very good. This is common for front-illuminated CCDs that have gate structures covering part of the active area and, in some cases, dead space in between each pixel to achieve other useful features. Relative QE values are not meaningful for comparison of different detector options unless the relative values can be converted into DQE values.

It is a common practice for scientific grade CCD manufacturers to report absolute QE values, and usually these values are equivalent to the DQE of the device. Many vacuum-based device manufacturers also report QE in absolute units, but they are often vague about whether it represents photocathode QE (more common) or detection QE (very uncommon). Because not all electrons liberated from the photocathode result in output signal, it is important to make sure that the figures being considered for comparison with the performance of solid-state devices represent DQE. Some typical response curves for a variety of detectors are shown in Fig. 11.9. It is important to recognize that while the

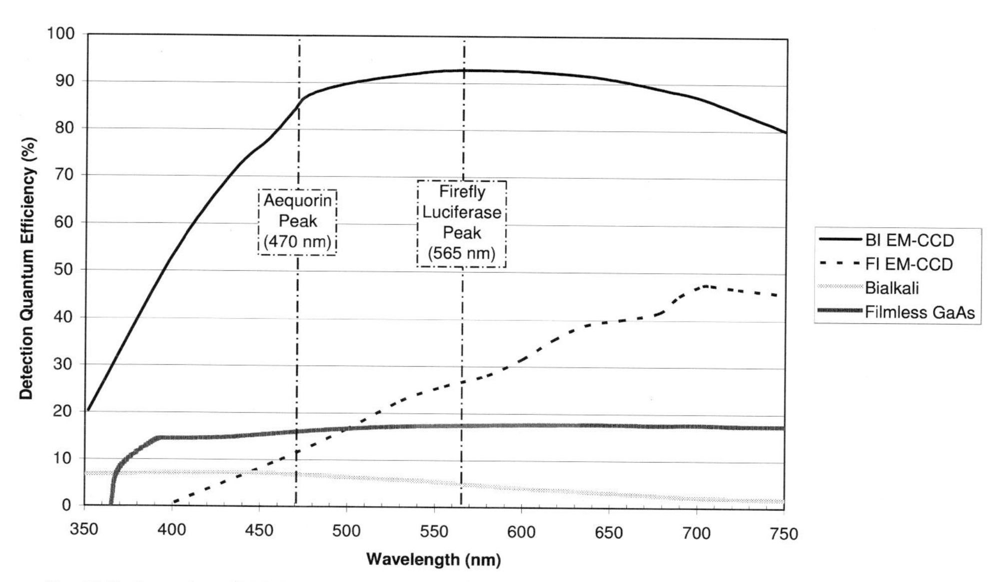

Fig. 11.9 Examples of DQE response curves with markers showing peak emission wavelengths for two traditional bioluminescent reporters.

data shown in Fig. 11.9 represent some of the best-published characteristics of real detectors, not all products from all vendors will perform according to the characteristics shown.

It is important to realize that very few biological and biochemical reporters emit light at a single wavelength. Normally the emission is spread over a range of tens of nanometers full width at half maximum (FWHM). For example, aequorin and firefly luciferase have a FWHM of around 70 nm. It is a good first-order approximation to evaluate detector response at the peak wavelength, but if the peak falls on a steep edge of the response curve, it can be critical to more carefully evaluate the impact of the range of emission wavelengths on the detected signal.

11.8 Detector Noise

As mentioned previously, the output of a detection system always consists of an inseparable combination of signal and noise. It is possible to make a measurement of the average noise level of a detector and subtract this average noise level from each measurement to estimate the net true signal. However, a signal cannot be reliably detected unless it is sufficiently strong relative to the noise level, so it is important for the noise level of a detector to be as low as possible.

Noise is a concern for both vacuum-based and solid-state detectors. In vacuum-based and hybrid-style detectors that have internal gain, the primary source of noise that must be taken into consideration is spuriously generated electrons in the photocathode material. These are electrons that become free inside the photocathode when there is no light incident on it. Thus, this noise is often called "dark noise" or "dark signal", because it represents the response of the device when it is in the complete dark. Some photocathode materials have a dark noise response that exhibits strong temperature dependence; therefore, there can be a significant benefit to cooling the photocathode material. The temperature dependence is typically exponential, so even a small amount of cooling can provide a tremendous benefit.

In solid-state detectors, there are two main types of noise that can cause trouble for low-light-level detection. The first is dark noise, which results from spuriously generated electrons that collect in the pixel wells even when there is no incident light. This dark noise has a variety of sources within the CCD structure. Ultimately, it exhibits exponential temperature dependence, and thus cooling the detector can be a tremendous benefit. An important method for reducing dark noise in CCDs by several orders of magnitude is called inverted-mode operation (IMO) or multiphase pinning (MPP). This method involves reverse biasing certain parts of the device structure during signal collection. These biases cannot be maintained during the readout cycle, and in some cases the contribution of dark noise from the readout cycle can be significant, even when the dark noise from the signal collection period is low.

Another kind of noise that can affect a solid-state detector is readout noise, which is introduced by the readout amplifier. Generally this noise can be reduced by cooling the readout amplifier and slowing down the readout process. However, the lowest read noises that have been achieved using these methods require very slow readout on the order of 20 000 pixels per second, which means it will take about 13 seconds of dead time to read out a 512×512 array. In addition, the noise level even at this low speed is still in the range of three electrons per pixel (rms), which can overwhelm even a modest amount of other dark noise sources. One approach that can be helpful is to bin together adjacent pixels, which reduces the impact of the read noise on the overall signal-to-noise ratio but also decreases the effective spatial resolution of the detector. A better solution to this problem has been achieved by electron-multiplying CCD technology, which can amplify the electrons in each pixel by a nearly noiseless gain of up to 1000 before sending the signal to the readout amplifier. With this technology, the read noise of even an uncooled readout amplifier operating at high speed can be easily overcome.

Another noise source that affects low-light-level detectors is cosmic rays, a name given to naturally occurring high-energy particles that are periodically absorbed by the detector material, usually producing thousands of free electrons in a very small space over a very short period of time. Cosmic rays are thought to arise from a variety of sources, and to date there are no known technologies for shielding detectors from this type of radiation. In many vacuum-based detectors it is possible to screen out such high-energy events by monitoring the energy level of each detected photon event. However, in a traditional CCD, it is not possible to directly distinguish free electrons produced by a cosmic ray from free electrons produced by visible light.

Fortunately, the flux of cosmic rays is relatively low, and the probability of one being intercepted by a detector is also relatively low, on the order of one to two events per minute per square centimeter. Because of the infrequency of these events, and because they deposit a very high amount of energy in a very small area over a very short period of time, such events normally appear as very small, intense, bright spots superimposed on a much lower intensity image. A variety of image-processing methods have been developed to identify cosmic ray spots in the image and estimate the "true" signal based on the signal level in surrounding pixels. In practice, cosmic ray noise is usually noticeable only in very low-light-level imaging with solid-state detectors when observation occurs over a continuous period of several minutes.

Based on the above discussion, the limiting factor for both vacuum-based and solid-state detectors is ultimately the dark noise-generated electrons that look the same as photoelectrons in the device. In order to make comparisons between different detectors, it is important to normalize the dark noise specification by the unit area it covers, since the signal from a sample will be spread out over a fixed area on the detector regardless of the pixel size. For example, a noise specification of 0.005 electrons per pixel per second in a 24-micron square pixel is actually lower than a noise specification of 0.0035 electrons per pixel per second in a 16-micron square pixel (8.7 events s^{-1} mm^{-2} for the former vs. 14 events s^{-1} mm^{-2} for the

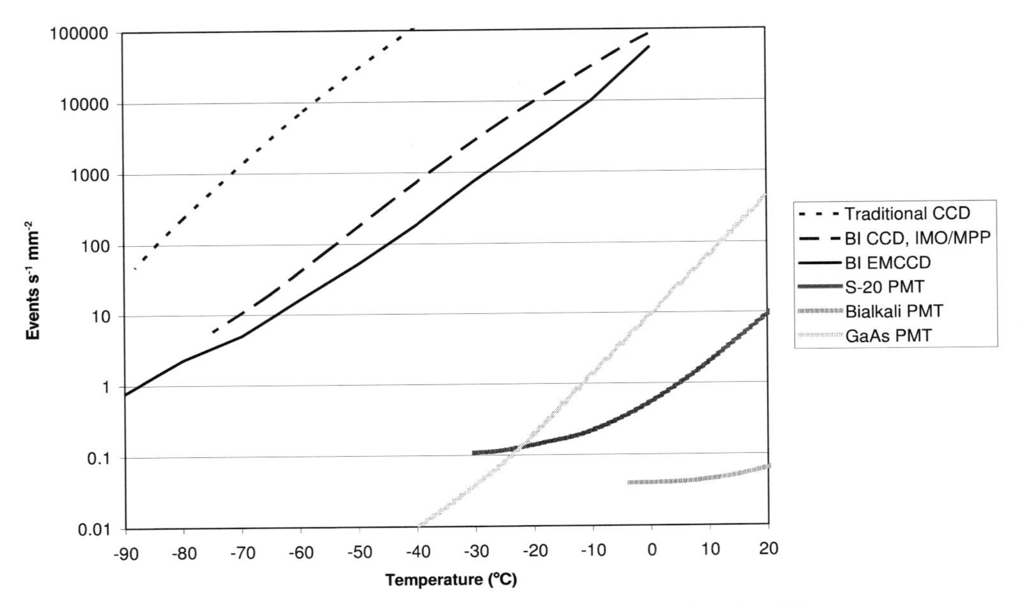

Fig. 11.10 Examples of dark noise as a function of temperature for selected detectors.

latter). A graph of common dark noise characteristics as a function of temperature for a variety of detectors is shown in Fig. 11.10.

As can be seen clearly from Fig. 11.10, there can be a tremendous benefit from cooling many types of detectors. It is important to recognize that while the data shown in Fig. 11.10 represent some of the best-published characteristics for real detectors, not all products from all vendors will perform according to the characteristics shown. For example, EMCCDs from different vendors will not all have the same specification for dark noise per unit area at –70 °C. When evaluating detector options for an application, it is important to obtain accurate detector performance data for each of the detectors being considered.

There are two additional considerations worth noting when a detector is to be cooled. One is that the cooling system will almost always require mounting the detector inside a sealed chamber in order to keep moisture from condensing on the cooled surfaces of the detector. This requires the use of an input window on the cooling chamber, which attenuates the input signal as a result of reflections at the window surfaces and the transmission characteristics of the window material. Thus, it is best to use only one highly transmitting, antireflection-coated window on the cooling chamber and have no protective window over the detector inside the cooling chamber. The cooling-system approach shown in Fig. 11.11b also prevents the use of a proximity-focused optical system and decreases the useable physical distance behind the mounting flange of the detector housing to from D_1 to D_2, as shown in Fig. 11.11b, which can preclude the use of certain macro lenses.

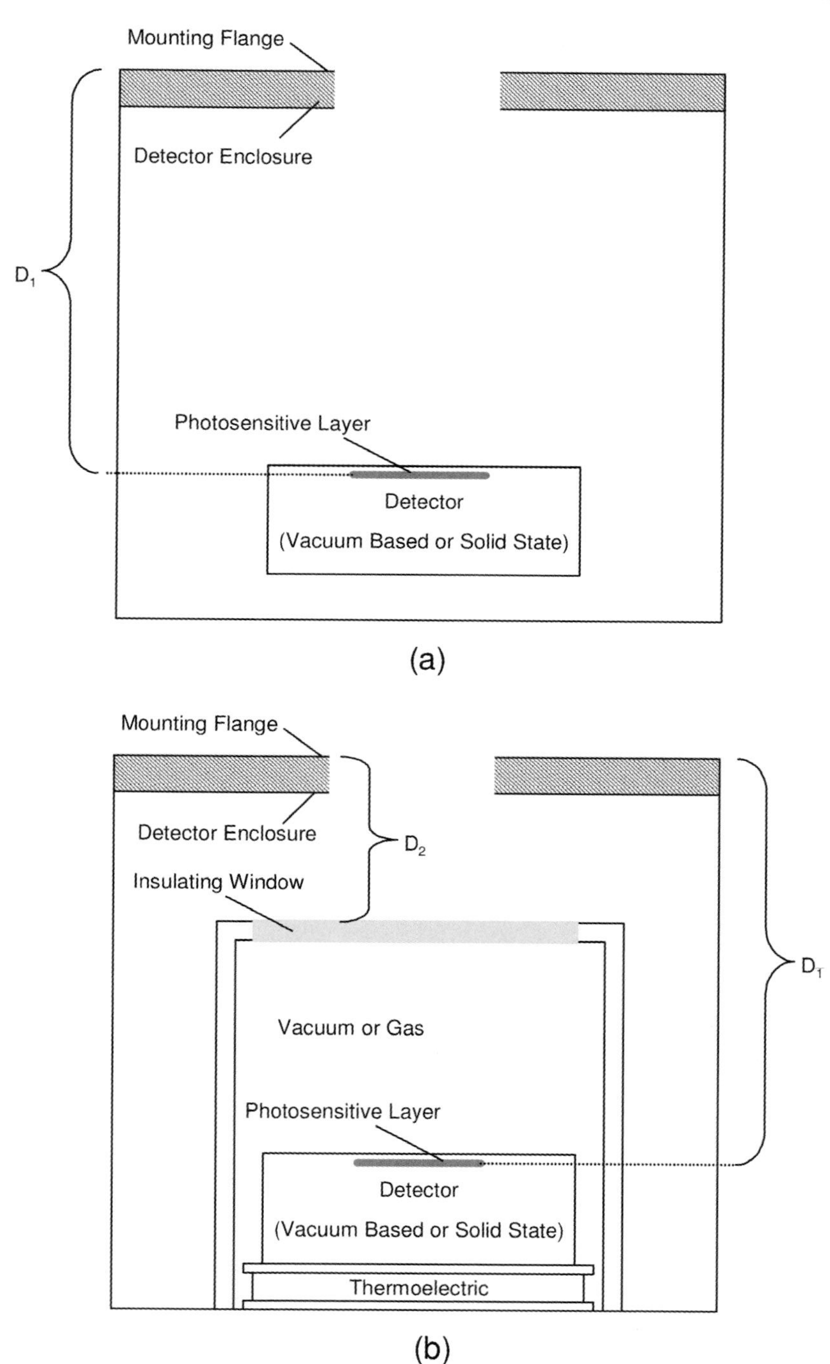

Fig. 11.11 (a) Uncooled and (b) cooled detector enclosures.

Another consideration when a detector is cooled is that the quantum efficiency of the detector is normally temperature dependent. For most common visible light detectors, quantum efficiency at wavelengths above 700 nm decreases, sometimes rapidly, as temperature decreases. At wavelengths below 700 nm there is often a slight increase in quantum efficiency as temperature decreases. If a detector is to be cooled, it is important to make sure that the detection quantum efficiency curve that is used to evaluate the detector's performance has been obtained at, or corrected to account for, the intended operation temperature.

11.9 Statistics of Photon Counting

As mentioned previously, visible light occurs in discrete units called photons. The smallest detectable amount of light is a single photon. In many situations, only a very small number of the total photons emitted from a biological or biochemical process are reported as signal by the detector being used. This is due to the limited ability of the optical system to relay the emitted light to the detector (see Fig. 11.4) and because of the limited ability of the detector to convert incident light into true signal (see Fig. 11.9).

Because low-level light detection involves counting discrete photon events, it is important to consider how counting statistics apply. This can be a very complex subject, and only a simple example will be considered here for illustration. Consider a situation where a 100-micron diameter spherical sample of aequorin with peak emission at 470 nm emits 1000 photons over a period of 100 seconds. For convenience, assume that these photons are emitted uniformly throughout space and time. If this sample is viewed with a 10× / 0.5 NA lens having 97% transmission efficiency, then the number of photons reaching the detector will be

$$\text{number of photons emitted} \times \text{collection efficiency} \times \text{transmission efficiency}$$
$$= 1000 \times 12.5\% \times 97\% = 121 \text{ photons.}$$

If these photons are incident on a bialkali photocathode operating at room temperature (20 °C), then the number of detected photons should be

$$\text{number of photons incident on the detector} \times \text{detection quantum efficiency}$$
$$= 121 \times 6.8\% = 8 \text{ photons.}$$

Surprisingly, this is less than 1% of the total signal! When it comes to counting eight photons in a row, even if the distribution is uniform through time and space, there is a finite probability that all eight detected photons will have been emitted in such a way that they land on the detector and are reported one after the other. If the signal is emitted at 10 photons per second and eight of them are detected in a row, they will be collected in a period of 0.08 seconds. If this signal of eight photons in 0.8 seconds is detected and then corrected for the 0.8% overall efficiency of the system, the data would suggest that the emission rate was 1250 photons per second over a 0.8-s period rather than 10 photons per second over a 100-s period. This extreme case is presented to illustrate why the issue of counting statistics is important. It is necessary to know something about the statistical variation of an observed signal before statements can be made about how representative the measured signal is of the true process.

In many cases it is sufficient to assume a Gaussian process that has a Poisson distribution of the form

$$P(x) = (\mu^x / x!)\, e^{-\mu} \tag{12}$$

where μ is the average value of a set of measurements and $P(x)$ is the probability that the measured value will be equal to x. It is possible to define a value called a standard deviation,

$$\sigma = \sqrt{\mu} \tag{13}$$

such that there is a 68% probability that any given measurement x will fall in the range of $\mu \pm \sigma$ and a 95% probability that it will fall in the range of $\mu \pm 2\,\sigma$.

Applying Eqs. (12) and (13) to the previous example of eight detected photons gives a standard deviation of $\sigma = \sqrt{8} = 2.8$ photons, so the measured signal can be reported as 8 ± 2.8 photons (±35%) with 68% certainty or as 8 ± 5.6 photons (±70%) with 95% certainty.

11.10 Summary

The uncertainties in the above example are quite high because of the small number of photons detected. In summary, some illustrations of how the measurement might be improved will be presented.

One option would be to use a higher numerical aperture objective, such as a 40× / 0.95 NA lens with 97% transmission, which would direct a larger number of emitted photons to the detector:

number of photons emitted × collection efficiency × transmission efficiency
= 1000 × 45% × 97% = 438 photons,

number of photons incident on the detector × detection quantum efficiency
= 438 × 6.8%
= 30 photons $\sigma = \sqrt{30} = 5.5$ photons
= 30 ± 5.5 photons (±18%) with 68% certainty
= 30 ± 11 photons (±37%) with 95% certainty.

Another option would be to use a higher efficiency detector such as a back-illuminated electron-multiplying CCD. In this case, the number of detected photons should be

number of photons incident on the detector $\times$ detection quantum efficiency
$= 121 \times 84\%$
$= 102$ photons $\sigma = \sqrt{102} = 10$ photons
$= 102 \pm 10$ photons ($\pm 10\%$) with 68% certainty
$= 102 \pm 20$ photons ($\pm 20\%$) with 95% certainty.

Even though these uncertainties may be more acceptable, there is still no guarantee that the detected signal will be discernable over the dark noise of the detector. In order to calculate the dark noise of the detector, it is necessary to know how much area of the detector is covered by the image of the sample. In the case of a 10× objective, a 100-micron diameter sample will be projected onto a 1-mm diameter circle on the detector having a 0.79-mm^2 area. For a bialkali detector, the amount of dark signal generated over the 100-s period in this circle will be

dark signal (counts mm^{-2} s^{-1}) $\times$ image area $\times$ time
$= 0.062$ counts mm^{-2} $s^{-1} \times 0.79$ $mm^2 \times 100$ s $= 5$ photons.

For a back-illuminated electron-multiplying CCD operating at −90 °C, the amount of dark signal generated would be

dark signal (counts mm^{-2} s^{-1}) $\times$ image area $\times$ time
$= 0.78$ counts mm^{-2} $s^{-1} \times 0.79$ $mm^2 \times 100$ s $= 61$ photons.

In the case of a 40× lens, the signal is projected onto a 4-mm diameter circle having an area of 12.6 mm^2, so the amount of noise generated in the region covered by the sample will be 16× higher, while the signal increase that is due to a larger NA will be only 3.6× higher.

A good way to assess whether the signal can be discerned over the noise is to calculate the signal-to-noise ratio (S/N). Generally an S/N of 3 or more is desirable, although it is possible to make some useful observations with signals having lower S/N. The S/N expectations for the example above are shown in Table 11.4.

Table 11.4 Summary of S/N results for an example application.

Lens	*Bialkali photocathode* *T = 20 °C*	*Back-illuminated EM-CCD* *T = −90 °C*
10× / 0.5 NA	8/5 = 1.6 (±35%) with 68% certainty	102/61 = 1.7 (±10%) with 68% certainty
40× / 0.95 NA	30/78 = 0.4 (±18%) with 68% certainty	368/940 = 0.4 (±5%) with 68% certainty

In this situation it appears that the best selection is a back-illuminated electron-multiplying CCD operating at −90 °C, as this will give the best S/N and best statistical certainty for the measurement.

There are many issues that must be taken into consideration that may limit the range of configurations worth considering. For example, a 40× objective is not a reasonable option for a sample larger than about 200 microns, as the projected image may not land on the physical detector unless it is a large-format detector. On the other hand, a 10× objective is generally not very useful for viewing samples with 10-micron features.

Other issues affecting detector selection include cost of equipment, software capabilities, and level of support available from the vendor. Ultimately, it may be necessary to settle for less than optimal signal-to-noise performance in order to accomplish the intended research. However, even in such situations it is best to select equipment that will optimize the signal-to-noise ratio and maximize the statistical certainty of the measurements being made.

References

Microscopy and Optics
http://micro.magnet.fsu.edu/index.html
http://www.microscopyu.com/
http://www.olympusmicro.com/

Detector Product Information
Andor Technologies (www.andor.com, www.emccd.com)
Burle Technologies, Inc. (www.burle.com)
Delft Electronic Products B. V. (www.dep.nl)
e2v Technologies, Ltd (www.e2v.com)
Electron Tubes, Ltd (www.electrontubes.com)
Hamamatsu Photonics, KK (www.hamamatsu.com)
Photek, Ltd (www.photek.com)
Roper Scientific, Inc. (www.roperscientific.com)

12 Photoproteins and Instrumentation: Their Availability and Applications in Bioanalysis

Leslie Doleman, Stephanie Bachas-Daunert, Logan Davies, Sapna K. Deo, and Sylvia Daunert

Aequorin

Protein	*Description/Properties*	*Applications*	*Commercial sources*
Recombinant aequorin	Emission λ_{max} 469 nm Detection limit: 1×10^{-18} M	Calcium detection, label in immunoassays	Senseomics www.senseomics.com MP Biomedical www.mpbio.com Invitrogen www.invitrogen.com LUX Biotechnology www.luxbiotech.com Chemicon www.chemicon.com
ChemFLASH™ Streptavidin Conjugate Pack	Aequorin conjugated with streptavidin	Detection of biotinylated target	Chemicon www.chemicon.com
DemoLite™ Plate (flash luminescence test plate)	Serial dilutions of lyophilized aequorin	Demonstration of instrument linearity	Chemicon www.chemicon.com
Universal Amplicon Detection AssayChem FLASH™ AquaLite™	Contains anti-FITC monoclonal antibody conjugated to aequorin	Detection of amplified DNA	Chemicon www.chemicon.com
XpressPack™ Luminescent Amplification Detection System, required for luminescent XpressPack™	Contains streptavidin conjugated to aequorin	Detection of amplified DNA	Chemicon www.chemicon.com

Photoproteins in Bioanalysis. Edited by Sylvia Daunert and Sapna K. Deo

ISBN: 3-527-31016-9

Protein	*Description/Properties*	*Applications*	*Commercial sources*
Aequorin biotin	Aequorin conjugated with biotin	Avidin-binding immunoassays	Senseomics www.senseomics.com MP Biomedical www.mpbio.com
Aquamax-Cortisol	Aequorin conjugated with cortisol	Cortisol detection	Senseomics www.senseomics.com
Aquamax-Digoxin	Aequorin conjugated with digoxigenin	Digoxin detection	Senseomics www.senseomics.com
Aquamax-6-keto-prostaglandin $F_{1\alpha}$	Aequorin conjugated with 6-keto prostaglandin F1α	Detection of 6-keto prostaglandin	Senseomics www.senseomics.com
Aquamax–HIV protease substrate	Genetically engineered fusion protein between aequorin and HIV-1 protease substrate	Sensing HIV-1 protease and its drug inhibitors	Senseomics www.senseomics.com
Aquamax-Leu-Enkephalin	Genetically engineered fusion protein between aequorin and Leu-enkephalin	Leu-enkephalin detection	Senseomics www.senseomics.com
Aequorin–Protein C conjugate	Aequorin conjugated with protein C	Detection of inflammatory marker protein C	Senseomics www.senseomics.com
Aequorin anti-salmonella conjugate	Genetically engineered fusion protein between aequorin and anti-salmonella ScFv	Salmonella detection	Senseomics www.senseomics.com

Obelin

Protein	*Description/Properties*	*Applications*	*Commercial sources*
Obelin	Emission λ_{max} 470 nm	Label	Metachem www.metachem.com Assay Designs www.assaydesigns.com
Obelin humanized photoprotein in PCRScript	Emission λ_{max} 470 nm	Obelin expression in mammalian cells	Lux Biotechnologies www.luxbiotech.com
Obelin native photo-protein in pUC19 vector	Emission λ_{max} 470 nm	Obelin expression in bacterial cells	Lux Biotechnologies www.luxbiotech.com

Protein	Description/Properties	Applications	Commercial sources
Obelin GxM IgG	Recombinant obelin conjugated to goat anti-mouse IgG Emission λ_{max} 470 nm	Immunoassay	Metachem www.metachem.com
Obelin GxR IgG	Recombinant obelin conjugated to goat anti-rabbit IgG Emission λ_{max} 470 nm	Immunoassay	Metachem www.metachem.com Assay Designs www.assaydesigns.com
Obelin biotin	Obelin conjugated with biotin Emission λ_{max} 470 nm	Avidin detection	Metachem www.metachem.com Assay Designs www.assaydesigns.com

Luciferases

Protein	Description/Properties	Applications	Commercial sources
Luciferase reporter assay kit	Kit includes firefly luciferase substrate and cell lysis buffer	Detection of luciferase activity	BD Biosciences www.bdbiosciences.com
Gaussia princeps luciferase	Spectral peak at 480 nm	Reporter	LUX Biotechnologies www.luxbiotech.com Nanolight Technolgoy www.nanolight.com
p-UC19, pcDNA3	Vectors of *Gaussia princeps* luciferase, *Renilla* luciferase, and *Pleuromamma*	Reporter	LUX Biotechnologies www.luxbiotech.com
Biotinylated luciferase	2× brighter than normal luciferase	Sensitive detection of avidin conjugated to proteins, DNA, etc.	LUX Biotechnologies www.luxbiotech.com
Renilla luciferase	Spectral peak at 480 nm	Reporter	LUX Biotechnologies www.luxbiotech.com
Pleuromamma luciferase	Spectral peak at 495 nm	Reporter	LUX Biotechnologies www.luxbiotech.com
Humanized *Gaussia* luciferase	Sodium-dependent expression in mammalian cells	Reporter	Nanolight Technology www.nanolight.com
QuantiLum® recombinant luciferase	2.0×10^{10} light units per mg protein	Reporter	Promega www.promega.com
Sodium and potassium salt D-luciferin	99% pure	Luciferase substrate	BD Biosciences www.bdbiosciences.com Gold Biotechnology Inc. Goldbio.com

Aequorea and Anthozoa Fluorescent Proteins

Protein	*Description/Properties*	*Applications*	*Commercial sources*
Ds-Red Express	Excitation λ_{max} 556 nm Emission λ_{max} 586 nm	Reporter	Clontech www.clontech.com
DsRed2	Excitation λ_{max} 563 nm Emission λ_{max} 582 nm	Reporter	Clontech www.clontech.com Evrogen www.evrogen.com
DsRed Monomer	Excitation λ_{max} 557 nm Emission λ_{max} 579 nm	Reporter	Clontech www.clontech.com
AsRed2	Excitation λ_{max} 588 nm Emission λ_{max} 618 nm	Reporter	Clontech www.clontech.com
HcRed1	Excitation λ_{max} 584 nm Emission λ_{max} 610 nm	Reporter	Clontech www.clontech.com
JRed	Excitation λ_{max} 584 nm Emission λ_{max} 610 nm	Reporter	Evrogen www.evrogen.com
KFP-Red	Excitation λ_{max} 580 nm Emission λ_{max} 600 nm	Reporter	Evrogen www.evrogen.com
GFP (wt)	Excitation λ_{max} 395, 488 nm Emission λ_{max} 509 nm Quantum yield 80%	Reporter	Biotek www.biotek.com Clontech www.clontech.com Invitrogen www.invitrogen.com
EGFP	Red-shifted excitation spectra; 4–35× brighter than GFP (wt) when excited at 488 nm; Excitation λ_{max} 488 nm Emission λ_{max} 507 nm Quantum yield 60%	Reporter	Biotek www.biotek.com Clontech www.clontech.com Invitrogen www.invitrogen.com Promega www.promega.com
EYFP	Excitation λ_{max} 513 nm Emission λ_{max} 527 nm Quantum yield 40%	Reporter	Biotek www.biotek.com Invitrogen www.invitrogen.com
ECFP	Excitation λ_{max} 433,453 nm Emission λ_{max} 475,501 nm Quantum yield 61%	Reporter	Biotek www.biotek.com Invitrogen www.invitrogen.com
GFPuv	18× brighter than wt-GFP Excitation λ_{max} 395 nm Emission λ_{max} 509 nm	Reporter	Biotek www.biotek.com Clontech www.clontech.com
AcGFP	Excitation λ_{max} 475 nm Emission λ_{max} 505 nm	Reporter	Clontech www.clontech.com

Coelenteraziness

Analogue	Description/Properties	Applications	Commercial sources
Coelenterazine native 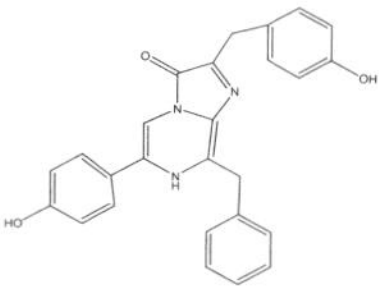	Emission λ_{max} 466 nm Relative luminescent capacity 1.00[a] Relative intensity 1[b] Half-rise time 6–30 s[c]	Aequorin chromophore	Anaspec www.anaspec.com Biotium www.biotium.com Biosynth www.biosynth.com Lux Biotechnologies www.luxbiotech.com Sigma www.sigmaaldrich.com Promega www.promega.com Chemicon www.chemicon.com Nanolight www.nanolight.com Invitrogen www.invitrogen.com
Coelenterazine *cp* 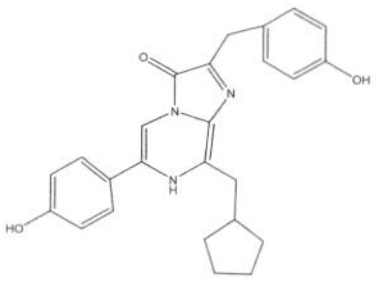	Emission λ_{max} 442 nm Relative luminescent capacity 0.95 Relative intensity 15 Half-rise time 2–5 s	Analogue with properties including a 15-fold increase in the luminescence intensity and quicker response time	Anaspec www.anaspec.com Biotium www.biotium.com Sigma www.sigmaaldrich.com Promega www.promega.com Invitrogen www.invitrogen.com
Coelenterazine *e* 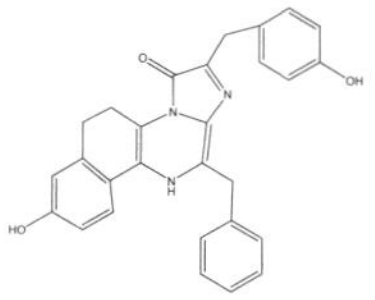	Emission λ_{max} 405, 465 nm Relative luminescent capacity 0.50 Relative intensity 4 Half-rise time 0.15–0.30 s	Analogue with properties including a 7-fold increase in luminescence intensity and quicker response	Biotium www.biotium.com Biosynth www.biosynth.com
Coelenterazine *f* 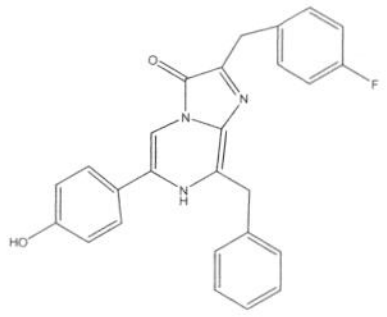	Emission λ_{max} 472 nm Relative luminescent capacity 0.80 Relative intensity 18 Half-rise time 6–30 s	Analogue with properties including a 20-fold increase in luminescence intensity and an 8-nm longer emission maximum	Anaspec www.anaspec.com Biotium www.biotium.com Sigma www.sigma-aldrich.com Invitrogen www.invitrogen.com

Analogue	Description/Properties	Applications	Commercial sources
Coelenterazine *fcp* 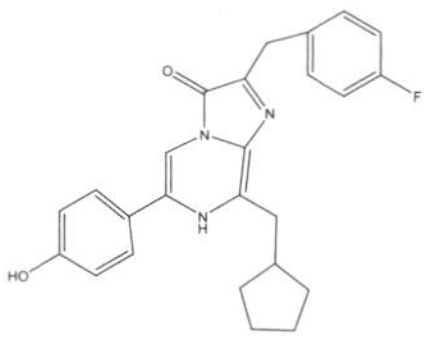	Emission λ_{max} 452 nm Relative luminescent capacity 0.57 Relative intensity 135 Half-rise time 0.4–0.8 s	Analogue with properties including a 135-fold increase in luminescence intensity	Biotium www.biotium.com Sigma www.sigma-aldrich.com Promega www.promega.com
Coelenterazine *h* 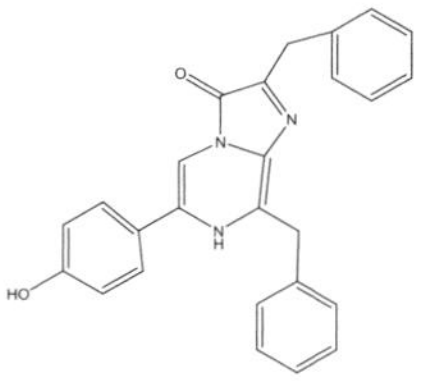	Emission λ_{max} 475 nm Relative luminescent capacity 0.82 Relative intensity 10 Half-rise time 6–30 s	Analogue with properties including a 10-fold increase in luminescence intensity	Anaspec www.anaspec.com Biotium www.biotium.com Biosynth www.biosynth.com Lux Biotechnologies www.luxbiotech.com Sigma www.sigma-aldrich.com Nanolight www.nanolight.com Invitrogen www.invitrogen.com
Coelenterazine *i* 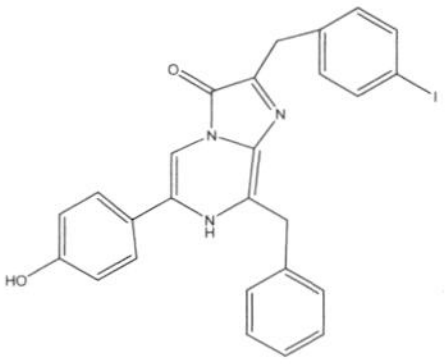	Emission λ_{max} 476 nm Relative luminescent capacity 0.70 Relative intensity 0.03 Half-rise time 8 s	Analogue with properties including the slowest response time and displays the capacity of 3% of native luminescence	Biotium www.biotium.com Sigma www.sigma-aldrich.com
Coelenterazine *ip* 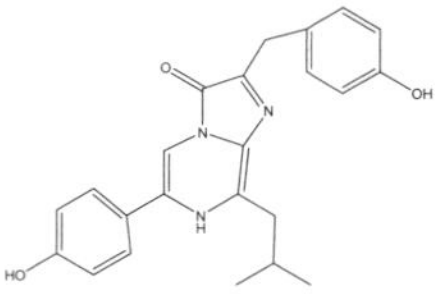	Emission λ_{max} 441 nm Relative luminescent capacity 0.54 Relative intensity 47 Half-rise time 1 s	Analogue with properties including a 50-fold increase in luminescence intensity slower response time than the native.	Biotium www.biotium.com Sigma www.sigma-aldrich.com
Coelenterazine *hcp* 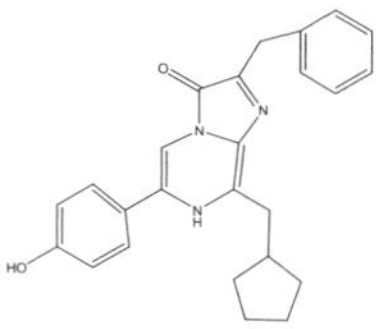	Emission λ_{max} 444 nm Relative luminescent capacity 0.67 Relative intensity 190 Half-rise time 2–5 s	Analogue with properties including a 190-fold increase in luminescence intensity and quicker response.	Anaspec www.anaspec.com Biotium www.biotium.com Sigma www.sigma-aldrich.com Promega www.promega.com Invitrogen www.invitrogen.com

Analogue	Description/Properties	Applications	Commercial sources
Coelenterazine *n* 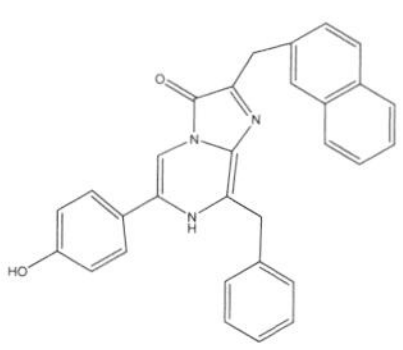	Emission λ_{max} 467 nm Relative luminescent capacity 0.26 Relative intensity 0.01 Half-rise time 6–30 s	Analogue with properties including the lowest cence intensity and a slow response time.	Anaspec www.anaspec.com Biotium www.biotium.com Sigma www.sigma-aldrich.com Invitrogen www.invitrogen.com

a) Relative luminescence capacity = total time-integrated emission of aequorin in saturating Ca^{2+} relative to native aequorin = 1.0.
b) Relative intensity = ratio of the luminescence of aequorin reconstituted with coelenterazine analogue relative to native aequorin.
c) Half-rise time = time for the luminescence signal to reach 50% of the maximum after addition of 1 mM Ca^{2+} to a standard of aequorin reconstituted with the coelenterazine analogue of interest. (http://probes.invitrogen.com/handbook/tables/0356.html).

Luminometers

Luminometer	Description/Properties	Commercial sources
Veritas™ Microplate Luminometer w/Single Reagent Injector	Detection limit: 3×10^{-21} moles luciferase Platform: 96-well microplate	Promega www.promega.com
Veritas™ Microplate Luminometer w/Dual Reagent Injectors	Detection limit: 3×10^{-21} moles luciferase Platform: 96-well microplate	Promega www.promega.com
BD Monolight 3096 Microplate Luminometer	Sensitivity: < 50 amol ATP Wavelength: 300–650 nm	BD Biosciences www.bdbiosciences.com
Luminoskan Ascent, one dispenser 100–240 V	Sensitivity: < 1 fmol ATP Platform: 96- and 384-well plates	Thermo Electron Corporation www.thermo.com
Fluoroskan Ascent FL	Sensitivity: 5 fmol ATP Platform: 96- and 384-well plates Absorbance, fluorescence, and luminescence detection	Thermo Electron Corporation www.thermo.com
LMax II 384 Luminometer	Wavelength: 380–630 nm Sensitivity: < 20 amol ATP per well Platform: 96- and 384-well microplates	Molecular Devices www.moleculardevices.com
FB12 – Single Tube Luminometer	Wavelength: 300–600 nm Sensitivity: Better than 1000 molecules firefly luciferase Platform: Various formats	Berthold Detection Systems GmbH www.berthold-ds.com

Luminometer	Description/Properties	Commercial sources
Sirius – Single Tube Luminometer	Wavelength: 300–600 nm Sensitivity: better than 1000 molecules firefly luciferase	Berthold Detection Systems GmbH www.berthold-ds.com
Smart line TL	Wavelength: 370–630 nm Up to two sample injectors	Berthold Detection Systems GmbH www.berthold-ds.com
LUMIstar Optima	Wavelength: 290–700 nm Sensitivity: 5×10^{-19} mole per well recombinant aequorin Platform: up to 384-well plate Absorbance, fluorescence, luminescence detection	BMG LABTECH www.bmglabtech.com
Chameleon: Multilabel Microplate Reader	Wavelength: 230–900 nm Platform: 6- to 384-well Two injectors Luminescence, fluorescence detection	Bioscan, Inc. www.bioscan.com
MLX™ Microplate Luminometer	Sensitivity: Alkaline phosphatase (10^{-21} moles) Dynamic range: 0.0008–10,000 RLU	Dynex Technologies, Inc. www.dynextechnologies.com
Microbeta Jet	Microtiter plate luminescence reader	PerkinElmer www.perkinelmer.com
MPL1 Luminometer	Spectral range: 300–650 nm Sensitivity: 1 fmole ATP Platform: 96-well microplate	Zylux www.zylux.com
MPL2 Luminometer	Spectral range: 300–650 nm Sensitivity: better than 1000 molecules of luciferase Platform: 96-well microplate	Zylux www.zylux.com
Orion I Luminometer	Spectral range: 300–630 nm Sensitivity: 20 attomole ATP Platform: 96-well microplate	Zylux www.zylux.com
Spectro System Luminometer	Spectral range: 400–700 nm Platform: up to 384-well	www.sciencewares.com

Fluorometers

Fluorometer	*Description/Properties*	*Commercial Sources*
Gemini XS Spectrofluoro-meter	Excitation wavelength: 250–850 nm Emission wavelength: 360–850 nm Detection limit: 3 fmol/well FITC Platform: 6-, 12-, 24-, 48-, 96-, and 384-well Dual monochromators	Molecular Devices www.moleculardevices.com
Cary Eclipse	Monochromator based Microtiter plate adaptor, fiber optic adaptor Flash xenon lamp	Varian Inc. www.varianinc.com
FluoroMax-3	Monochromator based Detection in cuvette UV-Vis: near IR range Xenon-arc lamp	Horiba Jovin Yvon www.jyinc.com
Fluorolog-3	Monochromator based Detection in microtiter plates, fiber optic UV-Vis: IR range Xenon lamp	Horiba Jovin Yvon www.jyinc.com
RF-5301	Monochromator-based Detection in cuvette Wavelength range 200–900 nm Xenon lamp High sensitivity, high throughput	Shimadzu www.ssi.shimadzu.com
RF-1501	Monochromator based Detection in cuvette Wavelength range 200–900 nm Xenon lamp Compact, small sample size	Shimadzu www.ssi.shimadzu.com
ND 3300 Fluorospectro-meter	Emission wavelength: 400–750 nm Platform: Droplet measurement pedestal Measurement of fluorophores, dyes, quantum dots	Nanodrop www.nanodrop.com

Portable Luminometers

Recently, many exciting advancements have been made in the field of bioluminescence and chemiluminescence detection. Practical field applications with portable luminometers are innovative methods of applying chemistry knowledge to everyday life. These small, portable, and battery-powered instruments are packaged with all the necessary equipment and reagents in a convenient travel suitcase, and they provide an innovative solution to a myriad of problems. For example, they can measure levels of somatic and microbial ATP on surfaces, thus providing an instant assessment of general cleanliness in the area. Consequently, this type of luminometer is useful in settings where hygiene is essential, such as food-preparation areas.

The applications for this device are endless: it can also be used in water-treatment facilities, healthcare centers, and in the manufacturing of paper. Many user-friendly features enhance BioTrace's portable luminometer, the Uni-*Lite* NG. Not only does it have an internal self-calibration device and a special docking station, but also the software can be upgraded remotely. This upgrading process saves much lost time attributed to mailing the instrument to its manufacturer, having the system upgraded, and waiting for its return.

Just a few years ago, a scant number of companies offered portable luminometers. However, nowadays this instrument has become mainstream. Currently, companies such as BioFix, Hygiena International, EG&G Berthold, CheckLight, Bio-Orbit, and Biotrace Ltd. manufacture this device. For more information, please see the websites below.

http://www.biotrace.co.uk/
https://macherey-nagel.com/web%5CMN-WEB-TestenKatalog.nsf/WebE/LUMINOMETER
http://www.hygiena.net/
http://www.berthold-us.com/
http://www.checklight.co.il/

Acknowledgment

This work was supported by the National Institutes of Health (NIH).

Disclaimer

The information in this chapter was compiled from various Internet resources to the best of our knowledge. This may not be a complete list of available products, and because of companies' constantly changing inventory, some products listed may no longer be available.

Subject Index

Photoproteins in Bioanalysis. Edited by Sylvia Daunert and Sapna K. Deo

ISBN: 3-527-31016-9